Springer Proceedings in Mathematics & Statistics

Volume 225

Springer Proceedings in Mathematics & Statistics

This book series features volumes composed of selected contributions from workshops and conferences in all areas of current research in mathematics and statistics, including operation research and optimization. In addition to an overall evaluation of the interest, scientific quality, and timeliness of each proposal at the hands of the publisher, individual contributions are all refereed to the high quality standards of leading journals in the field. Thus, this series provides the research community with well-edited, authoritative reports on developments in the most exciting areas of mathematical and statistical research today.

More information about this series at http://www.springer.com/series/10533

Samarjit Kar · Ujjwal Maulik
Xiang Li
Editors

Operations Research and Optimization

FOTA 2016, Kolkata, India, November 24–26

Springer

Editors
Samarjit Kar
Department of Mathematics
National Institute of Technology, Durgapur
Durgapur, West Bengal
India

Xiang Li
School of Economics and Management
Beijing University of Chemical Technology
Beijing
China

Ujjwal Maulik
Department of Computer Science
and Engineering
Jadavpur University
Kolkata, West Bengal
India

ISSN 2194-1009 ISSN 2194-1017 (electronic)
Springer Proceedings in Mathematics & Statistics
ISBN 978-981-10-7813-2 ISBN 978-981-10-7814-9 (eBook)
https://doi.org/10.1007/978-981-10-7814-9

Library of Congress Control Number: 2017962987

Mathematics Subject Classification (2010): 90B05, 90B06, 90B10, 90B22, 90B50, 90B99

Printed on acid-free paper

This Springer imprint is published by the registered company Springer Nature Singapore Pte Ltd.
part of Springer Nature
The registered company address is: 152 Beach Road, #21-01/04 Gateway East, Singapore 189721, Singapore

Foreword

The international conference on "*Frontiers in Optimization: Theory and Applications*" (FOTA 2016) from which this volume arose is a good example of interdisciplinary approach for optimization research.

The key questions that were in our mind while organizing the conference were: (i) What are the important problem areas that are faced by the industries and business sectors where optimization plays or could play a key role? (ii) What are the tools developed by the optimization research community over the decades to address the problems of industries and business sectors? Although these questions were never formally posed to the speakers, partial answers were offered by the twenty-six papers that were presented.

Durgapur, India Samarjit Kar

Preface

The first international conference on "*Frontiers in Optimization: Theory and Applications*" (FOTA 2016) was jointly organized by the Operational Research Society of India (ORSI) and the Department of Mathematics, Heritage Institute of Technology, Kolkata, during 24–26 November 2016 at Heritage Institute of Technology, Kolkata, India. Several distinguished speakers from academia, industry and business houses presented the state of the art in modelling and solved a variety of problems that were key to optimization and its applications.

The aim of the conference was to highlight the current advances in the optimization theory and its applications as well as to promote the research in this area to encourage the researchers, scientists, engineers and practitioners from both academia and industries through exchange of their ideas, problems and solutions. Representatives from industries like heads of research from Strand Life Sciences, Dastur Business and Technology Consulting, Microsoft Research Lab, Department of Atomic Energy (DAE), Government of India, participated and presented their research works in the conference. Speakers from academia including faculties from National Graduate Institute for Policy Studies, Tokyo, Japan; Tsinghua University, Beijing, China; Saint Mary's University, Halifax, Canada; Indian leading business schools as well as faculties from engineering and operations research departments also shared their research ideas.

FOTA 2016 received overwhelming submissions covering different areas related to optimization theory and its applications. With the help of our program committee and reviewers, these submissions went through an extensive peer review process. This volume comprises twenty-six of the accepted papers, which provides a comprehensive overview of the current research and future scope in optimization theory.

The papers collated in this volume reflect both current applications of optimization techniques by practitioners as well as development of new methodologies by academic researchers. The volume is organized in three parts:

- Optimization: Theory and modelling
- Engineering and Biomedical applications
- Management applications

Durgapur, India Samarjit Kar
Kolkata, India Ujjwal Maulik
Beijing, China Xiang Li

Acknowledgements

The conference was supported by the Board of Research in Nuclear Science (BRNS), Department of Atomic Energy (DAE), Government of India; Council of Scientific and Industrial Research, Government of India; National Aluminium Company (NALCO), India; Rashtriya Chemicals & Fertilizers Ltd. (RCFL), India. We are grateful to these organizations for their very generous support. No such event can be successfully organized without active help and support from many. We would like to acknowledge several individuals who assisted us in the presentation of this volume: Shyama Prasad Mukherjee, Rathindarnath Mukherjee, Somjit Dutta and Sandeep Chatterjee. We would like to thank all participants for their contributions to the conference. We would like to thank all keynote speakers, invited speakers, session chairs and authors for their excellent support to make FOTA 2016 a grand success. We are indebted to the program committee members and the reviewers for providing timely and quality reviews. Finally, we would like to thank all the volunteers for their tireless efforts contributing in making the conference run smoothly.

November 2016

Samarjit Kar
Ujjwal Maulik
Xiang Li

Contents

Editors and Contributors

About the Editors

Samarjit Kar received his B.Sc. in Mathematics from Presidency College, M.Sc. in Applied Mathematics from Calcutta University and Ph.D. in Mathematics from Vidyasagar University, India. He is presently an Associate Professor in the Department of National Institute of Technology, Durgapur, India. His research interest includes inventory management, financial modelling, fuzzy optimization and evolutionary computing. He has published 2 textbooks, 5 edited volume and more than 125 research articles in peer-reviewed journals and International conferences. He serves as Associate Editor of the Journal of Uncertainty Analysis and Applications, Springer. He has supervised 13 Ph.D. students and currently guiding 8 scholars for Ph.D. He is also a council member of Operational Research Society of India.

Ujjwal Maulik is a Professor and Head of the Department of Computer Science and Engineering, Jadavpur University, Kolkata, India, since 2004. Presently, he is a Visiting Professor at "International Center for Theoretical Physics (ICTP)", Italy. He did his Bachelors in Physics and Computer Science in 1986 and 1989, respectively. Subsequently, he did his Masters and Ph.D. in Computer Science in 1992 and 1997, respectively. He chaired the Department of Computer Science and Technology, Kalyani Government Engineering College, Kalyani, India, during 1996–1999. He has worked in Los Alamos National Laboratory, Los Alamos, New Mexico, USA, in 1997; University of New South Wales, Sydney, Australia, in 1999; University of Texas, Arlington, USA, in 2001; University of Maryland Baltimore County, USA, in 2004; Fraunhofer Institute AiS, St. Augustin, Germany, in 2005; Tsinghua University, China, in 2007; University of Rome, Italy, in 2008; University of Heidelberg, Germany, in 2009; German Cancer Research Center (DKFZ) in 2010–2012; Grenoble INP, France, in 2010, 2013 and 2014; ICM,

University of Warsaw, Poland, in 2013; International Center of Theoretical Physics (ICTP), Trieste, Italy, in 2014; University of Padoa in 2014 and 2016; Corvinus University of Budapest, Hungary, in 2015 and 2016; and University of Ljubljana, Slovenia, in 2015. He has also visited many institutes/universities around the world for invited lectures and collaborative research. He has been invited to supervise doctoral student in well-known university in France. He is a co-author of 7 books and more than 300 research publications. He is the recipient of Government of India BOYSCAST fellowship in 2001, Alexander Von Humboldt Fellowship for Experienced Researchers in 2010, 2011 and 2012 and Senior Associate of ICTP, Italy, in 2012. He has coordinated five Erasmus Mundus Mobility with Asia (EMMA) programs (European-Asian mobility program). He has been the Program Chair, Tutorial Chair and Member of the program committee of many international conferences and workshops. He is the Associate Editor of "IEEE Transaction on Fuzzy Systems" and "Information Sciences" and also in the editorial board of many journals including "Protein & Peptide Letters". Moreover, he has also served as guest co-editor of special issues of journals including "IEEE Transaction on Evolutionary Computation". He is the founder member of IEEE Computational Intelligence Society (CIS) Chapter, Kolkata section, and has worked as Secretary and Treasurer in 2011, as Vice Chair in 2012 and as the Chair in 2013, 2014 as well as 2016. He is a Fellow of Indian National Academy of Engineering (INAE), West Bengal Association of Science and Technology, Institution of Engineering and Telecommunication Engineers, Institution of Engineers and Senior Member of IEEE. His research interest includes Computational Intelligence, Computational Biology, Combinatorial Optimization, Pattern Recognition, Data Mining and Social Network.

Xiang Li received B.Sc. from the Jilin University, Changchun, China, in 2004 and Ph.D. in Operations Research and Control from Tsinghua University, Beijing, China, in 2008. He is currently a Professor in School of Economics and Management Science, Beijing University of Chemical Technology. He has published 1 book and more than 60 articles on international journals including Transportation Research Part B, Transportation Research Part C, European Journal of Operational Research, Information Sciences, IEEE Transactions on Fuzzy Systems, IEEE Transactions on Intelligent Transportation Systems, IEEE Transactions on Systems, Man, and Cybernetics: Systems and so on. He served as the Associate Editor of Information Sciences (SCI), Transportmetrica B: Transport Dynamics (SCI/SSCI), Journal of Ambient Intelligence and Humanized Computing (SCI) and Editorial Board Member of Sustainability (SCI/SSCI), International Journal of General Systems (SCI). He has also served as guest editor of Soft Computing, Applied Soft Computing, International Journal of Intelligent System, Flexible Service and Manufacturing Journal, International Journal of Uncertainty, Fuzziness and Knowledge-Based Systems and so on.

Contributors

Amr Alzouhri Alyafi GSCOP Laboratory, Grenoble Institute of Technology, Grenoble, France; LIG Laboratory, Grenoble Institute of Technology, Grenoble, France

Shalini Arora Indira Gandhi Delhi Technical University For Women, New Delhi, India

Hongfang Bai Department of Mathematics, Shijiazhuang Mechanical Engineering College, Shijiazhuang, People's Republic of China

Sanghamitra Bandyopadhyay Machine Intelligence Unit, Indian Statistical Institute, Kolkata, India

A. Banerjee Department of Mathematical Sciences, Indian Institute of Technology (BHU), Varanasi, India

Subrata Banerjee Department of Electrical Engineering, National Institute of Technology, Durgapur, West Bengal, India

A. D. Banik School of Basic Sciences, Indian Institute of Technology Bhubaneswar, Bhubaneswar, India

Debasis Basu School of Infrastructure, Indian Institute of Technology Bhubaneswar, Bhubaneswar, India

Uttam Kumar Bera Department of Mathematics, National Institute of Technology, Agartala, Barjala, Jirania, West Tripura, India

Nan-Bin Cao School of Mathematics and Science, Hebei GEO University, Shijiazhuang, People's Republic of China

Dipankar Chakraborty Department of Mathematics, Heritage Institute of Technology, Anandapur, Kolkata, West Bengal, India

Goutam Chakraborty Faculty of Software and Information Science, Iwate Prefectural University, Takizawa, Japan

Anupama Chanda Burdwan University, Bardhaman, India

Amitava Chatterjee Department of Electrical Engineering, Jadavpur University, Kolkata, India

Kajal Chatterjee Department of Mathematics, National Institute of Technology, Durgapur, India

M. L. Chaudhry Department of Mathematics and Computer Science, Royal Military College of Canada, Kingston, ON, Canada

Kaushik Das Sharma Department of Applied Physics, University of Calcutta, Kolkata, India

A. K. Das Indian Statistical Institute, Kolkata, India

Nibaran Das CMATER Laboratory, Department of Computer Science and Engineering, Jadavpur University, Kolkata, India

Tanmay De Department of Computer Science and Engineering, National Institute of Technology, Durgapur, India

Deepmala Indian Statistical Institute, Kolkata, India; Mathematics Discipline, PDPM Indian Institute of Information Technology, Design and Manufacturing, Jabalpur, India

Oshmita Dey Department of Mathematics, Techno India University, Kolkata, West Bengal, India

Asish Dhara Department of Computer Science and Engineering, National Institute of Technology, Durgapur, India

Pankaj Dutta Shailesh J. Mehta School of Management, Indian Institute of Technology Bombay, Mumbai, Maharashtra, India

Totan Garai Department of Mathematics, Indian Institute of Engineering Science and Technology, Shibpur, Howrah, West Bengal, India

Souvik Ghosh School of Basic Sciences, Indian Institute of Technology Bhubaneswar, Bhubaneswar, India

Dipayan Guha Department of Electrical Engineering, National Institute of Technology, Durgapur, West Bengal, India

G. K. Gupta Department of Mathematical Sciences, Indian Institute of Technology (BHU), Varanasi, India

R. Jana Jadavpur University, Kolkata, India

Shantanu Jana CMATER Laboratory, Department of Computer Science and Engineering, Jadavpur University, Kolkata, India

Samarjit Kar Department of Mathematics, National Institute of Technology, Durgapur, India

Bindu Kaushal Indira Gandhi Delhi Technical University For Women, New Delhi, India

Indadul Khan Department of Computer Science, Vidyasagar University, Medinipur, W.B, India

Mohan Krishnamoorthy Department of Mechanical and Aerospace Engineering, Monash University, Melbourne, VIC, Australia

M. Maiti Department of Applied Mathematics, Vidyasagar University, Midnapore, WB, India

Manoranjan Maiti Department of Applied Mathematics with Oceanology and Computer Programming, Vidyasagar University, Medinipur, W.B, India

Samir Maity Department of Computer Science, Vidyasagar University, Medinipur, W.B, India

Monalisa Mandal Department of Computer and Information Science, University of Science and Technology (NTNU), Trondheim, Norway

Ujjwal Maulik Department of Computer Science and Engineering, Jadavpur University, Kolkata, West Bengal, India

S. K. Mishra Department of Mathematics, Institute of Science, Banaras Hindu University, Varanasi, India

B. Mondal Raja N.L.K. Womens College, Midnapore, WB, India

Anindita Mukherjee Department of Mathematics, Techno India University, Kolkata, West Bengal, India

R. N. Mukherjee Jadavpur University, Kolkata, India

Anirban Mukhopadhyay Department of Computer Science and Engineering, University of Kalyani, Kalyani, West Bengal, India

Deepa Naik Department of Computer Science and Engineering, National Institute of Technology, Durgapur, India

Naorem Nalini Devi Department of CSE, NIT Manipur, Imphal, India

Mita Nasipuri CMATER Laboratory, Department of Computer Science and Engineering, Jadavpur University, Kolkata, India

Suvasis Nayak School of Basic Sciences, Indian Institute of Technology Bhubaneswar, Bhubaneswar, India

A. K. Ojha School of Basic Sciences, Indian Institute of Technology Bhubaneswar, Bhubaneswar, India

Tatsuo Oyama National Graduate Institute for Policy Studies, Tokyo, Japan

Monalisa Pal Machine Intelligence Unit, Indian Statistical Institute, Kolkata, India

Gopinath Panda School of Basic Sciences, Indian Institute of Technology Bhubaneswar, Bhubaneswar, India

Stéphane Ploix GSCOP Laboratory, Grenoble Institute of Technology, Grenoble, France

Anjan Rakshit Department of Electrical Engineering, Jadavpur University, Kolkata, India

Himansu Rathi Department of Computer Science and Engineering, National Institute of Technology, Durgapur, India

Patrick Reignier University of Grenoble Alpes, CNRS, INRIA, LIG, Grenoble, France

Arindam Roy Department of Computer Science, Prabhat Kumar College, Purba Medinipur, W.B, India

Jagannath Roy Department of Mathematics, National Institute of Technology, Durgapur, India

Provas Kumar Roy Department of Electrical Engineering, Kalyani Government Engineering College, Kalyani, West Bengal, India

Tapan Kumar Roy Department of Mathematics, Indian Institute of Engineering Science and Technology, Shibpur, Howrah, West Bengal, India

D. Sadhukhan Haldia Government College, Haldia, Purba Midnapore, WB, India

Bijan Sarkar Jadavpur University, Kolkata, India

Jyotirmoy Sarkar Indiana University-Purdue University Indianapolis, Indianapolis, USA

Ram Sarkar CMATER Laboratory, Department of Computer Science and Engineering, Jadavpur University, Kolkata, India

Dipanjana Sengupta Department of Mathematics, National Institute of Technology, Agartala, Barjala, Jirania, West Tripura, India

Himanshu Shrivastava IIT Bombay, IITB-Monash Research Academy, Mumbai, Maharashtra, India

Patrick Siarry Lab. LiSSi, Université Paris-Est Créteil, Vitry-sur-Seine, France

K. Sikdar Department of Mathematics, BMS Institute of Technology & Management, (Affiliated to VTU, Belgaum-18), Yelahanka, Bengaluru, India

Khundrakpam Johnson Singh Department of CSE, NIT Manipur, Imphal, India

Yadvendra Singh Department of Mathematics, Institute of Science, Banaras Hindu University, Varanasi, India

Pravin Suryawanshi Shailesh J. Mehta School of Management, Indian Institute of Technology Bombay, Mumbai, Maharashtra, India

Rui Xu Department of Mathematics, Shijiazhuang Mechanical Engineering College, Shijiazhuang, People's Republic of China

Edmundas Kazimieras Zavadskas Department of Construction Technology and Management, Vilnius Gediminas Technical University, Vilnius, Lithuania

Yu-Ping Zhang School of Mathematics and Science, Hebei GEO University, Shijiazhuang, People's Republic of China

FOTA 2016

Use of optimization techniques in different engineering and management applications has become common practices in the last few decades. This book presents some recent advances in optimization and its applications. Prominent researchers from academic institutions have presented some state-of-the-art techniques in optimization including theoretical and simulation aspects, while the contributors from the practitioners in industry and business sectors comprise the latest techniques practiced by them.

The book covers a wide range of topic: engineering optimization, logistic management, financial optimization, and models for risk management and diversification.

Not only the standard mathematical models were presented and analysed, but also different problems of engineering and management applications were thoroughly deliberated upon to suggest mathematical models and practices for evaluation and review in real-life situation.

Part I
Theory and Modelling

On Generalized Positive Subdefinite Matrices and Interior Point Algorithm

A. K. Das, R. Jana and Deepmala

Abstract In this paper, we propose an iterative and descent type interior point method to compute solution of linear complementarity problem LCP(q, A) given that A is real square matrix and q is a real vector. The linear complementarity problem includes many of the optimization problems and applications. In this context, we consider the class of generalized positive subdefinite matrices (GPSBD) which is a generalization of the class of positive subdefinite (PSBD) matrices. Though Lemke's algorithm is frequently used to solve small and medium-size LCP(q, A), Lemke's algorithm does not compute solution of all problems. It is known that Lemke's algorithm is not a polynomial time bound algorithm. We show that the proposed algorithm converges to the solution of LCP(q, A) where A belongs to GPSBD class. We provide the complexity analysis of the proposed algorithm. A numerical example is illustrated to show the performance of the proposed algorithm.

Keywords Interior point algorithm · Generalized positive subdefinite matrices (GPSBD) · Positive subdefinite matrices (PSBD) · Linear complementarity problem

A. K. Das · Deepmala (✉)
Indian Statistical Institute, 203 B. T. Road, Kolkata 700108, India
e-mail: dmrai23@gmail.com

A. K. Das
e-mail: akdas@isical.ac.in

R. Jana
Jadavpur University, Kolkata 700032, India
e-mail: rwitamjanaju@gmail.com

Deepmala
Mathematics Discipline, PDPM Indian Institute of Information Technology, Design and Manufacturing, Jabalpur 482005, India

S. Kar et al. (eds.), *Operations Research and Optimization*, Springer Proceedings in Mathematics & Statistics 225, https://doi.org/10.1007/978-981-10-7814-9_1

1 Introduction

We introduce a different approach to compute the solution of linear complementarity problem LCP(q, A) where A is real square matrix and q is a real vector. The proposed algorithm is formulated based on iterative descent method. The linear complementarity problem contains many of the optimization problems and applications. The class of generalized positive subdefinite matrices (GPSBD) is a generalization of the class of positive subdefinite (PSBD) matrices [7]. We study this algorithm for solving LCP(q, A) under $A \in$ GPSBD class. The algorithm proposed by Lemke to solve an LCP(q, A) contributed significantly to the development of the linear complementarity theory. However, Lemke's algorithm does not consider all problems. Interior point method is another approach to solve linear complementarity problem. If A is positive semidefinite matrix, then LCP(q, A) is solvable in polynomial time by ellipsoid method [4], path-following method [8], projective method [5]. Fathi [3] showed the computational complexity of LCP(q, A) related to symmetric positive definite matrix. For details, see [2, 12] and references cited therein. We suggest an interior point method in line with Pang [10] to compute solution of LCP(q, A) where A belongs to GPSBD class. We claim that the proposed algorithm is useful to compute solution of a large linear complementarity problem.

In Sect. 2, some results are presented that are used in the next sections. In Sect. 3, we proposed an algorithm based on interior point method to solve linear complementarity problem. We establish some new results related to the proposed interior point method. We prove that the proposed algorithm converges to the solutions of the problem. We include the complexity analysis of the proposed algorithm. Finally, we consider a numerical example to illustrate the performance of the proposed algorithm.

2 Preliminaries

R^n_{++} denotes the positive orthant in R^n. For any matrix $A \in R^{n\times n}$, A^T denotes its transpose. x_i denotes the ith coordinate of the vector x. Also, x^T denotes the transpose of x. $\|x\|$ denotes the norm of the vector x.

Now we start with the definition of linear complementarity problem. Suppose that a square matrix A of order n and an n dimensional vector q, we have to find n dimensional vectors u and v satisfying

$$v - Au = q, \quad u \geq 0, \ v \geq 0 \tag{1}$$

$$u^T v = 0. \tag{2}$$

Equation (1) indicates the feasibility of the problem, and (1), (2) jointly indicate the solution of the problem.

Martos [6] proposed positive subdefinite (PSBD) matrices to address pseudo-convex functions. The nonsymmetric PSBD matrices were studied to connect generalized monotonicity and the linear complementarity problem. Later, Crouzeix and Komlósi [1] enlarged PSBD class by introducing the class of GPSBD matrices. This class was studied in the context of the processability of linear complementarity problem by Lemke's algorithm. We recall that A is called *PSBD matrix* if for all $u \in R^n$, $u^T Au < 0$ implies $A^T u$ is unisigned. A matrix $A \in R^{n\times n}$ is called GPSBD [1, 9] if $\exists\ e_i \geq 0$ and $f_i \geq 0$ with $e_i + f_i = 1,\ i = 1, 2, \dots, n$ such that

$$\forall\ u \in R^n,\ \ u^T Au < 0 \Rightarrow \begin{cases} \text{either } & -e_i u_i + f_i (A^T u)_i \geq 0 \text{ for all } i, \\ \text{or} & -e_i u_i + f_i (A^T u)_i \leq 0 \text{ for all } i. \end{cases} \tag{3}$$

A matrix $A \in R^{n\times n}$ is called GPSBD if $\exists$ two diagonal matrices $E \geq 0$ and $F \geq 0$ with $E + F = I$ such that

$$\forall\ u \in R^n,\ \ u^T Au < 0 \Rightarrow \begin{cases} \text{either } & -Eu + FA^T u \geq 0 \\ \text{or} & -Eu + FA^T u \leq 0. \end{cases} \tag{4}$$

When $E = 0$, A is PSBD. A is called *merely generalized positive subdefinite* (MGPSBD) *matrix* when A is GPSBD but not PSBD matrix.

Theorem 2.1 *[10] Suppose* $u > 0$ *such that* $v = q + Au > 0$, $\kappa > n$ *and* $\psi : R^n_{++} \times R^n_{++} \to R$ *such that* $\psi(u, v) = \kappa\, log(u^T v) - \sum_{i=1}^n log(u_i v_i)$. *Then,* $\psi(u, v) \geq (\kappa - n) log(u^T v)$.

Proof Note that

$$\begin{aligned} \psi(u, v) &= \kappa \log(u^T v) - \textstyle\sum_{i=1}^n \log(u_i v_i) \\ &\geq \kappa \log(u^T v) - \log\left(\tfrac{1}{n} \textstyle\sum_{i=1}^n (u_i v_i)\right)^n \\ &= (\kappa - n)\log(u^T v) + n \log n \\ &\geq (\kappa - n)\log(u^T v) \end{aligned}$$

■

Theorem 2.2 *[10] Suppose* $u > 0$ *such that* $v = q + Au > 0$, $\kappa > n$ *and* $\psi : R^n_{++} \times R^n_{++} \to R$ *such that* $\psi(u, v) = \kappa\, log(u^T v) - \sum_{i=1}^n log(u_i v_i)$. *Then,* $(\nabla_u \psi(u, v))_i (\nabla_v \psi(u, v))_i = u_i v_i \left(\frac{\kappa}{u^T v} - \frac{1}{u_i v_i}\right)^2\ \forall i$.

Proof Note that

$$\begin{aligned} \left(\nabla_u \psi(u, v)\right)_i &= \tfrac{\kappa}{u^T v} v_i - \tfrac{1}{u_i v_i} v_i \\ &= v_i \left[\tfrac{\kappa}{u^T v} - \tfrac{1}{u_i v_i}\right]. \end{aligned}$$

Similarly, we show

$$\left(\nabla_v \psi(u, v)\right)_i = u_i \left[\tfrac{\kappa}{u^T v} - \tfrac{1}{u_i v_i}\right].$$

Hence, $(\nabla_u \psi(u, v))_i (\nabla_v \psi(u, v))_i = u_i v_i \left(\frac{\kappa}{u^T v} - \frac{1}{u_i v_i}\right)^2\ \forall i$. ■

Theorem 2.3 *[10] Suppose $u > 0$ such that $v = q + Au > 0$, $\kappa > n$ and $\psi : R^n_{++} \times R^n_{++} \to R$ such that $\psi(u,v) = \kappa\, log(u^T v) - \sum_{i=1}^n log(u_i v_i)$. Then, $(\nabla_u \psi(u,v))^T \nabla_v \psi(u,v) > 0$.*

3 Results

Let $u > 0$, $v = q + Au > 0$, $\kappa > n$ and $\psi : R^n_{++} \times R^n_{++} \to R$ such that $\psi(u,v) = \kappa \log(u^T v) - \sum_{i=1}^n \log(u_i v_i) \geq 0$. Todd et al. [11] considered this function in the context of linear programming. We propose an interior point algorithm in line with Pang [10] for finding solution of LCP (q,A) given that A is a MGPSBD $\cap\, C_0$ with $0 < f_i < 1\ \forall\, i$. We prove the following results which are required for the proposed algorithm.

Theorem 3.1 *Suppose $A \in MGPSBD \cap C_0$ with $0 < f_i < 1\ \forall\, i$. Then for $u, v > 0$, $\nabla_u \psi(u,v) + A^T \nabla_v \psi(u,v) \neq 0$.*

Proof Suppose $\nabla_u \psi(u,v) + A^T \nabla_v \psi(u,v) = 0$. It follows that

$$(\nabla_v \psi(u,v))_i (A^T \nabla_v \psi(u,v))_i = -(\nabla_u \psi(u,v))_i (\nabla_v \psi(u,v))_i \leq 0\ \forall\, i.$$

Let $I_1 = \{i\ :\ (\nabla_v \psi(u,v))_i > 0\}$ and $I_2 = \{i\ :\ (\nabla_v \psi(u,v))_i < 0\}$. We consider following three cases (**C1, C2, C3**).

C1: $I_2 = \emptyset$. Then,

$$\begin{aligned}(\nabla_v \psi(u,v))^T A (\nabla_v \psi(u,v)) &= (\nabla_v \psi(u,v))^T A^T (\nabla_v \psi(u,v)) \\ &= \sum_i (\nabla_v \psi(u,v))_i (A^T (\nabla_v \psi(u,v)))_i \\ &\leq 0.\end{aligned}$$

Since $A \in C_0$, $[(\nabla_v \psi(u,v))_i (A^T (\nabla_v \psi(u,v)))_i] = 0,\ \forall\, i$.

C2: $I_1 = \emptyset$. Then,

$$\begin{aligned}(-(\nabla_v \psi(u,v)))^T A^T (-(\nabla_v \psi(u,v))) &= (\nabla_v \psi(u,v))^T A^T (\nabla_v \psi(u,v)) \\ &= \sum_i (\nabla_v \psi(u,v))_i (A^T (\nabla_v \psi(u,v)))_i \\ &\leq 0.\end{aligned}$$

Since $A \in C_0$, $[(\nabla_v \psi(u,v))_i (A^T (\nabla_v \psi(u,v)))_i] = 0,\ \forall\, i$.

C3: Suppose $\exists\, (\nabla_v \psi(u,v))$ such that $(\nabla_v \psi(u,v))_i (A^T (\nabla_v \psi(u,v)))_i \leq 0$ for $i = 1, 2, \ldots, n$ and $(\nabla_v \psi(u,v))_k (A^T (\nabla_v \psi(u,v)))_k < 0$ for at least one $k \in \{1, 2, \ldots, n\}$. Let $I_1 \neq \emptyset$ and $I_2 \neq \emptyset$.

$$(\nabla_v \psi(u,v))^T A^T (\nabla_v \psi(u,v)) = \sum_i [(\nabla_v \psi(u,v))_i (A^T(\nabla_v \psi(u,v)))_i] < 0.$$

This implies

$$-e_i(\nabla_v \psi(u,v))_i + f_i(A^T(\nabla_v \psi(u,v)))_i \geq 0, \ \forall\, i \text{ or}$$

$$-e_i(\nabla_v \psi(u,v))_i + f_i(A^T(\nabla_v \psi(u,v)))_i \leq 0, \ \forall\, i.$$

We assume $-e_i(\nabla_v \psi(u,v))_i + f_i(A^T(\nabla_v \psi(u,v)))_i \geq 0, \ \forall\, i.$
Then, for all $i \in I_1$, $-e_i(\nabla_v \psi(u,v))_i^2 + f_i(\nabla_v \psi(u,v))_i(A^T(\nabla_v \psi(u,v)))_i \geq 0.$
This implies $[(\nabla_v \psi(u,v))_i(A^T(\nabla_v \psi(u,v)))_i] \geq \frac{e_i}{f_i}(\nabla_v \psi(u,v))_i^2 > 0, \ \forall\, i \in I_1.$
Therefore, $\sum_{i \in I_1} [(\nabla_v \psi(u,v))_i(A^T(\nabla_v \psi(u,v)))_i] > 0.$
Since $(\nabla_v \psi(u,v))_i(A^T(\nabla_v \psi(u,v)))_i \leq 0$ for $i = 1, \ldots, n$,
therefore, $[(\nabla_v \psi(u,v))_i(A^T(\nabla_v \psi(u,v)))_i] = 0, \ \forall\, i$. This contradicts Theorem 2.3.
Hence, $\nabla_u \psi(u,v) + A^T \nabla_v \psi(u,v) \neq 0$, for $u, v > 0$. ■

Theorem 3.2 *Suppose u and v are two positive n-vectors. $U = diag(u)$ and $V = diag(v)$ and $B = (U)^{-2} + A^T(V)^{-2}A$ where $A \in R^{n \times n}$. Then, B is symmetric positive definite matrix.*

Proof Note that

$$\begin{aligned} B^T = [(U)^{-2} + A^T(V)^{-2}A]^T &= (U)^{-2} + [A^T(V)^{-2}A]^T \\ &= (U)^{-2} + A^T(V)^{-2}A \\ &= (U)^{-2} + A^T(V)^{-2}A = B. \end{aligned}$$

Hence, B is symmetric. Again,

$$\begin{aligned} x^T A^T (V)^{-2} A x &= (Ax)^T (V)^{-2} A x \\ &= (y)^T (V)^{-2} y. \end{aligned}$$

Since $(y)^T(V)^{-2}y \geq 0, \ \forall\, y \in R^n$, $A^T(V)^{-2}A$ is positive semidefinite. Hence, B is positive definite. ■

We describe an interior point algorithm for solving LCP(q,A) where $A \in$ MGPSBD $\cap\, C_0$.

Algorithm

Step 1: Let $\beta, \gamma \in (0,1)$ and $\sigma \in (0, \frac{1}{2})$ following line search step and u^0 be a strictly feasible point of LCP(q,A) and $v^0 = q + Au^0 > 0$.

$$\nabla_u \psi_k = \nabla_u \psi(u^k, v^k), \qquad \nabla_v \psi_k = \nabla_v \psi(u^k, v^k)$$

and

$$U^k = diag(u^k), \qquad V^k = diag(v^k).$$

Step 2: Now to find the search direction, consider the following problem

$$\text{minimize} \quad (\nabla_u \psi_k)^T d_u + (\nabla_v \psi_k)^T d_v$$

$$\text{subject to} \quad d_v = A d_u, \qquad \|(U^k)^{-1} d_u\|^2 + \|(V^k)^{-1} d_v\|^2 \le \beta^2.$$

We apply scaled gradient reduction method to obtain search direction (d_u, d_v).

Step 3: Find m_k to be the smallest $m \ge 0$ integer such that

$$\psi(u^k + \gamma^m d_u^k, v^k + \gamma^m d_v^k) - \psi(u^k, v^k) \le \sigma \gamma^m [(\nabla_u \psi_k)^T d_u^k + (\nabla_v \psi_k)^T d_v^k].$$

Step 4: Set

$$(u^{k+1}, v^{k+1}) = (u^k, v^k) + \gamma^{m_k} (d_u^k, d_v^k).$$

Step 5: If (u^{k+1}, v^{k+1}) satisfies the termination criterion, i.e., $(u^{k+1})^T v^{k+1} \le \epsilon$, where $\epsilon > 0$ is a very small quantity, stop else $k = k + 1$.

Now to show (d_u^k, d_v^k) as descent direction for the merit function, we prove the following lemma.

Lemma 3.1 *Suppose* $A \in MGPSBD \cap C_0$ *with* $0 < f_i < 1 \; \forall \; i$, $u > 0$, $v = q + Au > 0$, $\kappa > n$ *and* $\psi : R_{++}^n \times R_{++}^n \to R$ *such that* $\psi(u, v) = \kappa \, log(u^T v) - \sum_{i=1}^n log(u_i v_i)$. *If there is a pair of vectors* (d_u^k, d_v^k) *such that* $(\nabla_u \psi_k)^T d_u^k + (\nabla_v \psi_k)^T d_v^k < 0$, *then* $\exists$ *a* $\gamma \in (0, 1)$ *such that* $\psi(u^k + \gamma^m d_u^k, v^k + \gamma^m d_v^k) - \psi(u^k, v^k) < 0$ *where* m *is a nonnegative integer and* (d_u^k, d_v^k) *is said to be the descent direction.*

Proof We have $d_u^k = -\frac{(A^k)^{-1} r^k}{\tau_k}$, $d_v^k = A d_u^k$ from the algorithm. According to Theorem 3.1, $r^k = \nabla_u \psi_k + A^T \nabla_v \psi_k \ne 0$ and $A^k = (U^k)^{-2} + A^T (V^k)^{-2} A$ is positive definite by Theorem 3.2. So $\tau_k = \dfrac{\sqrt{(r^k)^T (A^k)^{-1} r^k}}{\beta}$ is positive. Now we show that $(\nabla_u \psi_k)^T d_u^k + (\nabla_v \psi_k)^T d_v^k < 0$. We derive

$$\begin{aligned}(\nabla_u \psi_k)^T d_u^k + (\nabla_v \psi_k)^T d_v^k &= \left[\nabla_u \psi_k + M^T \nabla_v \psi_k\right]^T d_u^k \\ &= -\frac{1}{\tau_k} (\sqrt{(r^k)^t (M^k)^{-1} r^k})^2 \\ &= -\tau_k \beta^2 < 0.\end{aligned}$$

Now we consider $\psi(u^k + \gamma^m d_u^k, v^k + \gamma^m d_v^k) - \psi(u^k, v^k) \le \sigma \gamma^m [(\nabla_u \psi_k)^T d_u^k + (\nabla_v \psi_k)^T d_v^k]$. Since $0 < \beta, \gamma, \sigma < 1$, we say $\psi(u^k + \gamma^m d_u^k, v^k + \gamma^m d_v^k) - \psi(u^k, v^k) < 0$. ■

We prove the following theorem to show that the proposed algorithm converges to the solution under some defined condition.

Theorem 3.3 *If $A \in MGPSBD \cap C_0$ with $0 < f_i < 1 \; \forall \, i$ and LCP(q, A) has a strictly feasible solution, then every accumulation point of $\{u^k\}$ is the solution of LCP(q, A), i.e., algorithm converges to the solution.*

Proof Let us consider the subsequences $\{u^k : k \in \omega\}$. Suppose $\tilde{u}$ is the limit of the subsequence and $\tilde{v} = q + A\tilde{u}$. Again, we know $\psi(\tilde{u}, \tilde{v}) < \infty$. So either $\tilde{u}^T\tilde{v} = 0$ or $(\tilde{u}, \tilde{v}) > 0$. If the first case happen, then we are done. So let us consider that $(\tilde{u}, \tilde{v}) > 0$. Also, suppose $\tilde{r}$ and $\tilde{A}$ are the limits of the subsequences $\{r^k : k \in \omega\}$ and $\{A^k : k \in \omega\}$, respectively. Consider τ^k converges to $\tilde{\tau} = \frac{\sqrt{\tilde{r}^T\tilde{A}^{-1}\tilde{r}}}{\beta} (> 0)$, where $\tilde{A}$ remains positive definite. $(\tilde{d}_u, \tilde{d}_v)$ be the limits of the sequence of direction (d_u^k, d_v^k). So from the algorithm, we get

$$\tilde{d}_u = -\frac{\tilde{A}^{-1}\tilde{r}}{\tilde{\tau}}, \qquad \tilde{d}_v = A\tilde{d}_u.$$

Now as $\{\psi(u^{k+1}, v^{k+1}) - \psi(u^k, v^k)\}$ converges to zero and since $\lim m_k = \infty$ as $k \to \infty$, $\{(u^{k+1}, v^{k+1}) : k \in \omega\}$ and $\{(u^k + \gamma^{m_k-1}d_u^k, v^k + \gamma^{m_k-1}d_v^k) : k \in \omega\}$ converges to $(\tilde{u}, \tilde{v})$. As m_k is the smallest nonnegative integers, we have,

$$\frac{\psi(u^k + \gamma^{m_k-1}d_u^k, v^k + \gamma^{m_k-1}d_v^k) - \psi(u^k, v^k)}{\gamma^{m_k-1}} > -\sigma\beta^2\tau_k.$$

Again, on the other hand, from the algorithm,

$$\frac{\psi(u^{k+1}, v^{k+1}) - \psi(u^k, v^k)}{\gamma^{m_k}} \leq -\sigma\beta^2\tau_k.$$

Now taking limit $k \to \infty$, From the last two inequalities, we can write,

$$\nabla_u(\psi_(\tilde{u}, \tilde{v}))^T\tilde{d}_u + \nabla_v(\psi_(\tilde{u}, \tilde{v}))^T\tilde{d}_v = -\sigma\tilde{\tau}\beta^2.$$

Again from Lemma 3.1, we know

$$(\nabla_u\psi_k)^T d_u^k + (\nabla_v\psi_k)^T d_v^k = -\tau_k\beta^2.$$

Hence, by taking limit $k \to \infty$, we get

$$\nabla_u(\psi_(\tilde{u}, \tilde{v}))^T\tilde{d}_u + \nabla_v(\psi_(\tilde{u}, \tilde{v}))^T\tilde{d}_v = -\tilde{\tau}\beta^2.$$

This is a contradiction. So our proposed algorithm converges to the solution. ■

3.1 Complexity Analysis

In our proposed algorithm, we start with a interior feasible point (u^0, v^0) such that $\psi(u^0, v^0) \leq O(\kappa L)$, where L is the size of input data A and q and generates a sequence of interior feasible points $\{u^k, v^k : k \in \omega\}$ so that $\psi(u^k, v^k) \leq -(\kappa - n)L$. In the algorithm, (d_u^k, d_v^k) be the descent direction in the k-th iteration. Now to find the search direction, consider the following problem

$$\text{minimize} \quad (\nabla_u \psi_k)^T d_u + (\nabla_v \psi_k)^T d_v$$

$$\text{subject to} \quad d_v = A d_u, \qquad \|(U^k)^{-1} d_u\|^2 + \|(V^k)^{-1} d_v\|^2 \leq \beta^2.$$

According to Ye [14], we can rewrite,

$$\begin{bmatrix} (U^k)^{-1} d_u^k \\ (V^k)^{-1} d_v^k \end{bmatrix} = -\beta \frac{\alpha^k}{\|\alpha^k\|},$$

where

$$\alpha^k = \begin{bmatrix} \alpha_u^k \\ \alpha_v^k \end{bmatrix} = \begin{bmatrix} \frac{\kappa}{(u^T v)^k} U^k(v^k + A^T \pi^k) - e \\ \frac{\kappa}{(u^T v)^k} V^k(u^k - \pi^k) - e \end{bmatrix},$$

$\pi^k = ((V^k)^2 + A(U^k)^2 A^T)^{-1}(V^k - AU^k)(U^k v^k - \frac{(u^T v)^k}{\kappa} e)$ and e be the vector of all 1. Now for any $u, v > 0$, we define

$$h(u, v) = \frac{\kappa}{u^T v} Uv - e \text{ and}$$

$$H(u, v) = 2I - (UA^T - V)(V^2 + AU^2 A^T)^{-1}(AU - V).$$

$\|h(u, v)\|_H^2$ denotes the H-norm so that $\|h(u, v)\|_H^2 = h^T(u, v) H(u, v) h(u, v)$. Now we show that $H(u, v)$ is positive semidefinite (PSD).

Theorem 3.4 *Suppose $H(u, v) = 2I - (UA^T - V)(V^2 + AU^2 A^T)^{-1}(AU - V)$. Then, $H(u, v)$ is positive semidefinite matrix.*

Proof We define $\mathfrak{I} = \begin{pmatrix} 2I & UA^T - V \\ AU - V & V^2 + AU^2 A^T \end{pmatrix}$. $\mathfrak{I}_{11}$ and $\mathfrak{I}_{22}$ are symmetric and positive definite. This follows that $\mathfrak{I}$ is symmetric. Now we consider the Schur complement $2I - (UA^T - V)(V^2 + AU^2 A^T)^{-1}(AU - V)$ of $\mathfrak{I}_{22}$ in $\mathfrak{I}$. According to Schur complement lemma $2I - (UA^T - V)(V^2 + AU^2 A^T)^{-1}(AU - V)$ is positive semidefinite. This follows $H(u, v)$ is positive semidefinite. ∎

Now $\|\alpha^k\|^2 = h^T(u^k, v^k) H(u^k, v^k) h(u^k, v^k)$. We define the condition number $\zeta(q, A)$ for the LCP(q, A) as

$$\zeta(q,A) = inf\{\, \|h(u,v)\|_H^2 \;:\; u^T v \geq 2^{-L}, \psi(u,v) \leq O(\kappa L), u > 0, v = q + Au > 0\}.$$

The condition number $\zeta(q,A)$ represents the degree of difficulty for the proposed algorithm. Now we prove the following theorem.

Theorem 3.5 *Let $\kappa > n$ be fixed. Then, for $A \in MGPSBD \cap C_0$ with $0 < f_i < 1 \;\forall\, i$ and $\{u, v > 0, v = q + Au \,:\, \psi(u,v) \leq O(\kappa L)\}$ is bounded, $\zeta(q,A) > 0$.*

Proof We show that $\|h(u,v)\|_H^2 > 0$. If $\pi = 0$, then it is easy to verify that $\|h(u,v)\|_H^2 > 0$. Now if $\pi \neq 0$, we show that $\|h(u,v)\|_H^2 > 0$. If not, then $\|h(u,v)\|_H^2 = 0$. This follows

$$UA^T\pi + V\pi = 0. \tag{5}$$

However, as $A \in$ MGPSBD $\cap\, C_0$ with $0 < f_i < 1 \;\forall\, i$, According to Theorem 3.1, Eq. 5 can not be true. Hence, $\|h(u,v)\|_H^2 > 0$. Now $u^T v \geq 2^{-L}$ and $\forall\, u, v > 0, v = q + Au$ we have,

$$\begin{aligned}
O(\kappa L) \geq \psi(u,v) &= \kappa \log(u^T v) - \textstyle\sum_{i=1}^n \log(u_i v_i) \\
&= (\kappa - n + 1)\, \log(u^T v) + (n-1) \log(u^T v) \\
&\quad - \textstyle\sum_{i \neq j} \log(u_i v_i) - \log(u_j v_j) \\
&\geq (\kappa - n + 1)\, \log(u^T v) + (n-1) \log(u^T v - u_j v_j) \\
&\quad - \textstyle\sum_{i \neq j} \log\, (u_i v_i) - \log(u_j v_j) \\
&\geq (\kappa - n + 1)\, \log(u^T v) + (n-1) \log(n-1) - \log(u_j v_j) \\
&\geq -(\kappa - n + 1) L + (n-1) \log(n-1) - \log(u_j v_j).
\end{aligned}$$

Therefore, $\log(u_j v_j) \geq -O(\kappa L)$ for $j \in \{1, 2, \ldots, n\}$. Hence, $u_j v_j$ is bounded away from zero by $e^{-O(\kappa L)}\;\forall\, j$. Now there exists an $\bar{\epsilon}$ independent of (u,v) such that $u_j \geq \bar{\epsilon}$ and $v_j \geq \bar{\epsilon}$ following the proof line of Proposition 4 given in [15] Therefore,

$$\begin{aligned}
\zeta(q,A) &= inf\{\, \|h(u,v)\|_H^2 \;:\; u^T v \geq 2^{-L}, \psi(u,v) \leq O(\kappa L), u > 0, v = q + Au > 0\}, \\
&\geq inf\{\, \|h(u,v)\|_H^2 \;:\; u \geq \bar{\epsilon}e, v \geq \bar{\epsilon}e, \psi(u,v) \leq O(\kappa L), u \geq 0, v - q + Au \geq 0\}.
\end{aligned}$$

Now the set related to $\zeta(q,A)$ is closed and bounded and $\|h(u,v)\|_H^2 > 0$. Hence, $\zeta(q,A) > 0$. ■

Theorem 3.6 *The proposed algorithm with $\kappa > n$ and $\zeta(q,A) > 0$ solves the LCP(q,A) in $O(\frac{nL}{\xi(\zeta(q,A))})$ iterations.*

Proof We consider the merit function $\psi(u,v) = \kappa \log(u^T v) - \sum_{i=1}^n \log(u_i v_i)$. Based on the concavity of log function, Lemma 3.1 of [14] and Sect. 3 of [13], we have

$$\psi(u^k + d_u^k, v^k + d_v^k) - \psi(u^k, v^k) \leq -\beta \left\| \alpha^k \right\| + \frac{\beta^2}{2}(\kappa + \frac{1}{1-\beta}).$$

Now letting, $\beta = \min\{\frac{\|\alpha^k\|}{\kappa+2}, \frac{1}{\kappa+2}\} \leq 1/2$. So we can write,

$$\psi(u^k + d_u^k, v^k + d_v^k) - \psi(u^k, v^k) \leq -\xi(\left\|\alpha^k\right\|^2), \tag{6}$$

where

$$\xi(\left\|\alpha^k\right\|^2) = \begin{cases} \frac{\|\alpha^k\|^2}{2(\kappa+2)}, & \text{if } \|\alpha^k\|^2 \leq \frac{(\kappa+2)^2}{4} \\ \frac{\kappa+2}{8}, & \text{otherwise.} \end{cases} \tag{7}$$

Here, $\|\alpha^k\|^2$ is used as the amount of reduction of the potential function at k-th iteration. Now we find a interior feasible point for which each component is less than 2^L. The resulting point has a potential value less than $O(nL)$. Now from Eq. 6, we say that the potential function is reduced by $O(\xi(\zeta(q,A)))$ at every step of iteration. Hence, in total of $O(\frac{nL}{\xi(\zeta(q,A))})$ iterations, we have $\psi(u^k, v^k) < -(\kappa - n)L$ and $(u^k)^T v^k < 2^{-L}$. ■

3.2 Numerical Illustration

A numerical example is considered to demonstrate the effectiveness and efficiency of the proposed algorithm. We consider the following example of LCP(q, A), where

$$A = \begin{pmatrix} 1 & 0 & 0 & 0 \\ -2 & 1 & 0 & 0 \\ 4 & 0 & 1 & 0 \\ 10 & 0 & 0 & 1 \end{pmatrix} \text{ and } q = \begin{pmatrix} -5 \\ -4 \\ 2 \\ 8 \end{pmatrix}.$$

It is easy to show that A is not a PSD matrix. However, A satisfies the definitions of MGPSBD $\cap$ C_0-matrix. We apply proposed algorithm to find solution of the given problem. According to Theorem 3.3 algorithm converges to solution with $u^0, v^0 > 0$. To start with, we initialize $\beta = 0.5$, $\gamma = 0.5$, $\sigma = 0.2$, $\kappa = 5$, and $\epsilon = 0.00001$. We set $u^0 = \begin{pmatrix} 10 \\ 100 \\ 100 \\ 10 \end{pmatrix}$ and obtain $v^0 = \begin{pmatrix} 5 \\ 76 \\ 142 \\ 118 \end{pmatrix}$. We define *diff* $[\psi(u^k, v^k)] = \psi(u^k + \gamma^m d_u^k, v^k + \gamma^m d_v^k) - \psi(u^k, v^k)$.

Table 1 summarizes the computations for the first 10 iterations, 20th, 21st iteration, 45th iteration, and 46st iteration. It is clear that the sequence $\{u^k\}$ and $\{v^k\}$ produced by the proposed algorithm converges to the solution of the given LCP(q, A), i.e., $u^* = \begin{pmatrix} 5 \\ 14 \\ 0 \\ 0 \end{pmatrix}$ and $v^* = \begin{pmatrix} 0 \\ 0 \\ 22 \\ 58 \end{pmatrix}$.

Table 1 Summary of computation for the proposed algorithm

Iteration (k)	u^k	v^k	$\psi(u^k, v^k)$	d_u^k	d_v^k	$diff[\psi(u^k, v^k)]$
1	$\begin{pmatrix} 10.9 \\ 93.1 \\ 64.5 \\ 11.1 \end{pmatrix}$	$\begin{pmatrix} 5.85 \\ 67.39 \\ 109.88 \\ 127.62 \end{pmatrix}$	20.74057	$\begin{pmatrix} 0.852 \\ -6.908 \\ -35.523 \\ 1.1 \end{pmatrix}$	$\begin{pmatrix} 0.852 \\ -8.612 \\ -32.115 \\ 9.62 \end{pmatrix}$	−1.73264
2	$\begin{pmatrix} 11.9 \\ 79.5 \\ 44.7 \\ 12.1 \end{pmatrix}$	$\begin{pmatrix} 6.93 \\ 51.66 \\ 94.44 \\ 139.4 \end{pmatrix}$	19.00793	$\begin{pmatrix} 1.08 \\ -13.58 \\ -19.75 \\ 1.02 \end{pmatrix}$	$\begin{pmatrix} 1.08 \\ -15.73 \\ -15.44 \\ 11.78 \end{pmatrix}$	−1.42043
3	$\begin{pmatrix} 13.2 \\ 67.9 \\ 31.1 \\ 12.5 \end{pmatrix}$	$\begin{pmatrix} 8.19 \\ 37.55 \\ 85.88 \\ 152.35 \end{pmatrix}$	17.58749	$\begin{pmatrix} 1.258 \\ -11.587 \\ -13.595 \\ 0.368 \end{pmatrix}$	$\begin{pmatrix} 1.258 \\ -14.1 \\ -8.56 \\ 12.95 \end{pmatrix}$	−1.12179
4	$\begin{pmatrix} 14.6 \\ 60.3 \\ 21.3 \\ 11.2 \end{pmatrix}$	$\begin{pmatrix} 9.59 \\ 27.1 \\ 81.7 \\ 165.08 \end{pmatrix}$	16.46571	$\begin{pmatrix} 1.4 \\ -7.64 \\ -9.79 \\ -1.31 \end{pmatrix}$	$\begin{pmatrix} 1.4 \\ -10.45 \\ -4.17 \\ 12.73 \end{pmatrix}$	−0.85296
5	$\begin{pmatrix} 15.88 \\ 56.3 \\ 15.36 \\ 8.05 \end{pmatrix}$	$\begin{pmatrix} 10.9 \\ 20.5 \\ 80.9 \\ 174.9 \end{pmatrix}$	15.61275	$\begin{pmatrix} 1.29 \\ -3.98 \\ -5.99 \\ -3.13 \end{pmatrix}$	$\begin{pmatrix} 1.29 \\ -6.57 \\ -0.819 \\ 9.794 \end{pmatrix}$	−0.74863

(continued)

Table 1 (continued)

Iteration (k)	u^k	v^k	$\psi(u^k, v^k)$	d_u^k	d_v^k	$diff[\psi(u^k, v^k)]$
6	$\begin{pmatrix} 16.91 \\ 53.59 \\ 11.22 \\ 5.42 \end{pmatrix}$	$\begin{pmatrix} 11.9 \\ 15.8 \\ 80.9 \\ 182.5 \end{pmatrix}$	14.86412	$\begin{pmatrix} 1.02 \\ -2.71 \\ -4.13 \\ -2.63 \end{pmatrix}$	$\begin{pmatrix} 1.02 \\ -4.7589 \\ -0.0337 \\ 7.6215 \end{pmatrix}$	−0.67971
7	$\begin{pmatrix} 17.57 \\ 51.05 \\ 8.07 \\ 3.68 \end{pmatrix}$	$\begin{pmatrix} 12.6 \\ 11.9 \\ 80.3 \\ 187.3 \end{pmatrix}$	14.18441	$\begin{pmatrix} 0.66 \\ -2.54 \\ -3.16 \\ -1.74 \end{pmatrix}$	$\begin{pmatrix} 0.66 \\ -3.857 \\ -0.519 \\ 4.853 \end{pmatrix}$	−0.59796
8	$\begin{pmatrix} 17.65 \\ 48.23 \\ 5.76 \\ 2.51 \end{pmatrix}$	$\begin{pmatrix} 12.65 \\ 8.93 \\ 78.36 \\ 187.6 \end{pmatrix}$	13.58644	$\begin{pmatrix} 0.0831 \\ -2.8262 \\ -2.3069 \\ -1.1711 \end{pmatrix}$	$\begin{pmatrix} -0.0831 \\ -2.9925 \\ -1.9743 \\ -0.3396 \end{pmatrix}$	−0.51583
9	$\begin{pmatrix} 16.96 \\ 44.58 \\ 4.14 \\ 1.75 \end{pmatrix}$	$\begin{pmatrix} 11.96 \\ 6.67 \\ 73.96 \\ 179.3 \end{pmatrix}$	13.07061	$\begin{pmatrix} -0.694 \\ -3.647 \\ -1.618 \\ -0.76 \end{pmatrix}$	$\begin{pmatrix} -0.694 \\ -2.258 \\ -4.395 \\ -7.703 \end{pmatrix}$	−0.44552
10	$\begin{pmatrix} 15.49 \\ 40.04 \\ 3.06 \\ 1.27 \end{pmatrix}$	$\begin{pmatrix} 10.49 \\ 5.05 \\ 67.04 \\ 164.22 \end{pmatrix}$	12.62509	$\begin{pmatrix} -1.461 \\ -4.536 \\ -1.077 \\ -0.479 \end{pmatrix}$	$\begin{pmatrix} -1.461 \\ -1.61 \\ -6.92 \\ -15.09 \end{pmatrix}$	−0.40277
⋮	⋮	⋮	⋮	⋮	⋮	⋮

(continued)

Table 1 (continued)

Iteration (k)	u^k	v^k	$\psi(u^k, v^k)$	d_u^k	d_v^k	$diff[\psi(u^k, v^k)]$
20	$\begin{pmatrix} 5.8634 \\ 16.0745 \\ 0.2179 \\ 0.0842 \end{pmatrix}$	$\begin{pmatrix} 0.863 \\ 0.348 \\ 25.671 \\ 66.718 \end{pmatrix}$	8.96025	$\begin{pmatrix} -0.2643 \\ -0.6453 \\ -0.0721 \\ -0.0284 \end{pmatrix}$	$\begin{pmatrix} -0.2643 \\ -0.117 \\ -1.129 \\ -2.672 \end{pmatrix}$	−0.32641
21	$\begin{pmatrix} 5.6594 \\ 15.579 \\ 0.1635 \\ 0.0629 \end{pmatrix}$	$\begin{pmatrix} 0.659 \\ 0.26 \\ 24.801 \\ 64.657 \end{pmatrix}$	8.63385	$\begin{pmatrix} -0.2039 \\ -0.4955 \\ -0.0545 \\ -0.0213 \end{pmatrix}$	$\begin{pmatrix} -0.2039 \\ -0.0877 \\ -0.8701 \\ -2.0605 \end{pmatrix}$	−0.31937
⋮	⋮	⋮	⋮	⋮	⋮	⋮
45	$\begin{pmatrix} 5.000 \\ 14.000 \\ 0.0002 \\ 0.0001 \end{pmatrix}$	$\begin{pmatrix} 0.0007 \\ 0.0003 \\ 22.000 \\ 58.000 \end{pmatrix}$	1.58469	$\begin{pmatrix} -0.0002 \\ -0.0006 \\ -0.0001 \\ -0.0000 \end{pmatrix}$	$\begin{pmatrix} -0.0002 \\ -0.0001 \\ -0.0010 \\ -0.0024 \end{pmatrix}$	−0.28773
46	$\begin{pmatrix} 5.000 \\ 14.000 \\ 0.0001 \\ 0.0001 \end{pmatrix}$	$\begin{pmatrix} 0.0005 \\ 0.0002 \\ 22.000 \\ 58.000 \end{pmatrix}$	1.29696	$\begin{pmatrix} -0.0002 \\ -0.0004 \\ -0.0000 \\ -0.0000 \end{pmatrix}$	$\begin{pmatrix} -0.0002 \\ -0.0001 \\ -0.0007 \\ -0.0018 \end{pmatrix}$	−0.28771

Acknowledgements The second author R. Jana is thankful to the Department of Science and Technology, Govt. of India, INSPIRE Fellowship Scheme for financial support. The research work of the third author Deepmala is supported by the Science and Engineering Research Board (SERB), Government of India under SERB N-PDF scheme, File Number: PDF/2015/000799.

References

1. Crouzeix, J.P., Komlósi, S.: The linear complementarity problem and the class of generalized positive subdefinite matrices. In: Optimization Theory, pp. 45–63. Springer, US (2001)
2. Den Hertog, D.: Interior Point Approach to Linear, Quadratic and Convex Programming: Algorithms and Complexity, vol. 277. Springer Science & Business Media (2012)
3. Fathi, Y.: Computational complexity of LCPs associated with positive definite symmetric matrices. Math. Program. **17**(1), 335–344 (1979)
4. Khachiyan, L.G.: Polynomial algorithms in linear programming. USSR Comput. Math. Math. Phys. **20**(1), 53–72 (1980)
5. Kojima, M., Megiddo, N., Ye, Y.: An interior point potential reduction algorithm for the linear complementarity problem. Math. Program. **54**(1–3), 267–279 (1992)
6. Martos, B.: Subdefinite matrices and quadratic forms. SIAM J. Appl. Math. **17**(6), 1215–1223 (1969)
7. Mohan, S.R., Neogy, S.K., Das, A.K.: More on positive subdefinite matrices and the linear complementarity problem. Linear Algebra Appl. **338**(1), 275–285 (2001)
8. Monteiro, R.C., Adler, I.: An $O(n^3L)$ Primal-Dual Interior Point Algorithm for Linear Programming. Report ORC 87-4, Dept. of Industrial Engineering and Operations Research, University of California, Berkeley, CA (1987)
9. Neogy, S.K., Das, A.K.: Some properties of generalized positive subdefinite matrices. SIAM J. Matrix Anal. Appl. **27**(4), 988–995 (2006)
10. Pang, J.S.: Iterative descent algorithms for a row sufficient linear complementarity problem. SIAM J. Matrix Anal. Appl. **12**(4), 611–624 (1991)
11. Todd, M.J., Ye, Y.: A centered projective algorithm for linear programming. Math. Oper. Res. **15**(3), 508–529 (1990)
12. Wang, G.Q., Yu, C.J., Teo, K.L.: A full-Newton step feasible interior-point algorithm for $P_*(\kappa)$-linear complementarity problems. J. Glob. Optim. **59**(1), 81–99 (2014)
13. Ye, Y.: An $O(n^3L)$ potential reduction algorithm for linear programming. Math. Program. **50**(1–3), 239–258 (1991)
14. Ye, Y.: Interior Point Algorithms: Theory and Analysis, vol. 44. Wiley (2011)
15. Ye, Y., Pardalos, P.M.: A class of linear complementarity problems solvable in polynomial time. Linear Algebra Appl. **152**, 3–17 (1991)

Saddle Point Criteria for Semi-infinite Programming Problems via an η-Approximation Method

Yadvendra Singh and S. K. Mishra

Abstract In this paper, we consider a semi-infinite programming problem involving differentiable invex functions. We construct an η-approximated semi-infinite programming problem associated with the original semi-infinite programming problem and establish relationship between its saddle point and an optimal solution. We also establish relationship between an optimal solution of original semi-infinite programming problem and saddle point of η-approximated semi-infinite programming problem. Examples are given to illustrate the obtained results.

Keywords Semi-infinite programming · Generalized convexity · Optimality conditions

1 Introduction

In semi-infinite programming problems, the term semi-infinite means finitely many variables appear in infinitely many constraints. In recent years, semi-infinite programming problems have been an active field of research. Vaz et al. [1] have described how robot trajectory planning can be formulated as a semi-infinite programming problem. Tong et al. [2] have solved an optimal power flow problem with transient stability constraints by converting it to a semi-infinite programming problem. Vaz and Ferreira [3] have shown that air pollution control problems can be posed as semi-infinite programming problems. Winterfeld [4] has discussed semi-infinite programming problem in gemstone cutting industry.

Y. Singh
C.M.P. Degree College (A Constituent Postgraduate College of Central University of Allahabad), Allahabad 211002, India
e-mail: ysinghze@gmail.com

S. K. Mishra (✉)
Department of Mathematics, Institute of Science, Banaras Hindu University, Varanasi 221005, India
e-mail: bhu.skmishra@gmail.com

S. Kar et al. (eds.), *Operations Research and Optimization*, Springer Proceedings in Mathematics & Statistics 225, https://doi.org/10.1007/978-981-10-7814-9_2

In this paper, we consider the following semi-infinite programming problem:

$$\text{(SIP)} \qquad \min f(x)$$
$$\text{subject to} \qquad g_t(x) \leq 0, \quad t \in T,$$

where T is an infinite index set. Let $f : X \to \mathbb{R}$ and $g_t : X \to \mathbb{R},\ t \in T$ are differentiable functions on nonempty open set $X \subseteq \mathbb{R}^n$.

Optimality conditions for semi-infinite programming problems (SIP) have been widely studied. López et al. [5] have given optimality conditions for convex nondifferentiable semi-infinite programming problems which involve the notion of Lagrangian saddle point. Recently, Kanzi [6] established some constraint qualifications and necessary optimality conditions for semi-infinite programming problems. For more details, we refer the reader to the recent review papers by López and Still [7] and Shapiro [8] and reference therein.

Antczak [9] introduced η-approximation method for solving a nonlinear mathematical programming problems. Later, in [10], he extended η-approximation method and established equivalence between saddle point of associated η-approximated optimization problem and an optimal solution in original semi-infinite programming problem. Ratiu et al. [11] established connections between the feasible solutions of semi-infinite programming problems and its η-approximated semi-infinite programming problems and obtained some connections between the optimal solution of semi-infinite programming problems and related approximation problems.

Motivated by [6, 9–11], we construct an η-approximated semi-infinite programming problem associated with the original semi-infinite programming problem. The aim of our paper is to show how one can obtain optimality conditions of a semi-infinite programming problem by saddle point of a less complicated η-approximated semi-infinite programming problem.

The outline of this paper is as follows: In Sect. 2, we give some preliminary definitions and results which will be used in the sequel. In Sect. 3, we formulate an η-approximated semi-infinite programming problem associated with the original semi-infinite programming problem and establish relationship between its saddle point and an optimal solution. In Sect. 4, we establish relationship between an optimal solution of original semi-infinite programming problem and saddle point of η-approximated semi-infinite programming problem.

2 Basic Definition and Results

In this section, we recall some known definitions and results which will be used in the sequel. Given a nonempty set $D \subseteq \mathbb{R}^n$, we denote the closure of D by $\bar{D}$ and convex cone (containing origin) by cone(D). The polar cone is defined by

$$D^0 := \{d \in \mathbb{R}^n | \langle x, d\rangle \leq 0,\ \forall x \in D\},$$

where $\langle .,. \rangle$ exhibits the standard inner product in $\mathbb{R}^n$.
The contingent cone of attainable directions to D at $\bar{x}$ is defined by

$$A(D,\bar{x}) := \{d \in \mathbb{R}^n |\text{ for all } \{t_k\} \subset \mathbb{R}_+,\ t_k \to 0,\ \exists \{d_k\} \subset \mathbb{R}^n,\ d_k \to d : \\ \bar{x} + t_k d_k \in D\ \ \forall k \in \mathbb{N}\}.$$

Let $M := \{x \in X |\ g_t(x) \leq 0,\ t \in T\}$ denotes the set of all feasible solution of problem (SIP).

Definition 1 A feasible point $\bar{x} \in M$ is called an optimal solution of (SIP) if

$$f(\bar{x}) \leq f(x),\ \forall x \in M.$$

For $\bar{x} \in M$ we consider,

$$T(\bar{x}) := \{t : T |\ g_t(\bar{x}) = 0\} \text{ and } Z(\bar{x}) := \{\nabla g_t(\bar{x}) |\ t \in T\}.$$

Throughout the paper, we assume that the set $T(\bar{x})$, at $\bar{x}$, is nonempty and let

$$\mathbb{R}_+^T := \{\lambda \equiv (\lambda_t)_{t \in T} |\ \lambda_t \geq 0 \text{ and } \lambda_t \neq 0 \text{ for finitely many } t \in T\}.$$

The following Karush-Kuhn-Tucker type necessary optimality condition is a direct consequence of Theorem 4.3(b) of [6].

Theorem 1 *Suppose $\bar{x}$ is an optimal solution of (SIP) and assume the suitable constraint [6] qualification holds at $\bar{x}$ and cone($Z(\bar{x})$) is closed, then there exists $\bar{\lambda} \in \mathbb{R}_+^T$, such that*

$$\nabla f(\bar{x}) + \sum_{t \in T(\bar{x})} \bar{\lambda}_t \nabla g_t(\bar{x}) = 0. \tag{1}$$

Definition 2 [12] Let $f : X \to \mathbb{R}$ be a differentiable function on a nonempty set $X \subseteq \mathbb{R}^n$, then f is said to be invex at $\bar{x}$ on X with respect to η if the following inequality holds:

$$f(x) - f(\bar{x}) \geq \nabla f(\bar{x}) \eta(x, \bar{x}),\ \forall x \in X. \tag{2}$$

The following Lagrangian function is associated with the problem (*SIP*):

$$L(x, \lambda) := f(x) + \sum_{t \in T} \lambda_t\ g_t(x),\ \forall\ (x, \lambda) \in X \times \mathbb{R}_+^T.$$

Now, we give the definition of saddle point of the Lagrangian for (SIP).

Definition 3 A point $(\bar{x}, \bar{\lambda}) \in M \times \mathbb{R}_+^T$ is said to be a saddle point in (SIP) if

$$L(\bar{x}, \lambda) \leq L(\bar{x}, \bar{\lambda}) \leq L(x, \bar{\lambda}), \forall \lambda \in \mathbb{R}_+^T, \forall x \in M.$$

3 η-Saddle Point Optimality Criteria in the η-Approximated Semi-infinite Optimization Problem

Let $\bar{x}$ be a feasible point in (SIP). We consider the η-approximated semi-infinite optimization problem given by

$$(SIP_\eta(\bar{x})) \qquad \min\,(f(\bar{x}) + \nabla f(\bar{x})\eta(x,\bar{x})),$$

$$\text{subject to} \qquad g_t(\bar{x}) + \nabla g_t(\bar{x})\eta(x,\bar{x}) \leq 0, \quad t \in T,$$

where $f,\ g,\ X$ are define as in problem (SIP) and $\eta(.,\bar{x})$ is differentiable at the point $x = \bar{x}$ with respect to the first component.
Let $M(\bar{x}) := \{x \in X : g_t(\bar{x}) + \nabla g_t(\bar{x})\eta(x,\bar{x}) \leq 0\}$ denotes the set of all feasible solution in $(SIP_\eta)(\bar{x})$ and let

$$Z'(\bar{x}) := \{\nabla g_t(\bar{x})\eta_x(\bar{x},\bar{x}) |\ t \in T\}.$$

The following η-Lagrangian function is associated with the problem $(SIP_\eta(\bar{x}))$:

$$L_\eta(x,\lambda) := f(\bar{x}) + \sum_{t\in T} \lambda_t g_t(\bar{x}) + \nabla f(\bar{x})\eta(x,\bar{x}) + \sum_{t\in T} \lambda_t \nabla g_t(\bar{x})\eta(x,\bar{x}),$$
$$\forall\ (x,\lambda) \in X \times \mathbb{R}^T_+.$$

Now, we give the definition of η-saddle point of the η-Lagrangian for $(SIP_\eta(\bar{x}))$.

Definition 4 A point $(\bar{x},\bar{\lambda}) \in M(\bar{x}) \times \mathbb{R}^T_+$ is said to be an η-saddle point in the η-approximate semi-infinite programming problem $(SIP_\eta(\bar{x}))$ if

$$L_\eta(\bar{x},\lambda) \leq L_\eta(\bar{x},\bar{\lambda}) \leq L_\eta(x,\bar{\lambda}), \qquad \forall \lambda \in \mathbb{R}^T_+, \forall x \in M(\bar{x}). \tag{3}$$

The following constraint qualifications are generalization of constraint qualifications from [6] for η-approximated semi-infinite programming problem $(SIP_\eta(\bar{x}))$.

Definition 5 We say that the constraint qualification holds at $\bar{x} \in M(\bar{x})$ for η-approximated semi-infinite programming problem $(SIP_\eta(\bar{x}))$ if

$$Z'^{0}(\bar{x}) \subseteq A(M(\bar{x}),\bar{x}).$$

Now, we give the Karush-kuhn-Tucker necessary optimality condition for η-approximate semi-infinite programming problem $(SIP_\eta(\bar{x}))$.

Theorem 2 *Suppose $\bar{x}$ is optimal solution of $(SIP_\eta(\bar{x}))$ and assume that constraint qualification from Definition 5 holds at $\bar{x}$ and cone$(Z'(\bar{x}))$ is closed, then there exists $\bar{\lambda} \in \mathbb{R}^T_+$, such that*

$$\left[\nabla f(\bar{x}) + \sum_{t\in T(\bar{x})} \bar{\lambda}_t \nabla g_t(\bar{x})\right] \eta_x(x,\bar{x}) = 0. \tag{4}$$

Proof It follows from Theorem 1.

The following theorems give the equivalence between an η-saddle point of the η-Lagrangian and an optimal solution of problem $(SIP_\eta(\bar{x}))$.

Theorem 3 *If $(\bar{x}, \bar{\lambda}) \in M(\bar{x}) \times \mathbb{R}_+^T$ is an η-saddle point of $(SIP_\eta(\bar{x}))$, where $\eta(.,.) : X \times X \to X$ satisfies the condition $\eta(x,x) = 0,\ \forall x \in X$, then $\bar{x}$ is optimal in problem $(SIP_\eta(\bar{x}))$.*

Proof We proceed by contradiction. Suppose that $\bar{x}$ is not optimal in $(SIP_\eta(\bar{x}))$. Then, there exists $x^0 \in M(\bar{x})$ such that

$$f(\bar{x}) + \nabla f(\bar{x})\eta(x^0,\bar{x}) < f(\bar{x}) + \nabla f(\bar{x})\eta(\bar{x},\bar{x}), \tag{5}$$

$$g_t(\bar{x}) + \nabla g_t(\bar{x})\eta(x^0,\bar{x}) \leq 0,\ \forall\, t \in T. \tag{6}$$

Since $(\bar{x}, \bar{\lambda})$ is an η-saddle point of $(SIP_\eta(\bar{x}))$ and $\bar{\lambda} = (\bar{\lambda}_t)_{t\in T} \in \mathbb{R}_+^T$, then by (6), we get

$$\sum_{t\in T} \bar{\lambda}_t g_t(\bar{x}) + \sum_{t\in T} \bar{\lambda}_t \nabla g_t(\bar{x})\eta(x^0,\bar{x}) \leq 0. \tag{7}$$

By the Definition 4 and assumption $\eta(x,x) = 0$, the following inequality holds

$$f(\bar{x}) + \sum_{t\in T} \lambda_t g(\bar{x}) \leq f(\bar{x}) + \sum_{t\in T} \bar{\lambda}_t g(\bar{x}),\ \forall\, \lambda \in \mathbb{R}_+^T.$$

Let $\lambda_t = 0,\ \forall t \in T$, then we get

$$\sum_{t\in T} \bar{\lambda}_t g_t(\bar{x}) \geq 0. \tag{8}$$

Since $\bar{x} \in M$ and $\bar{\lambda} \in \mathbb{R}_+^T$, then we have

$$\sum_{t\in T} \bar{\lambda}_t g_t(\bar{x}) \leq 0. \tag{9}$$

Thus, from (8) and (9), we get

$$\sum_{t\in T} \bar{\lambda}_t g_t(\bar{x}) = 0. \tag{10}$$

From (5), (7), and (10), we get

$$f(\bar{x}) + \sum_{t\in T} \bar{\lambda}_t g_t(\bar{x}) + \nabla f(\bar{x})\eta(x^0, \bar{x}) + \sum_{t\in T} \bar{\lambda}_t \nabla_x g_t(\bar{x})\eta(x^0, \bar{x})$$
$$< f(\bar{x}) + \sum_{t\in T} \bar{\lambda}_t g_t(\bar{x}) + \nabla f(\bar{x})\eta(\bar{x}, \bar{x}) + \sum_{t\in T} \bar{\lambda}_t \nabla_x g_t(\bar{x})\eta(\bar{x}, \bar{x}).$$

Hence, we obtain the following inequality:

$$L_\eta(x^0, \bar{\lambda}) < L_\eta(\bar{x}, \bar{\lambda}),$$

which contradicts (3). This completes the proof.

Theorem 4 *Let $f(\bar{x}) + \nabla f(\bar{x})\eta(., \bar{x})$ and $g_t(\bar{x}) + \nabla g_t(\bar{x})\eta(., \bar{x})$, $\forall t \in T$ are invex at $\bar{x}$ on $M(\bar{x})$ with respect to the same function, but not necessary with respect to the function η and moreover we assume that constraints qualification from Definition 5 holds at $\bar{x}$. If $\bar{x}$ is an optimal solution in $(SIP_\eta(\bar{x}))$, then there exists $\bar{\lambda} \in \mathbb{R}^T_+$ such that $(\bar{x}, \bar{\lambda})$ is an η-saddle point in $(SIP_\eta(\bar{x}))$.*

Proof Since $f(\bar{x}) + \nabla f(\bar{x})\eta(., \bar{x})$ and $g_t(\bar{x}) + \nabla g_t(\bar{x})\eta(., \bar{x})\, \forall t \in T$ are invex at $\bar{x}$ on $M(\bar{x})$, then there exists a function $\varphi : M(\bar{x}) \times M(\bar{x}) \to M(\bar{x})$ such that, for all $x \in M(\bar{x})$, the following inequality holds

$$f(\bar{x}) + \nabla f(\bar{x})\eta(x, \bar{x}) \geq f(\bar{x}) + \nabla f(\bar{x})\eta(\bar{x}, \bar{x}) + \nabla f(\bar{x})\eta_x(\bar{x}, \bar{x})\varphi(x, \bar{x}), \tag{11}$$

$$g_t(\bar{x}) + \nabla g_t(\bar{x})\eta(x, \bar{x}) \geq g_t(\bar{x}) + \nabla g_t(\bar{x})\eta(\bar{x}, \bar{x}) + \nabla g_t(\bar{x})\eta_x(\bar{x}, \bar{x})\varphi(x, \bar{x}), \quad \forall t \in T. \tag{12}$$

Let $\bar{\lambda} \in \mathbb{R}^T_+$, therefore from (12), we get

$$\sum_{t\in T} \bar{\lambda}_t g_t(\bar{x}) + \sum_{t\in T} \bar{\lambda}_t \nabla g_t(\bar{x})\eta(x, \bar{x})$$
$$\geq \sum_{t\in T} \bar{\lambda}_t g_t(\bar{x}) + \sum_{t\in T} \bar{\lambda}_t \nabla g_t(\bar{x})\eta(\bar{x}, \bar{x}) + \sum_{t\in T} \bar{\lambda}_t \nabla g_t(\bar{x})\eta_x(\bar{x}, \bar{x})\varphi(x, \bar{x}). \tag{13}$$

Adding both sides of inequality (11) and (13) and using the necessary optimality condition for $(SIP_\eta(\bar{x}))$ and define $\bar{\lambda}_t = 0,\ t \notin T(\bar{x})$, we get

$$f(\bar{x}) + \sum_{t\in T} \bar{\lambda}_t g_t(\bar{x}) + \nabla f(\bar{x})\eta(x, \bar{x}) + \sum_{t\in T} \bar{\lambda}_t \nabla g(\bar{x})\eta(x, \bar{x})$$
$$\geq f(\bar{x}) + \sum_{t\in T} \bar{\lambda}_t g_t(\bar{x}) + \nabla f(\bar{x})\eta(\bar{x}, \bar{x}) + \sum_{t\in T} \bar{\lambda}_t \nabla g_t(\bar{x})\eta(\bar{x}, \bar{x}).$$

Thus,

$$L_\eta(x, \bar{\lambda}) \geq L_\eta(\bar{x}, \bar{\lambda}), \quad \forall x \in M(\bar{x}).$$

Since $\bar{x} \in M(\bar{x})$, then

$$g_t(\bar{x}) + \nabla g_t(\bar{x})\eta(\bar{x}, \bar{x}) \leq 0,\ \forall t \in T,$$

Therefore for all $\lambda \in \mathbb{R}^T_+$, we get

$$\sum_{t\in T} \lambda_t g_t(\bar{x}) + \sum_{t\in T} \lambda_t \nabla g_t(\bar{x})\eta(\bar{x}, \bar{x}) \leq 0,$$

Since $\bar{\lambda}_t = 0, t \notin T(\bar{x})$, thus we get

$$\begin{aligned} f(\bar{x}) + \nabla f(\bar{x})\eta(\bar{x}, \bar{x}) + \sum_{t\in T} \lambda_t g_t(\bar{x}) + \sum_{t\in T} \lambda_t \nabla g_t(\bar{x})\eta(\bar{x}, \bar{x}) \\ \geq f(\bar{x}) + \sum_{t\in T} \bar{\lambda}_t g_t(\bar{x}) + \nabla f(\bar{x})\eta(\bar{x}, \bar{x}) + \sum_{t\in T} \bar{\lambda}_t \nabla g_t(\bar{x})\eta(\bar{x}, \bar{x}), \end{aligned}$$

$$\begin{aligned} f(\bar{x}) + \nabla f(\bar{x})\eta(\bar{x}, \bar{x}) + \sum_{t\in T} \lambda_t g_t(\bar{x}) + \sum_{t\in T} \lambda_t \nabla g_t(\bar{x})\eta(\bar{x}, \bar{x}) \\ \geq f(\bar{x}) + \sum_{t\in T} \bar{\lambda}_t g_t(\bar{x}) + \nabla f(\bar{x})\eta(\bar{x}, \bar{x}) + \sum_{t\in T} \bar{\lambda}_t \nabla g_t(\bar{x})\eta(\bar{x}, \bar{x}). \end{aligned}$$

Hence, we obtain the following inequality:

$$L_\eta(\bar{x}, \lambda) \leq L_\eta(\bar{x}, \bar{\lambda}), \quad \forall \lambda \in \mathbb{R}^T_+.$$

4 η-Saddle Point Optimality Criteria in the Semi-infinite Optimization Problem

In this section, we show the equivalence between the semi-infinite programming problem (SIP) and its η-approximated semi-infinite problem. We shall use the following lemma to prove our main results.

Lemma 1 *[11] Let $\bar{x} \in M$ and assume that $g_t, \forall t \in T$ is differentiable invex function at $\bar{x}$ on $M$with respect to same η, then*

$$M \subseteq M(\bar{x}).$$

The following example shows that the invexity imposed in the above lemma is essential.

Example 1 Let $g_t : \mathbb{R} \to \mathbb{R}$, the function defined by

$$g_t(x) = x^3 - t, \quad t \in \mathbb{N}.$$

We observe that feasible set $M = \{x \in \mathbb{R} : g_t(x) \leq 0\} = (-\infty, 1]$.

It is easy to verify that for $\bar{x} = -1$ and the function $\eta : \mathbb{R} \times \mathbb{R} \to \mathbb{R}$, defined by $\eta(x, y) = x - y$, $g_t, t \in \mathbb{N}$ are not invex.
For $\bar{x} = -1$, we have $M(\bar{x}) = \{x \in \mathbb{R} : g_t(\bar{x}) + \nabla g_t(\bar{x})\eta(x, \bar{x}) \leq 0\} = (-\infty, -\frac{1}{3}]$.
Hence,

$$M \nsubseteq M(\bar{x}).$$

The following theorems give the equivalence between an η-saddle point of the η-Lagrangian and an optimal solution of the problem (SIP).

Theorem 5 *Let $\bar{x}$ be a feasible solution in semi-infinite programming problem (SIP). We assume that $f(\cdot)$ and $g_t(\cdot), \forall t \in T$ are invex at $\bar{x}$ on M with respect to η satisfying the condition $\eta(\bar{x}, \bar{x}) = 0$. If $(\bar{x}, \bar{\lambda}) \in M(\bar{x}) \times \mathbb{R}_+^T$ is an η-saddle point in the $(SIP_\eta(\bar{x}))$, then $\bar{x}$ is optimal in the original mathematical programming problem (SIP).*

Proof Since $(\bar{x}, \bar{\lambda}) \in M(\bar{x}) \times \mathbb{R}_+^T$ is an η-saddle point in the η-approximate semi-infinite programming problem $(SIP_\eta(\bar{x}))$, then by Definition 4, we have

$$L_\eta(\bar{x}, \lambda) \leq L_\eta(\bar{x}, \bar{\lambda}), \ \forall \lambda \in \mathbb{R}_+^T.$$

Using the definition of an η- Lagrange function in the problem $(SIP_\eta(\bar{x}))$, we have

$$\begin{aligned} &f(\bar{x}) + \sum_{t\in T} \lambda_t g_t(\bar{x}) + \nabla f(\bar{x})\eta(\bar{x}, \bar{x}) + \sum_{t\in T} \lambda_t \nabla g_t(\bar{x})\eta(\bar{x}, \bar{x}) \\ \geq\ &f(\bar{x}) + \sum_{t\in T} \bar{\lambda}_t g_t(\bar{x}) + \nabla f(\bar{x})\eta(\bar{x}, \bar{x}) + \sum_{t\in T} \bar{\lambda}_t \nabla g_t(\bar{x})\eta(\bar{x}, \bar{x}), \forall \lambda \in \mathbb{R}_+^T. \end{aligned}$$

Let $\lambda_t = 0, \ \forall t \in T$ and $\eta(\bar{x}, \bar{x}) = 0$, we get

$$\sum_{t\in T} \bar{\lambda}_t g_t(\bar{x}) \geq 0. \tag{14}$$

Since $\bar{x} \in M$ and $\bar{\lambda} \in \mathbb{R}_+^T$, then we have

$$\sum_{t\in T} \bar{\lambda}_t g_t(\bar{x}) \leq 0, \tag{15}$$

from (14) and (15), we get

$$\sum_{t\in T} \bar{\lambda}_t g_t(\bar{x}) = 0. \tag{16}$$

By the Definition 4 of η-saddle point of η-lagrangian, we have

$$L_\eta(\bar{x}, \lambda) \leq L_\eta(x, \lambda), \forall x \in M(x).$$

Using the definition of an η- Lagrange function in the problem $(SIP_\eta(\bar{x}))$, we have

$$f(\bar{x}) + \sum_{t \in T} \lambda_t g_t(\bar{x}) + \nabla f(\bar{x})\eta(\bar{x}, \bar{x}) + \sum_{t \in T} \lambda_t \nabla g_t(\bar{x})\eta(\bar{x}, \bar{x})$$
$$\leq f(\bar{x}) + \sum_{t \in T} \bar{\lambda}_t g_t(\bar{x}) + \nabla f(\bar{x})\eta(x, \bar{x}) + \sum_{t \in T} \bar{\lambda}_t \nabla g_t(\bar{x})\eta(x, \bar{x}), \forall x \in M(x),$$

and so,

$$\nabla f(\bar{x})\eta(x, \bar{x}) + \sum_{t \in T} \bar{\lambda}_t \nabla g_t(\bar{x})\eta(x, \bar{x}) \geq 0, \forall x \in M(\bar{x}). \tag{17}$$

Since $f,\ g_t(\cdot), \forall t \in T$ are invex with respect to η, then we have

$$f(x) - f(\bar{x}) \geq \nabla f(\bar{x})\eta(x, \bar{x}),\ \forall x \in M, \tag{18}$$

$$g_t(x) - g_t(\bar{x}) \geq \nabla g_t(\bar{x})\eta(x, \bar{x}),\ \forall x \in M, t \in T. \tag{19}$$

Hence, by $\bar{\lambda} \in \mathbb{R}^T_+$, we get

$$\sum_{t \in T} \bar{\lambda}_t g_t(x) - \sum_{t \in T} \bar{\lambda}_t g_t(\bar{x}) \geq \sum_{t \in T} \bar{\lambda}_t \nabla g_t(\bar{x})\eta(x, \bar{x}),\ \forall x \in M. \tag{20}$$

Therefore by (16), we get

$$\sum_{t \in T} \bar{\lambda}_t g_t(x) \geq \sum_{t \in T} \bar{\lambda}_t \nabla_x g_t(\bar{x})\eta(x, \bar{x}),\ \forall x \in M. \tag{21}$$

Since $x \in M$, then by feasibility condition of (SIP), we have

$$\sum_{t \in T} \bar{\lambda}_t g_t(x) \leq 0. \tag{22}$$

Therefore, by (22) and (21), we get

$$\sum_{t \in T} \lambda_t \nabla g_t(x)\eta(x, \bar{x}) \leq 0, \forall x \in M. \tag{23}$$

Since by Lemma 1, it follows that $M \subseteq M(\bar{x})$, then by (23) and (17), we get

$$\nabla f(x)\eta(x, \bar{x}) \geq 0,\ \forall x \in M. \tag{24}$$

Hence by (18) and (24)

$$f(x) \geq f(\bar{x}),\ \forall x \in M.$$

Therefor, $\bar{x}$ is optimality in (SIP).

Theorem 6 *Let $\bar{x}$ be an optimal solution for semi-infinite programming problem (SIP) and assume that a suitable constraint qualification [6] holds at $\bar{x}$ and cone($Z(\bar{x})$) is closed. Further, assume that $f(\cdot)$ and $g_t(\cdot), \forall t \in T$ are invex with respect to the* *same η at $\bar{x}$ on M. Then, there exists $\bar{\lambda} \in \mathbb{R}^T_+$ such that $(\bar{x}, \bar{\lambda})$ is an η-saddle point in its η-approximated semi-infinite programming problem $(SIP_\eta(\bar{x}))$.*

Proof Since $\bar{x}$ is optimal solution for SIP and a suitable constraint qualification holds and $cone(Z(\bar{x}))$ is closed, then by Theorem 1, the following inequality is obviously fulfilled

$$L_\eta(\bar{x}, \bar{\lambda}) \leq L_\eta(x, \bar{\lambda}), \ \forall x \in M(\bar{x}).$$

By optimality condition, we have

$$\sum_{t \in T} \bar{\lambda}_t g_t(\bar{x}) + \left[\nabla f(\bar{x}) + \sum_{t \in T} \bar{\lambda}_t \nabla g_t(\bar{x}) \right] \eta(\bar{x}, \bar{x}) = 0. \tag{25}$$

Since $g_t(\cdot), t \in T$ are invex at $\bar{x}$ on M with respect to η then by Lemma 1 $M \subseteq M(\bar{x})$, then the inequality (23) holds at $\bar{x} \in M$. Hence, we get

$$\sum_{t \in T} \lambda_t \nabla g_t(\bar{x}) \eta(\bar{x}, \bar{x}) \leq 0. \tag{26}$$

Since f is invex at $\bar{x}$ on M, then

$$f(x) - f(\bar{x}) \geq \nabla f(\bar{x}) \eta(x, \bar{x}), \ \forall x \in M. \tag{27}$$

Therefore by putting $x = \bar{x}$, we get

$$\nabla f(\bar{x}) \eta(\bar{x}, \bar{x}) \leq 0, \ \forall x \in M. \tag{28}$$

For $\bar{x} \in M$ and $\lambda \in \mathbb{R}^T_+$, we have

$$\sum_{t \in T} \lambda_t g_t(\bar{x}) \leq 0. \tag{29}$$

Hence, from (26), (28), and (29), we get

$$\sum_{t \in T} \lambda_t g_t(\bar{x}) + \sum_{t \in T} \lambda_t \nabla_x g_t(\bar{x}) \eta(\bar{x}, \bar{x}) + \nabla f(\bar{x}) \eta(\bar{x}, \bar{x}) \leq 0. \tag{30}$$

Hence, by (25) and (30), we get

$$\begin{aligned} & f(\bar{x}) + \sum_{t \in T} \lambda_t g_t(\bar{x}) + \nabla f(\bar{x}) \eta(\bar{x}, \bar{x}) + \sum_{t \in T} \lambda_t \nabla g_t(\bar{x}) \eta(\bar{x}, \bar{x}) \\ \leq & f(\bar{x}) + \sum_{t \in T} \bar{\lambda}_t g_t(\bar{x}) + \nabla f(\bar{x}) \eta(\bar{x}, \bar{x}) + \sum_{t \in T} \bar{\lambda}_t \nabla g_t(\bar{x}) \eta(\bar{x}, \bar{x}), \end{aligned}$$

holds for all $\lambda \in \mathbb{R}^T_+$. Then, we have

$$L_\eta(\bar{x}, \lambda) \leq L_\eta(\bar{x}, \bar{\lambda}), \ \forall \lambda \in \mathbb{R}^T_+.$$

Example 2 We consider the following mathematical programming problem:

$$(SIP) \qquad \min f(x) = (x-1)^2$$

$$\text{subject to} \qquad g_t(x) = tx \leq 0, \ \forall t \in \mathbb{N}.$$

The feasible solution for problem (SIP) is $M := \{x \in \mathbb{R} : -\infty < x \leq 0\}$. Let $\bar{x} = 0$ be any feasible point. For $\eta(x, \bar{x}) = x - \bar{x}$, it is easy that to prove that f and $g_t, \forall t \in T$ are invex with respect η at $\bar{x}$ on M.
Thus, we obtain the following linear optimization problem:

$$(SIP_\eta(0)) \qquad \min \ 1 - 2x$$

$$\text{subject to} \qquad tx \leq 0, \ \ t \in N.$$

The η-approximated Lagrangian L_η of problem $(SIP_\eta(0))$ is

$$L_\eta(x, \lambda) = 1 - 2x + \sum_{t \in N} \lambda_t tx = 1 - 2x + (\lambda_1 + 2\lambda_2 + \cdots)x.$$

It is easy to show by Definition that $(\bar{x}, \bar{\lambda})$, where $\bar{x} = 0$ and $\bar{\lambda} = (1, 0, \cdots)$ is an saddle point of η-approximated Lagrangian L_η in the problem $(SIP_\eta(0))$. Then, by Theorem 5, we conclude that $\bar{x}$ is optimal in considered semi-infinite programming problem (SIP).

Remark 1 Not that there exists more than one η satisfying all conditions of the Theorem 5. In another words, there exists more than one η-approximated semi-infinite programming problem $(SIP_\eta(\bar{x}))$ associated with original semi-infinite programming problem (SIP).

Let $\eta(x, \bar{x}) = \exp(x) - \exp(\bar{x})$ in the above example. It is easy to verify that functions f and $g_t, t \in \mathbb{N}$ are invex with respect to that η.
Then, we get the following associated nonlinear η-approximated semi-infinite programming problem

$$(SIP_\eta(0)) \qquad \min \ 3 - 2\exp(x)$$

$$\text{subject to} \qquad t(\exp(x) - 1) \leq 0, \ \ t \in N.$$

It is easy to show that $(\bar{x}, \bar{\lambda})$, where $\bar{x} = 0$ and $\bar{\lambda} = (1, 0, \cdots)$ is an saddle point of η-approximated Lagrangian L_η in the problem $(SIP_\eta(0))$. Then, by Theorem 5, we conclude that $\bar{x}$ is optimal in considered semi-infinite programming problem (SIP).

5 Conclusions

In this paper, we have constructed an η-approximated semi-infinite programming problem to solve original problem using saddle point criteria. We have established relationship among an optimal solution of semi-infinite programming problem and saddle point of associated η-approximated semi-infinite programming problem. In future, η-approximated method given in this paper can be extended to nonsmooth case.

Acknowledgements The first author is supported by the Council of Scientific and Industrial Research(CSIR), New Delhi, India, through grant no. 09/013(0474)/2012-EMR-1.

References

1. Vaz, A.I.F., Fernandes, E.M.G.P., Gomes, M.P.S.F.: Robot trajectory planning with semi-infinite programming. Eur. J. Oper. Res. **153**, 607–617 (2004)
2. Tong, X., Ling, C., Qi, L.: A semi-infinite programming algorithm for solving optimal power flow with transient stability constraints. J. Comput. Appl. Math. **217**(2), 432–447 (2008)
3. Vaz, A.I.F., Ferreira, E.C.: Air pollution control with semi-infinite programming. Appl. Math. Model. **33**(4), 1957–1969 (2009)
4. Winterfeld, A.: Application of general semi-infinite programming to lapidary cutting problems. Eur. J. Oper. Res. **191**(3), 838–854 (2008)
5. López, M.A., Vercher, E.: Optimality conditions for nondifferentiable convex semi-infinite programming. Math. Program. **27**, 307–319 (1983)
6. Kanzi, N.: Necessary optimality conditions for nonsmooth semi-infinite programming problems. J. Glob. Optim. **49**, 713–725 (2011)
7. López, M.A., Still, G.: Semi-infinite programming. Eur. J. Oper. Res. **180**, 491–518 (2007)
8. Shapiro, A.: Semi-infinite programming, duality, discretization and optimality conditions. Optimization **58**, 133–161 (2009)
9. Antczak, T.: An η-approximation approach to nonlinear mathematical programming involving invex functions. Numer. Funct. Anal. Optim. **25**, 423–438 (2004)
10. Antczak, T.: Saddle point criteria in an η-approximated method for nonlinear mathematical programming problem involving invex functions. J. Optim. Theory Appl. **132**(1), 71–87 (2007)
11. Ratiu, A., Duca, I.D.: Semi-infinite optimization problems and their approximations. Stud. Univ. Babes-Bolyai Math. **58**(3), 401–411 (2013)
12. Mishra, S.K., Giorgi, G.: Invexity and Optimization. Springer, Berlin, Heidelberg (2008)

A Solution Approach to Multi-level Nonlinear Fractional Programming Problem

Suvasis Nayak and A. K. Ojha

Abstract This paper studies multi-level nonlinear fractional programming problem (ML-NLFPP) of maximization type and proposes a solution approach which is based on the concept of fuzzy and simultaneous minimization, maximization of the objectives from their ideal, anti-ideal values, respectively. Nonlinear polynomial functions are considered as the numerators and denominators of the fractional objectives at each level. In the objective space, distance function or Euclidean metric is implemented to measure the distances between numerators, denominators and their ideal, anti-ideal values which need to be minimized and maximized. Goals for the controlled decision variables of upper levels are ascertained from the individual best optimal solutions of the corresponding levels, and tolerances are defined by decision makers to avoid the situation of decision deadlock. Fuzzy goal programming with reduction of only under-deviation from the highest membership value derives the best compromise solution of the concerned multi-level problem. An illustrative numerical example is discussed to demonstrate the solution approach and its effectiveness.

Keywords Multi-level programming · Fractional programming · Distance function · Fuzzy goal programming · Best compromise solution

1 Introduction

Multi-level programming problems(MLPP) arise in hierarchical organizations comprising multiple interactive decision-making units or decision makers (DM) to solve decentralized planning problems where DM at each level controls a set of decision variables independently. Some common characteristics [1] of MLPP are: DMs are

S. Nayak (✉) · A. K. Ojha
School of Basic Sciences, Indian Institute of Technology Bhubaneswar, Bhubaneswar, India
e-mail: sn14@iitbbs.ac.in

A. K. Ojha
e-mail: akojha@iitbbs.ac.in

S. Kar et al. (eds.), *Operations Research and Optimization*, Springer Proceedings in Mathematics & Statistics 225, https://doi.org/10.1007/978-981-10-7814-9_3

interactive in nature within the hierarchical structure; decisions are sequentially processed from upper to lower level; each level DM tries to maximize its own benefit but their decisions are affected by the actions and reactions of each other and sometimes causes a situation of decision deadlock. Thus, DMs must possess a compromising and cooperative motivation for determination of a solution for the overall benefit of the system at which each level DM gains a minimum standard of satisfaction. MLPP are extremely useful to decentralized systems [1] such as agriculture, transportation, network design, government policy, finance, economic system. In many real-world decision-making situations, objectives in form of fraction $(\frac{f^N(x)}{f^D(x)})$ of physical and/or economical quantities are encountered to be optimized [2], e.g., profit/cost, output/employ, debit/equity, risk assets/capital. Such mathematically modelled optimization problems belong to the class of fractional programming in which objectives are defined as ratio of linear or nonlinear functions. Multi-level fractional programming can be encountered [3, 4] in the fields of control theory, resource allocation, complex network design, pattern recognition and so forth.

Bi-level and tri-level problems belong to the class of multi-level programming, and bi-level programming problems are most considerably studied in the literature. Shih et al. [1] extended the concept of satisfactory solution of Lai [5] to solve MLPP using fuzzy membership functions and Zimmermanns [6] max-min operator technique. White [7] developed a penalty function approach to solve a tri-level programming problem. Osman et al. [8] implemented tolerance membership functions to solve a tri-level nonlinear multi-objective problem, whereas Zhang et al. [9] proposed Kth best algorithm to solve linear tri-level problem.

Pramanik and Roy [10] developed fuzzy goal programming (FGP) to solve MLPP by extending goal programming (GP) approach of Mohamed [11]. Baky [12] used FGP to solve a multi-level multi-objective programming problem (MLMOPP) and proposed technique for order preference by similarity to ideal solution (TOPSIS) [13] algorithm to solve multi-level nonlinear multi-objective decision-making problems. Abo-Sinna and Baky [14] proposed interactive balance space approach to solve MLMOPP. Sinha [15] used fuzzy mathematical programming to solve MLPP. Lachhwani [16] proposed FGP approach to solve MLMOPP with linear objectives and linear fractional objectives [4]. In literature, almost all works on MLPP deal with linear or non-linear objectives but the proposed method solves MLPP with the objectives are defined as fraction of non-linear polynomial functions which is not supposed to be studied earlier.

The paper is organized as follows: Following introduction, Sect. 2 interprets the mathematical formulation of ML-NLFPP and Sect. 3 incorporates some basic ideas about distance functions. The details of the proposed solution technique and a stepwise algorithm are explained in Sect. 4. A numerical example is worked out in Sect. 5 to demonstrate the proposed solution approach. Finally, some conclusions are incorporated in Sect. 6.

2 Mathematical Formulation of ML-NLFPP

Consider a q-level mathematical programming which comprises its objectives as nonlinear fractional functions at each level. DM$_i$ denotes the decision maker at ith level and controls a set of decision variables X_i independently. The ML-NLFPP of maximization type can be formulated as follows:

$$\text{Level-1: } \max_{X_1} f_1(x) = \frac{f_1^N(x)}{f_1^D(x)} = \frac{\sum_{j=1}^{r_1} c_{1j} x_1^{\alpha_{j1}^{(1)}} x_2^{\alpha_{j2}^{(1)}} \ldots x_n^{\alpha_{jn}^{(1)}}}{\sum_{j=1}^{r_1'} d_{1j} x_1^{\beta_{j1}^{(1)}} x_2^{\beta_{j2}^{(1)}} \ldots x_n^{\beta_{jn}^{(1)}}}$$

where $X_2, X_3, \ldots, X_q$ solve,

$$\text{Level-2: } \max_{X_2} f_2(x) = \frac{f_2^N(x)}{f_2^D(x)} = \frac{\sum_{j=1}^{r_2} c_{2j} x_1^{\alpha_{j1}^{(2)}} x_2^{\alpha_{j2}^{(2)}} \ldots x_n^{\alpha_{jn}^{(2)}}}{\sum_{j=1}^{r_2'} d_{2j} x_1^{\beta_{j1}^{(2)}} x_2^{\beta_{j2}^{(2)}} \ldots x_n^{\beta_{jn}^{(2)}}}$$

⋮

where X_q solves

$$\text{Level-}q\text{: } \max_{X_q} f_q(x) = \frac{f_q^N(x)}{f_q^D(x)} = \frac{\sum_{j=1}^{r_q} c_{qj} x_1^{\alpha_{j1}^{(q)}} x_2^{\alpha_{j2}^{(q)}} \ldots x_n^{\alpha_{jn}^{(q)}}}{\sum_{j=1}^{r_q'} d_{qj} x_1^{\beta_{j1}^{(q)}} x_2^{\beta_{j2}^{(q)}} \ldots x_n^{\beta_{jn}^{(q)}}}$$

subject to

$x = (X_1, X_2, \ldots, X_q) \in \Omega = \{g_i(x)(\leq, =, \geq)0, x > 0, i = 1, 2, \ldots, r\} \neq \phi$, or
$x = (X_1, X_2, \ldots, X_q) \in \Omega = \{A_1X_1 + A_2X_2 + \cdots + A_qX_q(\leq, =, \geq)b, x > 0\}$
$\neq \phi$
where,
$X_1 = (x_{11}, x_{12}, \ldots, x_{1n_1}) \in R^{n_1}$, $X_2 = (x_{21}, x_{22}, \ldots, x_{2n_2}) \in R^{n_2}$, $\ldots$, $X_q = (x_{q1}, x_{q2}, \ldots, x_{qn_q}) \in R^{n_q}$, $x = (X_1, X_2, \ldots, X_q) \in R^n$, i.e., $n = n_1 + n_2 + \cdots + n_q$ and $f_i^N, f_i^D : R^n \to R$, $c_{ij}, d_{ij} \in R$ and $\alpha_{jk}^{(i)}, \beta_{jk}^{(i)} \in R^+$ for $i = 1, 2, \ldots, q$, $k = 1, 2, \ldots, n$.

$g_i(x)$ are nonlinear constraints whereas a set of linear constraints is produced by $A_1 \in R^{m \times n_1}$, $A_2 \in R^{m \times n_2}, \ldots, A_q \in R^{m \times n_q}$, $b \in R^m$ and Ω is assumed to be a convex feasible region of constraints.

3 Some Basics of Distance Function

Consider a vector of objective functions $F(x) = (f_1(x), f_2(x), \ldots, f_k(x))$ in k-dimensional objective space of a multi-objective optimization problem (MOOP) of maximization type. Let $F^* = (f_1^*, f_2^*, \ldots, f_k^*)$ and $\bar{F} = (\bar{f}_1, \bar{f}_2, \ldots, \bar{f}_k)$ be the ideal or reference point and the anti-ideal or nadir point, respectively, where:

$$f_i^* = \max_{x \in \Omega} f_i(x), \bar{f}_i = \min_{x \in \Omega} f_i(x) \text{ for } i = 1, 2, \ldots, k$$

Define the following vectors F^{IP} and F^{NP} in the objective space.

$$F^{IP}(x) = (f_1^* - f_1(x), f_2^* - f_2(x), ..., f_k^* - f_k(x))$$
$$F^{NP}(x) = (f_1(x) - \bar{f}_1, f_2(x) - \bar{f}_2, ..., f_k(x) - \bar{f}_k)$$

As we consider a problem of maximization type, $(F^* - F(x))$ and $(F(x) - \bar{F})$ need to be minimized and maximized, respectively, to obtain the best compromise solution. Define the following distance functions or L_p-metrics in the objective space to optimize the differences.

$$d_p(F^{IP}(x)) = [\sum_{i=1}^{k} \gamma_i^p \{\frac{f_i^* - f_i(x)}{f_i^* - \bar{f}_i}\}^p]^{\frac{1}{p}} = \sum_{i=1}^{k} \gamma_i^p \{\mu_{f_i}^{min}(x)\}^p]^{\frac{1}{p}},\ 1 \leq p < \infty$$
$$d_p(F^{IP}(x)) = \max_{1 \leq i \leq k}[\gamma_i \{\frac{f_i^* - f_i(x)}{f_i^* - \bar{f}_i}\}] = \max_{1 \leq i \leq k}[\gamma_i \{\mu_{f_i}^{min}(x)\}],\ p = \infty$$
$$d_p(F^{NP}(x)) = [\sum_{i=1}^{k} \gamma_i^p \{\frac{f_i(x) - \bar{f}_i}{f_i^* - \bar{f}_i}\}^p]^{\frac{1}{p}} = \sum_{i=1}^{k} \gamma_i^p \{\mu_{f_i}^{max}(x)\}^p]^{\frac{1}{p}},\ 1 \leq p < \infty$$
$$d_p(F^{NP}(x)) = \max_{1 \leq i \leq k}[\gamma_i \{\frac{f_i(x) - \bar{f}_i}{f_i^* - \bar{f}_i}\}] = \max_{1 \leq i \leq k}[\gamma_i \{\mu_{f_i}^{max}(x)\}],\ p = \infty$$

where, γ_i $(i = 1, 2, ..., k)$ are the relative weights assigned to the objectives in order of their preference by the decision maker and also are usually assumed to be positive and normalized. Each of the deviations $\{f_i^* - f_i(x)\}$ and $\{f_i(x) - \bar{f}_i\}$, $i = 1, 2, ..., k$ is divided by the largest possible deviation $\{f_i^* - \bar{f}_i\}$ for achieving the values in $[0, 1]$. As $\bar{f}_i \leq f_i(x) \leq f_i^*$, $\mu_{f_i}^{min}(x)$ and $\mu_{f_i}^{max}(x)$ can be treated as fuzzy linear membership functions of the objectives $f_i(x)$ of minimization and maximization type, respectively. In our proposed method, we have considered the Euclidean metric, i.e., $p = 2$, for computations required in numerical problems. As we deal with the L_p-metric in finite dimensional objective space, other values of $p \neq 2$ can also be considered. More about these results can be found in [17, 18].

4 Proposed Method to Solve ML-NLFPP

As we consider a multi-level problem which involves fractional objectives of maximization type at each level, $\max_{x \in \Omega} \frac{f_i^N(x)}{f_i^D(x)}$, $i = 1, 2, ..., q$ occurs when $f_i^N(x)$ and $f_i^D(x)$ simultaneously attain their best possible maximum and minimum values, respectively, on the constrained feasible region. Thus at each level-i, the problem is reformulated as a multi-objective optimization problem with non-fractional objectives as: $\{\max f_i^N(x), \min f_i^D(x)\}$ subject to $x \in \Omega$

To obtain the best and worst values of $f_i^N(x)$ and $f_i^D(x)$, optimize them individually over the feasible region of constraints.

$$\max f_i^N(x) = f_i^{N^*},\ \min f_i^N(x) = \bar{f}_i^N,\ \min f_i^D(x) = f_i^{D^*},\ \max f_i^D(x) = \bar{f}_i^D$$

The range of variations for numerator and denominator functions at each level is computed as:

$$\bar{f}_i^N \leq f_i^N(x) \leq f_i^{N^*} \text{ and } f_i^{D^*} \leq f_i^D(x) \leq \bar{f}_i^D, \; i = 1, 2, ..., q$$

At each level-i, the ideal and anti-ideal objective points for $(f_i^N(x), f_i^D(x))$ are obtained as $(f_i^{N^*}, f_i^{D^*})$ and $(\bar{f}_i^N, \bar{f}_i^D)$, respectively. Construct the following distance functions separately for each level-$i = 1, 2, ..., q$.

$$d_p(F_i^*) = [\lambda_i^p \{\frac{f_i^{N^*} - f_i^N(x)}{f_i^{N^*} - \bar{f}_i^N}\}^p + \beta_i^p \{\frac{f_i^D(x) - f_i^{D^*}}{\bar{f}_i^D - f_i^{D^*}}\}^p]^{\frac{1}{p}}$$
$$d_p(\bar{F}_i) = [\lambda_i^p \{\frac{f_i^N(x) - \bar{f}_i^N}{f_i^{N^*} - \bar{f}_i^N}\}^p + \beta_i^p \{\frac{\bar{f}_i^D - f_i^D(x)}{\bar{f}_i^D - f_i^{D^*}}\}^p]^{\frac{1}{p}}$$

$d_p(F_i^*)$ and $d_p(\bar{F}_i)$ need to be simultaneously minimized and maximized, respectively, to maintain shortest and farthest distance from the best and worst values of the objectives so as to determine the best compromise solution of the ML-NLFPP. Thus, at each level-i, the problem is formulated as:

Problem-P$_i$: $\{\min d_p(F_i^*), \max d_p(\bar{F}_i)\}$ subject to $x \in \Omega$
Evaluate the aspired (best) and acceptable (worst) values of $d_p(F_i^*)$ and $d_p(\bar{F}_i)$ as follows:

$$d_p^{*min}(F_i) = \min_{x \in \Omega} d_p(F_i^*) \text{ which occurs at } x = x_{F_i^*}$$
$$\bar{d}_p^{\,max}(F_i) = \max_{x \in \Omega} d_p(\bar{F}_i) \text{ which occurs at } x = x_{\bar{F}_i}$$
$$d_p^{*max}(F_i) = \max_{x \in \Omega} d_p(F_i^*), \; \bar{d}_p^{\,min}(F_i) = \min_{x \in \Omega} d_p(\bar{F}_i)$$

This is suggested that $d_p^{*max}(F_i)$ and $\bar{d}_p^{\,min}(F_i)$ can also be evaluated as:

$$d_p^{*max}(F_i) = d_p(F_i^*(x_{\bar{F}_i})) \text{ and } \bar{d}_p^{\,min}(F_i) = d_p(\bar{F}_i(x_{F_i^*}))$$

Finally, the range of variations for $d_p(F_i^*)$ and $d_p(\bar{F}_i)$ is obtained as:

$$d_p^{*min}(F_i) \leq d_p(F_i^*) \leq d_p^{*max}(F_i), \; \bar{d}_p^{\,min}(F_i) \leq d_p(\bar{F}_i) \leq \bar{d}_p^{\,max}(F_i)$$

Construct the following fuzzy linear membership functions for $d_p(F_i^*)$ and $d_p(\bar{F}_i)$ at each level-$i = 1, 2, ..., q$ as follows:

$$\mu_{d_p(F_i^*)}(x) = \begin{cases} 1, & d_p(F_i^*) \leq d_p^{*min}(F_i) \\ \frac{d_p^{*max}(F_i) - d_p(F_i^*)}{d_p^{*max}(F_i) - d_p^{*min}(F_i)}, & d_p^{*min}(F_i) < d_p(F_i^*) \leq d_p^{*max}(F_i) \\ 0, & d_p(F_i^*) > d_p^{*max}(F_i) \end{cases}$$

$$\mu_{d_p(\bar{F}_i)}(x) = \begin{cases} 1, & d_p(\bar{F}_i) \geq \bar{d}_p^{\,max}(F_i) \\ \frac{d_p(\bar{F}_i) - \bar{d}_p^{\,min}(F_i)}{\bar{d}_p^{\,max}(F_i) - \bar{d}_p^{\,min}(F_i)}, & \bar{d}_p^{\,min}(F_i) \leq d_p(\bar{F}_i) < \bar{d}_p^{\,max}(F_i) \\ 0, & d_p(\bar{F}_i) < \bar{d}_p^{\,min}(F_i) \end{cases}$$

As decisions of upper-level DM_i are expressed through their respective controlled decision variables, the goals for X_i can be determined by solving the following problem separately for each $i = 1, 2, ..., q-1$ except the qth level.

Problem-P_i': max $\{\mu_{d_p(F_i^*)}(x), \mu_{d_p(\bar{F}_i)}(x)\}$ subject to $x \in \Omega$

To solve the above problem, fuzzy membership goals can be expressed as:

$\mu_{d_p(F_i^*)}(x) + d_i^{*-} - d_i^{*+} = 1$, $\mu_{d_p(\bar{F}_i)}(x) + \bar{d}_i^- - \bar{d}_i^+ = 1$

where, $d_i^{*-}, \bar{d}_i^-$ and $d_i^{*+}, \bar{d}_i^+$ are under and over deviational variables with $d_i^{*-}, \bar{d}_i^-$, $d_i^{*+}, \bar{d}_i^+ \geq 0$ and $d_i^{*-}.d_i^{*+} = 0$, $\bar{d}_i^-.\bar{d}_i^+ = 0$. As over deviation from the highest membership value unity(1) represents the state of complete achievement, only under deviational variables are considered to be minimized to solve the Problem-P_i' which can be reformulated as:

$$\frac{\text{Problem} - P_i''}{}$$

$$\min\ (d_i^{*-} + \bar{d}_i^-)$$

subject to

$$\mu_{d_p(F_i^*)}(x) = \frac{d_p^{*max}(F_i) - d_p(F_i^*)}{d_p^{*max}(F_i) - d_p^{*min}(F_i)} + d_i^{*-} \geq 1$$

$$\mu_{d_p(\bar{F}_i)}(x) = \frac{d_p(\bar{F}_i) - \bar{d}_p^{min}(F_i)}{\bar{d}_p^{max}(F_i) - \bar{d}_p^{min}(F_i)} + \bar{d}_i^- \geq 1$$

$$x \in \Omega, d_i^{*-}, \bar{d}_i^- \geq 0$$

On solving Problem-P_i'' separately for each $i = 1, 2, ..., (q-1)$ by DM_i, the corresponding solutions are obtained as $x_i^* = (X_1^{i*}, X_2^{i*}, ..., X_q^{i*})$. So, the goals for the controlled decision variables $X_1, X_2, ..., X_{q-1}$ are obtained as $X_1^{1*}, X_2^{2*}, ..., X_{q-1}^{(q-1)*}$, respectively. DM_i declares some positive and negative tolerances to the goals X_i^{i*} for its controlled decision variables X_i as relaxation to avoid the situation of decision deadlock. Thus, X_i can be treated as a triangular fuzzy number as $(X_i^{i*} - I_i^-, X_i^{i*}, X_i^{i*} + I_i^+)$. I_i^- and I_i^+ are not necessarily same and defined by DM_i considering the practical problem.

Construct the following fuzzy membership functions for the controlled decision variables X_i at each level-$i = 1, 2, ..., q-1$.

$$\mu_{\tilde{X}_i}(X_i) = \begin{cases} \frac{X_i - X_i^{i*} + I_i^-}{I_i^-}, & X_i^{i*} - I_i^- \leq X_i \leq X_i^{i*} \\ \frac{X_i^{i*} + I_i^+ - X_i}{I_i^+}, & X_i^{i*} \leq X_i \leq X_i^{i*} + I_i^+ \\ 0, & otherwise \end{cases}$$

Figures 1, 2, and 3 represent the linear membership functions for the distance functions $d_p(\bar{F}_i)$, $d_p(F_i^*)$ and the controlled decision variables X_i, respectively.

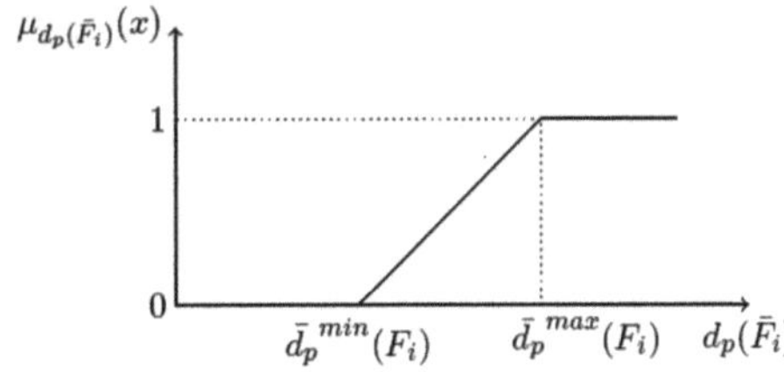

Fig. 1 Membership function for $d_p(\bar{F}_i)$

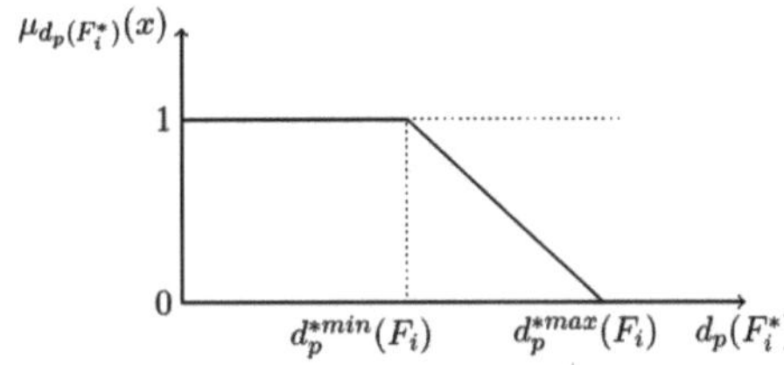

Fig. 2 Membership function for $d_p(F_i^*)$

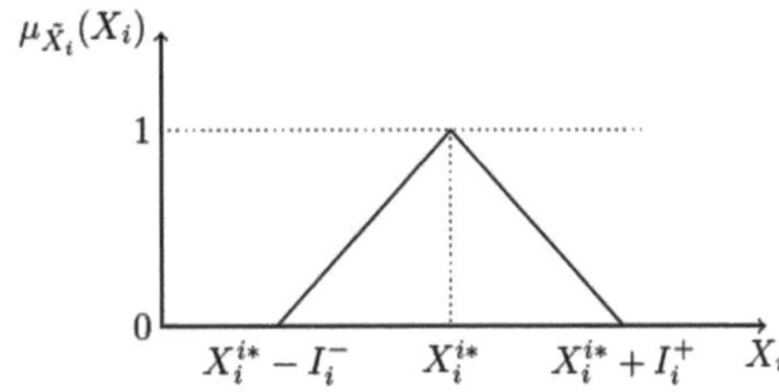

Fig. 3 Membership function for X_i

Finally to determine the best compromise solution of ML-NLFPP, all the above constructed membership functions are maximized simultaneously; i.e., the following problem-P_i^* is solved.

$$\underline{\text{Problem} - \text{P}_i^{''}}$$

$$\max\ \{\mu_{d_p(F_i^*)}(x), \mu_{d_p(\bar{F}_i)}(x)\ : i = 1, 2, ..., q,\ \mu_{\tilde{X}_i}(X_i)\ :\ i = 1, 2, ..., (q-1)\}$$

$$\text{subject to}$$

$$x \in \Omega$$

Using fuzzy goal programming introduced by Mohamed [11], Problem-P_i^* can be reformulated and expressed in generalized form as follows:

$$\underline{\text{Problem} - \text{P}_i^{**}}$$

$$\min\ \sum_{i=1}^{q}(d_i^{*-} + \bar{d}_i^-) + \sum_{i=1}^{q-1}(d_i^{l-} + d_i^{r-})$$

$$\text{subject to}$$

$$\mu_{d_p(F_i^*)}(x) = \frac{d_p^{*max}(F_i) - d_p(F_i^*)}{d_p^{*max}(F_i) - d_p^{*min}(F_i)} + d_i^{*-} \geq 1, \ i = 1, 2, ..., q$$

$$\mu_{d_p(\bar{F}_i)}(x) = \frac{d_p(\bar{F}_i) - \bar{d}_p^{\ min}(F_i)}{\bar{d}_p^{\ max}(F_i) - \bar{d}_p^{\ min}(F_i)} + \bar{d}_i^{-} \geq 1, i = 1, 2, ..., q$$

$$\mu_{\tilde{X}_i}(X_i) = \frac{X_i - X_i^{i*} + I_i^-}{I_i^-} + d_i^{l-} \geq 1, i = 1, 2, ..., (q-1)$$

$$\mu_{\tilde{X}_i}(X_i) = \frac{X_i^{i*} + I_i^+ - X_i}{I_i^+} + d_i^{r-} \geq 1, i = 1, 2, ..., (q-1)$$

$$x \in \Omega, d_i^{*-}, \bar{d}_i^-, d_i^{l-}, d_i^{r-} \geq 0$$

If all DMs get unsatisfied with the best compromise solution obtained by solving Problem-P_i^{**}, the tolerances for the goals of decision variables are modified and the compromise solution is re-evaluated. But such case rarely happens since the tolerances are defined by respective DMs after considerable observation towards the practical problem. To summarize the steps of the proposed solution approach, an algorithm is presented below.

4.1 Algorithm to Solve ML-NLFPP

The following steps are executed sequentially to determine the best compromise solution of the ML-NLFPP.

Step 1. Maximize and minimize $f_i^N(x)$ and $f_i^D(x)$ individually for each $i = 1, 2, ..., q$ to obtain $f_i^{N^*}, \bar{f}_i^N, f_i^{D^*}$, and $\bar{f}_i^D$.
Step 2. Construct the distance functions $d_p(F_i^*)$ and $d_p(\bar{F}_i)$ for each level-$i = 1, 2, ..., q$.
Step 3. Maximize and minimize $d_p(F_i^*)$ and $d_p(\bar{F}_i)$ individually for each $i = 1, 2, ..., q$ to obtain $d_p^{*min}(F_i), d_p^{*max}(F_i), \bar{d}_p^{\ min}(F_i)$ and $\bar{d}_p^{\ max}(F_i)$.
Step 4. Construct the fuzzy membership functions $\mu_{d_p(F_i^*)}(x)$ and $\mu_{d_p(\bar{F}_i)}(x)$ for each $i = 1, 2, ..., q$.
Step 5. Solve Problem-P_i'' for each $i = 1, 2, ..., (q-1)$ to obtain the solutions $x_i^* = (X_1^{i*}, X_2^{i*}, ..., X_q^{i*})$.
Step 6. Obtain the goals X_i^{i*} for each $X_i, i = 1, 2, ..., (q-1)$.
Step 7. Define the negative and positive tolerances I_i^- and I_i^+ for the controlled decision variables $X_i, i = 1, 2, ..., (q-1)$.
Step 8. Construct the fuzzy membership functions $\mu_{\tilde{X}_i}(X_i)$ for $i = 1, 2, ..., (q-1)$.
Step 9. Solve Problem-P_i^{**} to obtain the best compromise solution.
Step 10. If the DMs remain unsatisfied with the above-obtained solution, the tolerances for X_i are modified and the steps 8 and 9 are repeated to obtain a new best compromise solution of ML-NLFPP.

5 Illustrative Numerical Example

To demonstrate the proposed solution approach and justify its feasibility, the following ML-NLFPP is solved.

Level 1: $\max_{x_1} \frac{x_1^2+2x_1x_3+x_3^2}{x_1^2+x_2+x_3}$

where x_2, x_3 solve
Level 2: $\max_{x_2} \frac{2x_1x_2^2+x_2^2+x_3}{x_1+x_2^2+x_3}$
where x_3 solves
Level 3: $\max_{x_3} \frac{x_1^2-3x_1^2x_3+x_1x_2^2}{x_1+x_2+x_3^2}$
subject to

$$\Omega = \begin{cases} 2x_1 + x_2 + 2x_3 \le 4 \\ x_1 + x_2 + x_3 \ge 1 \\ x_1, x_2, x_3 \ge 0 \end{cases}$$

Individual maximal and minimal values of the numerator and denominator functions of all levels are computed as:

$\max_{x\in\Omega} f_1^N(x) = 4,\ \min_{x\in\Omega} f_1^N(x) = 0,\ \min_{x\in\Omega} f_1^D(x) = 0.75,\ \max_{x\in\Omega} f_1^D(x) = 4$
$\max_{x\in\Omega} f_2^N(x) = 18.5185,\ \min_{x\in\Omega} f_2^N(x) = 0,\ \min_{x\in\Omega} f_2^D(x) = 0.75,\ \max_{x\in\Omega} f_2^D(x) = 16$
$\max_{x\in\Omega} f_3^N(x) = 5.2526, \min_{x\in\Omega} f_3^N(x) = -2.0576,$
$\min_{x\in\Omega} f_3^D(x) = 0.75, \max_{x\in\Omega} f_3^D(x) = 4$

Using the concept of the proposed solution approach, construct the following distance functions at each level-1, 2, 3 as follows:

At level-1: Ideal and anti-ideal points are $(f_1^{N*}, f_1^{D*}) = (4, 0.75)$ and $(\bar{f}_1^N, \bar{f}_1^D) = (0, 4)$, respectively.

$d_p(F_1^*) = [\lambda_1^2(\frac{4-f_1^N(x)}{4})^2 + \beta_1^2(\frac{f_1^D(x)-0.75}{3.25})^2]^{\frac{1}{2}},\ d_p(\bar{F}_1) = [\lambda_1^2(\frac{f_1^N(x)}{4})^2 + \beta_1^2(\frac{4-f_1^D(x)}{3.25})^2]^{\frac{1}{2}}$

On substituting the values of $f_1^N(x), f_1^D(x)$ and $\lambda_1 = \beta_1 = 0.5$:
$d_p(F_1^*) = [0.0156(4 - x_1^2 - 2x_1x_3 - x_3^2)^2 + 0.0237(x_1^2 + x_2 + x_3 - 0.75)^2]^{\frac{1}{2}}$
$d_p(\bar{F}_1) = [0.0156(x_1^2 + 2x_1x_3 + x_3^2)^2 + 0.0237(4 - x_1^2 - x_2 - x_3)^2]^{\frac{1}{2}}$

At level-2: Ideal and anti-ideal points are $(f_2^{N*}, f_2^{D*}) = (18.5185, 0.75)$ and $(\bar{f}_2^N, \bar{f}_2^D) = (0, 16)$, respectively.

$d_p(F_2^*) = [\lambda_2^2(\frac{18.5185-f_2^N(x)}{18.5185})^2 + \beta_2^2(\frac{f_2^D(x)-0.75}{15.25})^2]^{\frac{1}{2}}$
$d_p(\bar{F}_2) = [\lambda_2^2(\frac{f_2^N(x)}{18.5185})^2 + \beta_2^2(\frac{16-f_2^D(x)}{15.25})^2]^{\frac{1}{2}}$

On substituting the values of $f_2^N(x), f_2^D(x)$ and $\lambda_2 = \beta_2 = 0.5$:

$$d_p(F_2^*) = [0.0007(18.5185 - 2x_1x_2^2 - x_2^2 - x_3)^2 + 0.0011(x_1 + x_2^2 + x_3 - 0.75)^2]^{\frac{1}{2}}$$
$$d_p(\bar{F}_2) = [0.0007(2x_1x_2^2 + x_2^2 + x_3)^2 + 0.0011(16 - x_1 - x_2^2 - x_3)^2]^{\frac{1}{2}}$$

At level-3: Ideal and anti-ideal points are $(f_3^{N*}, f_3^{D*}) = (5.2526, 0.75)$ and $(\bar{f}_3^N, \bar{f}_3^D) = (-2.0576, 4)$, respectively.

$$d_p(F_3^*) = [\lambda_3^2(\tfrac{5.2526 - f_3^N(x)}{7.3102})^2 + \beta_3^2(\tfrac{f_3^D(x) - 0.75}{3.25})^2]^{\frac{1}{2}}$$
$$d_p(\bar{F}_3) = [\lambda_3^2(\tfrac{f_3^N(x) + 2.0576}{7.3102})^2 + \beta_3^2(\tfrac{4 - f_3^D(x)}{3.25})^2]^{\frac{1}{2}}$$

On substituting the values of $f_3^N(x), f_3^D(x)$ and $\lambda_3 = \beta_3 = 0.5$:

$$d_p(F_3^*) = [0.0047(5.2526 - x_1^2 + 3x_1^2x_3 - x_1x_2^2)^2 + 0.0237(x_1 + x_2 + x_3^2 - 0.75)^2]^{\frac{1}{2}}$$
$$d_p(\bar{F}_3) = [0.0047(x_1^2 - 3x_1^2x_3 + x_1x_2^2 + 2.0576)^2 + 0.0237(4 - x_1 - x_2 - x_3^2)^2]^{\frac{1}{2}}$$

The aspired and acceptable values, i.e., the range of variations for $d_p(F_i^*)$ and $d_p(\bar{F}_i)$, $i = 1, 2, 3$, are obtained as:

$0.1468 \le d_p(F_1^*) \le 0.7071$, $\quad 0 \le d_p(\bar{F}_1) \le 0.6079$, $\quad 0.2107 \le d_p(F_2^*) \le 0.5102$, $0.3309 \le d_p(\bar{F}_2) \le 0.5228$, $0.2097 \le d_p(F_3^*) \le 0.6164$, $0.1411 \le d_p(\bar{F}_3) \le 0.5201$

Construct the following membership functions for $d_p(F_i^*)$ and $d_p(\bar{F}_i)$, $i = 1, 2, 3$ as follows:

$$\mu_{d_p(F_1^*)}(x) = \begin{cases} 1, & d_p(F_1^*) \le 0.1468 \\ \frac{0.7071 - d_p(F_1^*)}{0.5603}, & 0.1468 < d_p(F_1^*) \le 0.7071 \\ 0, & d_p(F_1^*) > 0.7071 \end{cases}$$

$$\mu_{d_p(\bar{F}_1)}(x) = \begin{cases} 1, & d_p(\bar{F}_1) \ge 0.6079 \\ \frac{d_p(\bar{F}_1)}{0.6079}, & 0 \le d_p(\bar{F}_1) < 0.6079 \\ 0, & d_p(\bar{F}_1) < 0 \end{cases}$$

$$\mu_{d_p(F_2^*)}(x) = \begin{cases} 1, & d_p(F_2^*) \le 0.2107 \\ \frac{0.5102 - d_p(F_2^*)}{0.2995}, & 0.2107 < d_p(F_2^*) \le 0.5102 \\ 0, & d_p(F_2^*) > 0.5102 \end{cases}$$

$$\mu_{d_p(\bar{F}_2)}(x) = \begin{cases} 1, & d_p(\bar{F}_2) \ge 0.5228 \\ \frac{d_p(\bar{F}_2) - 0.3309}{0.1919}, & 0.3309 \le d_p(\bar{F}_2) < 0.5228 \\ 0, & d_p(\bar{F}_2) < 0.3309 \end{cases}$$

$$\mu_{d_p(F_3^*)}(x) = \begin{cases} 1, & d_p(F_3^*) \le 0.2097 \\ \frac{0.6164 - d_p(F_3^*)}{0.4067}, & 0.2097 < d_p(F_3^*) \le 0.6164 \\ 0, & d_p(F_3^*) > 0.6164 \end{cases}$$

$$\mu_{d_p(\bar{F}_3)}(x) = \begin{cases} 1, & d_p(\bar{F}_3) \ge 0.5201 \\ \frac{d_p(\bar{F}_3) - 0.1411}{0.379}, & 0.1411 \le d_p(\bar{F}_3) < 0.5201 \\ 0, & d_p(\bar{F}_3) < 0.1411 \end{cases}$$

DM_1 solves the following problem to obtain the goal for x_1 by maximizing $\mu_{d_p(F_1^*)}(x)$ and $\mu_{d_p(\bar{F}_1)}(x)$ simultaneously.

$$\min\ (d_1^* + \bar{d}_1)$$

$$\text{subject to } \frac{0.7071 - d_p(F_1^*)}{0.5603} + d_1^* \ge 1,\ \frac{d_p(\bar{F}_1)}{0.6079} + \bar{d}_1 \ge 1$$
$$2x_1 + x_2 + 2x_3 \le 4,\ x_1 + x_2 + x_3 \ge 1,\ x_1, x_2, x_3, d_1^*, \bar{d}_1 \ge 0$$

The solution is obtained as $(x_1, x_2, x_3) = (0.5, 0, 1.4999)$. The goal for x_1 is ascertained as 0.5, and DM_1 defines $\tilde{x_1} = (0.2, 0.5, 0.7)$ with tolerances.

DM_2 solves the following problem to obtain the goal for x_2 by maximizing $\mu_{d_p(F_2^*)}(x)$ and $\mu_{d_p(\bar{F}_2)}(x)$ simultaneously.

$$\min\ (d_2^* + \bar{d}_2)$$

$$\text{subject to } \frac{0.5102 - d_p(F_2^*)}{0.2995} + d_2^* \ge 1,\ \frac{d_p(\bar{F}_2) - 0.3309}{0.1919} + \bar{d}_2 \ge 1$$

$$2x_1 + x_2 + 2x_3 \le 4,\ x_1 + x_2 + x_3 \ge 1,\ x_1, x_2, x_3, d_2^*, \bar{d}_2 \ge 0$$

The solution is obtained as $(x_1, x_2, x_3) = (0.7627, 2.4745, 0)$. The goal for x_2 is ascertained as 2.4745, and DM_2 defines $\tilde{x_2} = (0.5, 2.4745, 3)$ with tolerances.

Construct the fuzzy membership functions for x_1 and x_2 as follows:

$$\mu_{\tilde{x_1}}(x_1) = \begin{cases} \frac{x_1 - 0.2}{0.3}, & 0.2 \le x_1 \le 0.5 \\ \frac{0.7 - x_1}{0.2}, & 0.5 \le x_1 \le 0.7 \\ 0, & otherwise \end{cases} \quad \mu_{\tilde{x_2}}(x_2) = \begin{cases} \frac{x_2 - 0.5}{1.9745}, & 0.5 \le x_2 \le 2.4745 \\ \frac{3 - x_2}{0.5255}, & 2.4745 \le x_2 \le 3 \\ 0, & otherwise \end{cases}$$

The best compromise solution of the ML-NLFPP is obtained by solving the following problem where fuzzy goal programming reducing only negative deviational variables is applied to maximize $\mu_{\tilde{x_1}}(x_1)$, $\mu_{\tilde{x_2}}(x_2)$ and $\mu_{d_p(\bar{F}_i)}(x)$, $\mu_{d_p(\bar{F}_i)}(x)$, $i = 1, 2, 3$.

$$\min \sum_{i=1}^{3} d_i^* + \sum_{i=1}^{3} \bar{d}_i + \sum_{i=1}^{2} d_i^{l-} + \sum_{i=1}^{2} d_i^{r-}$$

subject to

$$\frac{0.7071 - d_p(F_1^*)}{0.5603} + d_1^* \geq 1, \frac{0.5102 - d_p(F_2^*)}{0.2995} + d_2^* \geq 1, \frac{0.6164 - d_p(F_3^*)}{0.4067} + d_3^* \geq 1$$
$$\frac{d_p(\bar{F}_1)}{0.6079} + \bar{d}_1 \geq 1, \frac{d_p(\bar{F}_2) - 0.3309}{0.1919} + \bar{d}_2 \geq 1, \frac{d_p(\bar{F}_3) - 0.1411}{0.379} + \bar{d}_3 \geq 1$$
$$\frac{x_1 - 0.2}{0.3} + d_1^{l-} \geq 1, \frac{x_2 - 0.5}{1.9745} + d_2^{l-} \geq 1$$
$$\frac{0.7 - x_1}{0.2} + d_1^{r-} \geq 1, \frac{3 - x_2}{0.5255} + d_2^{r-} \geq 1$$
$$2x_1 + x_2 + 2x_3 \leq 4, \; x_1 + x_2 + x_3 \geq 1$$
$$x_1, x_2, x_3, d_1^*, d_2^*, d_3^*, \bar{d}_1, \bar{d}_2, \bar{d}_3, d_1^{l-}, d_2^{l-}, d_1^{r-}, d_2^{r-} \geq 0$$

The solution is obtained as $x^* = (0.5, 0.5, 0)$, and the values of the numerator and denominator functions at this point are: $f_1^N(x^*) = 0.25, f_2^N(x^*) = 0.5, f_3^N(x^*) = 0.375, f_1^D(x^*) = 0.75, f_2^D(x^*) = 0.75, f_3^D(x^*) = 1$ which shows the feasibility of the proposed approach as the obtained values of the numerator and denominator functions of the objectives lie in their predefined range of variations.

6 Conclusion

This paper proposed a solution approach to multi-level fractional programming problem with the objectives as fraction of nonlinear polynomial functions. The numerator and denominator functions simultaneously maintained possible shortest and farthest distance from their respective ideal and anti-ideal points to derive the best compromise solution. Reduction of only negative deviations from the goals of fuzzy membership functions produced the optimal solution of ML-NLFPP. Numerical example illustrated the feasibility and effectiveness of the solution approach. LINGO software and MATLAB software were used for the computational works of the problem.

Acknowledgements Authors are grateful to the editor and anonymous referees for their valuable comments and suggestions to improve the quality of the paper.

References

1. Shih, H.-S., Lai, Y.-J., Lee, E.S.: Fuzzy approach for multi-level programming problems. Comput. Oper. Res. **23**, 73–91 (1996)
2. Stancu-Minasian, I.M.: Fractional programming: Theory, Methods and Applications. Kluwer Academic Publishers (1997)
3. Bellman, R.: Dynamic Programming. Princeton University Press (1957)
4. Lachhwani, K.: Modified FGP approach for multi-level multi objective linear fractional programming problems. Appl. Math. Comput. **266**, 1038–1049 (2015)

5. Lai, Y.-J.: Hierarchical optimization: a satisfactory solution. Fuzzy Sets Syst. **77**, 321–335 (1996)
6. Zimmermann, H.-J.: Fuzzy programming and linear programming with several objective functions. Fuzzy Sets Syst. **1**, 45–55 (1978)
7. White, D.J.: Penalty function approach to linear trilevel programming. J. Optim. Theory Appl. **93**, 183–197 (1997)
8. Osman, M.S., Abo-Sinna, M.A., Amer, A.H., Emam, O.E.: A multi-level non-linear multi-objective decision-making under fuzziness. Appl. Math. Comput. **153**, 239–252 (2004)
9. Zhang, G., Lu, J., Montero, J., Zeng, Y.: Model, solution concept, and Kth-best algorithm for linear trilevel programming. Inf. Sci. **180**, 481–492 (2010)
10. Pramanik, S., Roy, T.K.: Fuzzy goal programming approach to multilevel programming problems. Eur. J. Oper. Res. **176**, 1151–1166 (2007)
11. Mohamed, R.H.: The relationship between goal programming and fuzzy programming. Fuzzy Sets Syst. **89**, 215–222 (1997)
12. Baky, I.A.: Solving multi-level multi-objective linear programming problems through fuzzy goal programming approach. Appl. Math. Model. **34**, 2377–2387 (2010)
13. Baky, I.A.: Interactive TOPSIS algorithms for solving multi-level non-linear multi-objective decision-making problems. Appl. Math. Model. **38**, 1417–1433 (2014)
14. Abo-Sinna, M.A., Baky, I.A.: Interactive balance space approach for solving multi-level multi-objective programming problems. Inf. Sci. **177**, 3397–3410 (2007)
15. Sinha, S.: Fuzzy programming approach to multi-level programming problems. Fuzzy Sets Syst. **136**, 189–202 (2003)
16. Lachhwani, K.: On solving multi-level multi objective linear programming problems through fuzzy goal programming approach. Opsearch **51**, 624–637 (2014)
17. Baky, I.A., Abo-Sinna, M.A.: TOPSIS for bi-level MODM problems. Appl. Math. Model. **37**, 1004–1015 (2013)
18. Miettinen, K.M.: Nonlinear Multiobjective Optimization. Springer Science & Business Media (2012)

On Finite Buffer *BMAP/G/1* Queue with Queue Length Dependent Service

A. Banerjee, K. Sikdar and G. K. Gupta

Abstract This paper deals with the analysis of a finite buffer queueing system where customers are arriving according to the batch Markovian arrival process (*BMAP*). The service time is considered to be generally distributed and is dependent on the queue length at service initiation epoch. The stationary queue length distribution at various epoch is obtained using the embedded Markov chain technique and the supplementary variable technique. A computational procedure has been discussed by considering phase-type service time distribution. Finally, some numerical results are given to show the numerical compatibility of the analytical results. Also a comparative study is carried out to establish the fact that our model may help in optimizing system performance by controlling the service rate depending on the state of the system.

Keywords Batch Markovian arrival process · Blocking probability Congestion · Queue length dependent service

1 Introduction

Modern wireless/wired computer communication and telecommunication system have been designed to support a wide range of multimedia applications, such as voice, data and video, with expectation to maintain Quality of Service (QoS). The traffic flow in such communication networks is statistically multiplexed, highly

A. Banerjee (✉) · G. K. Gupta
Department of Mathematical Sciences, Indian Institute of Technology (BHU), Varanasi 221005, India
e-mail: anuradha.mat@iitbhu.ac.in

G. K. Gupta
e-mail: gopalgupta.bhu90@gmail.com

K. Sikdar
Department of Mathematics, BMS Institute of Technology & Management, (Affiliated to VTU, Belgaum-18), Post Box No. 6443, Doddaballapura Main Road, Yelahanka, Bengaluru 560064, India
e-mail: karabis06@gmail.com; karabi@bmsit.in

S. Kar et al. (eds.), *Operations Research and Optimization*, Springer Proceedings in Mathematics & Statistics 225, https://doi.org/10.1007/978-981-10-7814-9_4

irregular (e.g. bursty and correlated) and transmitted in superhigh speed. Hence, this type of traffic flow cannot be well approximated in terms of stationary Poisson process. The mathematical model for telecommunication networks is well approximated by the *BMAP* input flow as it is a stochastic point process that generalizes the standard Poisson process by allowing for batches of arrivals, dependent interarrival times, nonexponential interarrival time distributions and correlated batch sizes. For more detail on *BMAP*, readers are referred to see [1–10] and the references therein.

The purpose of studying any queueing model is to optimize its system performance. After arriving to the system, the customers have to wait in the queue if server is found unavailable. Long waiting line may cause queueing delay and increase in loss probability which ultimately results in poor system performance. Hence, to avoid this situation it is needed to develop control policy (service rate/arrival rate) to reduce congestion. The queueing models with queue length dependent service policy have been studied by Choi and Choi [11], Choi et al. [12], Banerjee [13] and the references therein. Choi and Choi [11] analysed a finite buffer queue, where customers arrive according to the Markov modulated Poisson process (*MMPP*), with queue length dependent service. *MMPP* is a special case of Markovian arrival process (*MAP*). Recently, Banerjee [13] considered a more general queue length dependent service policy than the one considered in [11], with *MAP*. No literature was found on *BMAP*/*G*/1/*N* queueing model with queue length dependent service policy and this motivated us to study this model theoretically.

In this paper, we deal with a finite buffer *BMAP*/*G*/1 queue whose service time distribution changes dynamically depending on the queue length at service initiation epoch. Also service time distribution is considered to be generally distributed. Following the approach of Banerjee [13], we obtain queue length distribution at service completion epoch (departure epoch) and arbitrary epoch. Then we present a computational procedure by considering phase-type service time distribution.

For use in sequel, let $\boldsymbol{e}(r)$, $\boldsymbol{e}_j(r)$ and I_r denote, respectively, the column vector of dimension r consisting of 1's, column vector of dimension r with 1 in the jth position and 0 elsewhere, and an identity matrix of dimension r. When there is no need to emphasize the dimension of these vectors, we will suppress the suffix. Thus, $\boldsymbol{e}$ will denote a column vector of 1's of appropriate dimension.

The rest of this paper is organized as follows. In Sect. 2, description of the model and its analysis at various epoch is given. The computational procedure when service time follows phase-type distribution is spelled out in Sect. 3. System performance measures and numerical results are given in Sections 4 and 5, respectively. The paper ends with some concluding remark in Sect. 6.

2 Model Description and Analysis

We consider a single-server queueing system in which the customers arrive according to a *BMAP* with matrix representation (D_k, $k = 0, 1, 2, \ldots$) of dimension m. The arrivals are governed by an underlying m-state Markov chain with transition rate

d_{ij}^0, $1 \le i,j \le m$, $i \ne j$, there is a transition from state i to state j in the underlying Markov chain without an arrival, and with transition rate d_{ij}^k, $1 \le i,j \le m$, $k \geqslant 1$, there is a transition from state i to state j in the underlying Markov chain with an arrival of batch size k. The matrix $D_0 = [d_{ij}^0]$ has nonnegative off-diagonal and negative diagonal elements, and the matrices $D_k = [d_{ij}^k]$, $k \ge 1$ have nonnegative elements. Hence, $\sum_{k=0}^{\infty} D_k = D$ is the infinitesimal generator of the underlying Markov chain $\{J(t)\}$, where $J(t)$ is the state of the underlying Markov chain at time t with state space $\{i : 1 \le i \le m\}$. For this arrival process, we have $D\mathbf{e} = 0$ and there exists a stationary probability vector $\boldsymbol{\delta}$ such that

$$\boldsymbol{\delta} D = \mathbf{0}, \boldsymbol{\delta}\boldsymbol{e} = 1. \tag{1}$$

The fundamental arrival rate (average arrival rate) and average batch arrival rate of the above Markov process are given by $\lambda^* = \boldsymbol{\delta}\sum_{k=0}^{\infty} kD_k\mathbf{e}$ and $\lambda_g = \boldsymbol{\delta}\sum_{k=1}^{\infty} D_k\mathbf{e}$, respectively. For more detail on this topic, see Lucantoni et al. [1].

Let $\widetilde{N}(t)$ denotes the number of customers arriving in $(0, t]$. Then $\{\widetilde{N}(t), J(t)\}$ is a two-dimensional Markov process of batch Markovian arrival process (*BMAP*), with state space $\{(n, i) : n \ge 0, 1 \le i \le m\}$. The infinitesimal generator of the above Markov process is given by

$$Q = \begin{pmatrix} D_0 & D_1 & D_2 & D_3 & \cdots \\ 0 & D_0 & D_1 & D_2 & \cdots \\ 0 & 0 & D_0 & D_1 & \cdots \\ \vdots & \vdots & \vdots & \vdots & \ddots \end{pmatrix}.$$

Let $\{\tilde{P}(n, t), n \ge 0, t \ge 0\}$ be the square matrices of order m whose (i, j)-th elements are the conditional probabilities defined as

$$\tilde{p}_{i,j}(n, t) = P\{\widetilde{N}(t) = n, J(t) = j|\widetilde{N}(0) = 0, J(0) = i\} \; n \ge 0, \; 1 \le i, j \le m.$$

These matrices satisfy the following system of differential–difference equations

$$\widetilde{P}'(0, t) = \widetilde{P}(0, t)D_0,$$

$$\widetilde{P}'(n, t) = \sum_{i=0}^{n} \widetilde{P}(n - i, t)D_i, \quad n \ge 1,$$

$$\text{with} \qquad \widetilde{P}(0, 0) = I \text{ and } \widetilde{P}(n, 0) = 0, \quad n \ge 1.$$

These matrices, associated with the counting process $\{\widetilde{N}(t), J(t); t \ge 0\}$, have been extensively studied in the literature, and an efficient procedure for computing them is given in Neuts [14]. Let us define the matrix-generating function of $\widetilde{P}(n,t)$ $(n \geq 0)$ as $\widetilde{P}^*(z,t) = \sum_{n=0}^{\infty} \widetilde{P}(n,t)z^n$, $\quad |z| \leq 1$. With usual process of generating function, we have $\widetilde{P}^{*'}(z,t) = \widetilde{P}^*(z,t)D(z)$, $\widetilde{P}^{*'}(z,0) = I$. Solving these matrix-differential equations, we get $\widetilde{P}^*(z,t) = e^{D(z)t}$, $\quad |z| \leq 1,\ t \geq 0$, where $D(z) = \sum_{i=0}^{\infty} D_i z^i$.

We are dealing with a finite buffer queue of buffer size $N > 1$, so at any time maximum $(N+1)$ customers can be present in the system. The batches which upon arrival are unable to find enough space in the buffer for all the members of the batch are, either fully rejected, or a part of the batch is rejected. These rejection rules are known as total batch rejection or partial batch rejection policy, respectively. Since the partial batch rejection policy utilizes the buffer space in an optimal manner, we consider only this policy in this paper.

The service times are assumed to be generally distributed and dependent on the queue length. Specifically, let T_n, $1 \leq n \leq N$, denote the service time with distribution function $H_n(.)$. Let $\tilde{h}_n(.)$ and $H_n^*(.)$, $1 \leq n \leq N$, denote, respectively, the probability density function and the Laplace–Stieltjes transform of T_n. Let h_n denote the mean service time.

The steady-state analysis of the model under consideration will be carried out using the embedded Markov chain approach and the supplementary variable technique since the service times are assumed to be generally distributed. First, we will look at the semi-Markov process embedded at the points of departure of a customer. Towards this end, let us define the square matrices $\mathcal{A}_{n,k}(x)$ and $\mathcal{B}_{0,k}(x)$ of order m for $x \geq 0$ whose (i,j)th elements $(\mathcal{A}_{n,k}(x))_{ij}$ and $(\mathcal{B}_{0,k}(x))_{ij}$, respectively, are defined as follows.

- $(\mathcal{A}_{n,k}(x))_{ij} = Pr\{$Given a departure at time 0, which left n $(1 \leq n \leq N)$ customer in the queue and the arrival process in phase i, the next departure occurs no later than time x with the arrival process in phase j, and during that service k $(k \geq 0)$ customers arrive$\}$,
- $(\mathcal{B}_{0,k}(x))_{ij} = Pr\{$Given a departure at time 0, which left 0 customer in the queue and the arrival process in phase i, the next departure occurs no later than time x with the arrival process in phase j, and during that service k $(k \geq 0)$ customers arrive$\}$.

$$\mathcal{A}_{n,k}(x) = \int_0^x \widetilde{P}(k,t)dH_n(t), \quad 1 \leq n \leq N,\ 0 \leq k \leq N-n,$$

$$\bar{\mathcal{A}}_{n,k}(x) = \sum_{l=k}^{\infty} \mathcal{A}_{n,l}(x), \quad 1 \leq n \leq N,\ k = N-n+1,$$

$$\mathcal{B}_{0,k}(x) = \sum_{i=1}^{k+1} \widetilde{D}_i \mathcal{A}_{i,k-i+1}, \quad 0 \leq k \leq N-1,$$

$$\bar{\mathcal{B}}_{0,N}(x) = \sum_{i=1}^{N} \widetilde{D}_i \bar{\mathcal{A}}_{i,N-i+1}.$$

For use in sequel, we define

$$\begin{aligned}
\mathcal{A}_{n,k} &= \mathcal{A}_{n,k}(\infty),\ 1 \leq n \leq N,\ 0 \leq k \leq N-n, \\
\bar{\mathcal{A}}_{n,k} &= \bar{\mathcal{A}}_{n,k}(\infty),\ 1 \leq n \leq N,\ k = N-n+1, \\
\mathcal{B}_{0,k} &= \mathcal{B}_{0,k}(\infty),\ 0 \leq k \leq N-1, \\
\bar{\mathcal{B}}_{0,N} &= \bar{\mathcal{B}}_{0,N}(\infty),
\end{aligned} \tag{2}$$

with $\widetilde{D}_i = (-D_0)^{-1}D_i$, $1 \leq i \leq N-1$ and $\widetilde{D}_N = (-D_0)^{-1}\widehat{D}_N$, where $\widehat{D}_k = \sum_{i=k}^{\infty} D_i$ for $k \geqslant 1$.

2.1 *Queue Length Distribution at Departure Epoch*

In this section, the joint distribution of the number of customers in the queue and phase of the arrival process at departure epoch has been obtained using the embedded Markov chain technique. Towards this end, let N_n^+ and J_n^+ denote, respectively, the number of customers in the queue and the phase of the arrival process immediately after the n-th departure of a customer. Then the discrete-time process $\{(N_n^+, J_n^+); n \geq 0\}$ constitutes a two-dimensional Markov chain with state space $\{(i,j); 0 \leq i \leq N,\ 1 \leq j \leq m\}$. Now observing the system immediately after a departure, the transition probability matrix (*TPM*) $\mathcal{P}$ of the above Markov process is obtained as

$$\mathcal{P} = \begin{bmatrix}
\mathcal{B}_{0,0} & \mathcal{B}_{0,1} & \mathcal{B}_{0,2} & \cdots & \mathcal{B}_{0,N-1} & \bar{\mathcal{B}}_{0,N} \\
\mathcal{A}_{1,0} & \mathcal{A}_{1,1} & \mathcal{A}_{1,2} & \cdots & \mathcal{A}_{1,N-1} & \bar{\mathcal{A}}_{1,N} \\
0 & \mathcal{A}_{2,0} & \mathcal{A}_{2,1} & \cdots & \mathcal{A}_{2,N-2} & \bar{\mathcal{A}}_{2,N-1} \\
0 & 0 & \mathcal{A}_{3,0} & \cdots & \mathcal{A}_{3,N-3} & \bar{\mathcal{A}}_{3,N-2} \\
\vdots & \vdots & \vdots & \ddots & \vdots & \vdots \\
0 & 0 & 0 & \cdots & \mathcal{A}_{N,0} & \bar{\mathcal{A}}_{N,1}
\end{bmatrix} \tag{3}$$

Let $\pi_i^+(n)$, $0 \leq n \leq N$, be the joint probability that there are n customers in the queue and the state of the arrival process is i $(1 \leq i \leq m)$ immediately after the departure of a customer. Further, define $\tilde{\boldsymbol{\pi}}^+(n) = (\pi_1^+(n), \pi_2^+(n), \ldots, \pi_m^+(n))$, $0 \leq n \leq N$. The unknown quantities $\tilde{\boldsymbol{\pi}}^+(n)$ can be obtained by solving the system of equations $\tilde{\boldsymbol{\pi}}^+\mathcal{P} = \tilde{\boldsymbol{\pi}}^+$ with $\tilde{\boldsymbol{\pi}}^+\mathbf{e} = 1$, where $\tilde{\boldsymbol{\pi}}^+ = (\tilde{\boldsymbol{\pi}}^+(0), \tilde{\boldsymbol{\pi}}^+(1), \ldots, \tilde{\boldsymbol{\pi}}^+(N))$ is a vector of dimension $(N+1)m$.

2.2 Queue Length Distribution at an Arbitrary Epoch

In this section, we obtain joint distribution of queue length and phase of the system at arbitrary epoch. Towards this end, we define the state of the system at time t as follows:

- $N_q(t)$ is the number of customers in the queue waiting for service
- $J(t)$ is the phase of the arrival process
- $U(t)$ is the remaining service time of a customer in service, if any
- $\xi(t)$ be the state of the server, that is,

$$\xi(t) = \begin{cases} 1, \text{ if server is busy,} \\ 0, \text{ if server is idle.} \end{cases}$$

Let us define for $1 \leq i \leq m$,

$$\begin{aligned} p_i(t) &= Pr\{N_q(t) = 0, J(t) = i, \xi(t) = 0\}, \\ \pi_i(n, u; t)du &= Pr\{N_q(t) = n, J(t) = i, u < U(t) \leq u + du, \xi(t) = 1\}, \quad 0 \leq n \leq N, \ u \geq 0. \end{aligned}$$

Define the steady-state probabilities, for $1 \leq i \leq m$, as

$$\begin{aligned} p_i &= \lim_{t \to \infty} p_i(t), \\ \pi_i(n, u) &= \lim_{t \to \infty} \pi_i(n, u; t), \ 0 \leq n \leq N \ u \geq 0. \end{aligned}$$

Let us define the vectors $\boldsymbol{p}$ and $\boldsymbol{\pi}(n, u)$, $0 \leq n \leq N$, $u \geq 0$, of order m whose jth components are given by p_j and $\pi_j(n, u)$, respectively. Then relating the state of the system at time t and $t + dt$ and using the supplementary variable technique, in steady state, we have the following (matrix) differential equations.

$$\mathbf{0} = \boldsymbol{p}D_0 + \boldsymbol{\pi}(0, 0), \tag{4}$$

$$-\frac{d}{du}\boldsymbol{\pi}(0, u) = \boldsymbol{\pi}(0, u)D_0 + h_1(u)\boldsymbol{\pi}(1, 0) + h_1(u)\boldsymbol{p}D_1, \tag{5}$$

$$-\frac{d}{du}\boldsymbol{\pi}(n, u) = \boldsymbol{\pi}(n, u)D_0 + h_{n+1}(u)\boldsymbol{p}D_{n+1} + h_{n+1}(u)\boldsymbol{\pi}(n + 1, 0) + \sum_{i=1}^{n} \boldsymbol{\pi}(n - i, u)D_i, \quad 1 \leq n \leq N - 2, \tag{6}$$

$$-\frac{d}{du}\boldsymbol{\pi}(N - 1, u) = \boldsymbol{\pi}(N - 1, u)D_0 + h_N(u)\boldsymbol{p}\sum_{i=N}^{\infty} D_i + h_N(u)\boldsymbol{\pi}(N, 0)\sum_{i=1}^{N-1} \boldsymbol{\pi}(N - 1 - i, u)D_i, \tag{7}$$

$$-\frac{d}{du}\boldsymbol{\pi}(N, u) = \boldsymbol{\pi}(N, u)\sum_{i=0}^{\infty} D_i + \sum_{j=1}^{N}\sum_{i=j}^{\infty} \boldsymbol{\pi}(N - j, u)D_i. \tag{8}$$

Define the Laplace transform of $\boldsymbol{\pi}(n, u)$ as $\boldsymbol{\pi}^*(n, s) = \int_0^\infty e^{-su}\boldsymbol{\pi}(n, u)du, \quad 0 \leq n \leq N, \ \mathbb{R}e\ s \geq 0$ and observe that $\boldsymbol{\pi}(n) \equiv \boldsymbol{\pi}^*(n, 0) = \int_0^\infty \boldsymbol{\pi}(n, u)du, 0 \leq n \leq N.$

Now multiplying (5)–(8) by e^{-su} and integrating with respect to u over 0 to ∞, we have

$$-s\boldsymbol{\pi}^*(0,s)+\boldsymbol{\pi}(0,0)=\boldsymbol{\pi}^*(0,s)D_0+H_1^*(s)\boldsymbol{\pi}(1,0)+H_1^*(s)\boldsymbol{p}D_1, \tag{9}$$

$$-s\boldsymbol{\pi}^*(n,s)+\boldsymbol{\pi}(n,0)=\boldsymbol{\pi}^*(n,s)D_0+H_{n+1}^*(s)\boldsymbol{p}D_{n+1}+H_{n+1}^*(s)\boldsymbol{\pi}(n+1,0)+\sum_{i=1}^{n}\boldsymbol{\pi}^*(n-i,s)D_i,\quad 1\le n\le N-2, \tag{10}$$

$$-s\boldsymbol{\pi}^*(N-1,s)+\boldsymbol{\pi}(N-1,0)=\boldsymbol{\pi}^*(N-1,s)D_0+H_N^*(s)\boldsymbol{p}\sum_{i=N}^{\infty}D_i+H_N^*(s)\boldsymbol{\pi}(N,0)+\sum_{i=1}^{N-1}\boldsymbol{\pi}^*(N-1-i,s)D_i, \tag{11}$$

$$-s\boldsymbol{\pi}^*(N,s)+\boldsymbol{\pi}(N,0)=\boldsymbol{\pi}^*(N,s)\sum_{i=0}^{\infty}D_i+\sum_{j=1}^{N}\sum_{i=j}^{\infty}\boldsymbol{\pi}^*(N-j,s)D_i. \tag{12}$$

Post multiplying (9)–(12) by the vector $\mathbf{e}$, adding them and using $\sum_{i=0}^{\infty}D_i\mathbf{e}=\mathbf{0}$, we obtain

$$\sum_{n=0}^{N}\boldsymbol{\pi}^*(n,s)\mathbf{e}=\sum_{n=1}^{N-1}\frac{1-H_n^*(s)}{s}\boldsymbol{p}D_n\mathbf{e}+\frac{1-H_N^*(s)}{s}\sum_{n=N}^{\infty}\boldsymbol{p}D_n\mathbf{e}+\sum_{n=1}^{N}\frac{1-H_n^*(s)}{s}\boldsymbol{\pi}(n,0)\mathbf{e}. \tag{13}$$

Taking limit $s\to 0$ in (13) yields

$$\sum_{n=0}^{N}\boldsymbol{\pi}(n)\mathbf{e}=\sum_{n=1}^{N-1}h_n\boldsymbol{p}D_n\mathbf{e}+h_N\sum_{n=N}^{\infty}\boldsymbol{p}D_n\mathbf{e}+\sum_{n=1}^{N}h_n\boldsymbol{\pi}(n,0)\mathbf{e},$$

$$\Rightarrow 1-\boldsymbol{p}\mathbf{e}=\sum_{n=1}^{N-1}h_n\boldsymbol{\pi}(0,0)\widetilde{D}_n\mathbf{e}+h_N\boldsymbol{\pi}(0,0)\widetilde{D}_N\mathbf{e}+\sum_{n=1}^{N}h_n\boldsymbol{\pi}(n,0)\mathbf{e}. \tag{14}$$

Using (4) and the normalizing condition $\boldsymbol{p}\mathbf{e}+\sum_{n=0}^{N}\boldsymbol{\pi}(n)\mathbf{e}=1$, the above relation has been obtained.

Before giving relations between $\{\boldsymbol{p},\boldsymbol{\pi}(n)\}$ and $\{\tilde{\boldsymbol{\pi}}^+(n)\}$, let us first derive the following results which will be used later.

It may be noted here that as $\tilde{\boldsymbol{\pi}}^{+}(n)$ and $\boldsymbol{\pi}(n,0)$ are proportional to each other, hence

$$\tilde{\boldsymbol{\pi}}^{+}(n) = d\boldsymbol{\pi}(n,0), \quad 0 \le n \le N, \tag{15}$$

where d is a proportionality constant. The following lemma gives an expression for d.

Lemma 1 *The value of d, as appeared in (15), is given by*

$$d^{-1} = \sum_{n=0}^{N} \boldsymbol{\pi}(n,0)\mathbf{e} = \frac{1-\boldsymbol{p}\mathbf{e}}{g}, \tag{16}$$

where $g = \sum_{n=1}^{N-1} h_n \tilde{\boldsymbol{\pi}}^{+}(0)\widetilde{D}_n\mathbf{e} + h_N \tilde{\boldsymbol{\pi}}^{+}(0)\widetilde{D}_N\mathbf{e} + \sum_{n=1}^{N} h_n \tilde{\boldsymbol{\pi}}^{+}(n)\mathbf{e}$.

Proof Summing both sides of (15) over the range of n, the desired result $d^{-1} = \sum_{n=0}^{N} \boldsymbol{\pi}(n,0)\mathbf{e}$ is obtained. Then dividing (14) by $\sum_{n=0}^{N} \boldsymbol{\pi}(n,0)\mathbf{e}$ and using (15), the desired result (16) is obtained after little algebraic operation.

Theorem 1 *The state probabilities* $\{\boldsymbol{p}, \boldsymbol{\pi}(n)\}$ *and* $\{\tilde{\boldsymbol{\pi}}^{+}(n)\}$ *are related by*

$$\boldsymbol{p} = \tau[\tilde{\boldsymbol{\pi}}^{+}(0)(-D_0)^{-1}] \tag{17}$$

$$\boldsymbol{\pi}(0) = \tau[\tilde{\boldsymbol{\pi}}^{+}(1) - \tilde{\boldsymbol{\pi}}^{+}(0) + \tilde{\boldsymbol{\pi}}^{+}(0)\widetilde{D}_1](-D_0)^{-1} \tag{18}$$

$$\boldsymbol{\pi}(n) = \tau[\tilde{\boldsymbol{\pi}}^{+}(n+1) - \tilde{\boldsymbol{\pi}}^{+}(n) + \tilde{\boldsymbol{\pi}}^{+}(0)\widetilde{D}_{n+1}](-D_0)^{-1} + \sum_{i=1}^{n} \boldsymbol{\pi}(n-i)D_i(-D_0)^{-1}, \quad 0 \le n \le N-2, \tag{19}$$

$$\boldsymbol{\pi}(N-1) = \tau[\tilde{\boldsymbol{\pi}}^{+}(N) - \tilde{\boldsymbol{\pi}}^{+}(N-1) + \tilde{\boldsymbol{\pi}}^{+}(0)\widetilde{D}_N](-D_0)^{-1} + \sum_{i=1}^{N-1} \boldsymbol{\pi}(N-1-i)D_i(-D_0)^{-1}, \tag{20}$$

$$\boldsymbol{\pi}(N) = \boldsymbol{\delta} - \boldsymbol{p} - \sum_{n=0}^{N-1} \boldsymbol{\pi}(n), \tag{21}$$

where $\tau^{-1} = g + \tilde{\boldsymbol{\pi}}^{+}(0)(-D_0)^{-1}\mathbf{e}$ *and g is given in Lemma 1.*

Proof Dividing (4) by $\sum_{n=0}^{N} \boldsymbol{\pi}(n,0)\mathbf{e}$ and using (15) and Lemma 1, after little manipulations the desired result (17) is obtained. Now setting $s = 0$ in (9)–(11), using (15) and following a recursive procedure, the desired result (18)–(20) is obtained with the help of Lemma 1. The last relation (21) follows immediately from the normalizing condition $\boldsymbol{p} + \sum_{n=0}^{N} \boldsymbol{\pi}(n) = \boldsymbol{\delta}$.

3 Computational Procedure

This section describes the necessary steps required for the computation of the elements and the matrices $\mathcal{A}_{n,k}$ of *TPM* $\mathcal{P}$ by considering phase-type (PH distribution) service time distribution. PH distribution can be completely represented by $(\boldsymbol{\beta}, \mathbf{S})$, where $\boldsymbol{\beta}$ and $\mathbf{S}$ are of dimension ν (i.e. $\boldsymbol{\beta}$ is an initial probability vector and $\mathbf{S}$ is a square matrix governing the transitions to various transition states). For more detail on PH distribution and their properties, see Neuts [15]. The following theorem gives a procedure for the computation of the matrices $\mathcal{A}_{n,k}$.

Theorem 2 *Let $H_n(\cdot)$ $(1 \le n \le N)$ follows a PH distribution with irreducible representation $(\boldsymbol{\beta}_n, \mathbf{S}_n)$, where $\boldsymbol{\beta}_n$ and $\mathbf{S}_n$ are of dimension ν, then the matrices $\mathcal{A}_{n,k}$ appearing in (3) are given by*

$$\mathcal{A}_{n,k} = (I_m \otimes \boldsymbol{\beta}_n)M_{n,k}(I_m \otimes \mathbf{S}_n^0), \quad 1 \le n \le N, \ 0 \le k \le N-n, \tag{22}$$

$$\bar{\mathcal{A}}_{n,k} = (I_m \otimes \boldsymbol{\beta}_n)\bar{M}_{n,k}(I_m \otimes \mathbf{S}_n^0), \quad 1 \le n \le N, \ k = N-n+1, \tag{23}$$

where

$$\mathbf{S}_n^0 = -\mathbf{S}_n\mathbf{e},$$

$$M_{n,k} = \sum_{i=1}^{k} M_{n,k-i}(D_i \otimes I_\nu)M_{n,0} \quad 1 \le n \le N, \ 1 \le k \le N-n$$

$$\bar{M}_{n,k} = \left[-\sum_{l=0}^{k-1} M_{n,l}(\widehat{D}_{k-l} \otimes I_\nu)(D \oplus \mathbf{S}_n)^{-1}\right] \quad 1 \le n \le N, \ k = N-n+1,$$

$$M_{n,0} = -(D_0 \oplus \mathbf{S}_n)^{-1}, \ 1 \le n \le N-1.$$

Proof Proof follows similar steps as described in Banerjee et al. [16] and Banerjee [13].

4 Performance Measures

The performance measures are key features of any queueing system as they reflect the efficiency of the queueing model under consideration. The average number of customers in the queue (L_q) and average waiting time of a customer in the queue (W_q) is given by $L_q = \sum_{n=0}^{N} n\boldsymbol{\pi}(n)\mathbf{e}$ and $W_q = \frac{L_q}{\lambda^*}$, respectively. Another important performance measure that is the loss probability, which is also termed as blocking probability, is discussed below.

4.1 Loss Probability

- The blocking probability of the first customer of an arriving batch is given by $PBL_F = \frac{\boldsymbol{\pi}(N)\widehat{D}_1\mathbf{e}}{\lambda_g}$
- The blocking probability of an arbitrary arriving customer is given by

$$PBL_A = \boldsymbol{p} \sum_{k=N+1}^{\infty} G_k\mathbf{e} + \sum_{n=0}^{N} \boldsymbol{\pi}(n) \sum_{k=N-n+1}^{\infty} G_k\mathbf{e}$$

where $G_k = \frac{\widehat{D}_k}{\lambda^*}$, $k = 1, 2, ...,$ is a matrix of dimension m and represents that the position of an arbitrary customer in a accepted batch is k.

- The blocking probability of the last customer is given by $PBL_L = \frac{\boldsymbol{p}\widehat{D}_{N+1}\mathbf{e}}{\lambda_g}$ $+\sum_{n=0}^{N} \sum_{i=N-n+1}^{\infty} \frac{\boldsymbol{\pi}(n)D_i\mathbf{e}}{\lambda_g}$.

5 Numerical Results

The numerical compatibility of the analytical results, as obtained in the previous sections, is illustrated in this section in self-explanatory tables and graphs. Towards this end, in Table 1 and Table 2, the queue length distributions at various epoch for $BMAP/PH_n/1/30$ and $BMAP/PH_n/1/20$ queue have been displayed, respectively. At the bottom of the table, various performance measures are also presented. In Table 1, the input parameters are considered as follows:
The *BMAP* representation is taken as

$$D_0 = \begin{bmatrix} -0.986 & 0.12 & 0.023 \\ 0.01 & -0.999 & 0.04 \\ 0.11 & 0.150 & -1.439 \end{bmatrix}, D_2 = \begin{bmatrix} 0.013 & 0.140 & 0.0 \\ 0.15 & 0.0 & 0.135 \\ 0.206 & 0.198 & 0.078 \end{bmatrix}, D_5 = \begin{bmatrix} 0.123 & 0.0 & 0.127 \\ 0.116 & 0.090 & 0.008 \\ 0.167 & 0.087 & 0.197 \end{bmatrix}$$

and $D_7 = \begin{bmatrix} 0.120 & 0.230 & 0.09 \\ 0.45 & 0.0 & 0.0 \\ 0.0 & 0.119 & 0.127 \end{bmatrix}$. For service time, PH-distribution is taken as $\boldsymbol{\beta}_n = \begin{pmatrix} 0.4 & 0.6 \end{pmatrix}$ and $\mathbf{S}_n = \begin{pmatrix} -\mu_n & \mu_n \\ 0 & -\mu_n \end{pmatrix}$ for $1 \leq n \leq N$, where $\mu_n = \mu_{n-1} + 0.3$ for $1 \leq n \leq N$ and $\mu_0 = 5.0$. In Table 2, the input parameters are considered as follows:

Table 1 Queue length distribution at departure epoch and arbitrary epoch for $BMAP/PH_n/1/30$ queue with $m = 3$, $\lambda^* = 4.74360$ and $\lambda_g = 0.93874$

	$\tilde{\boldsymbol{\pi}}^+(n)$				$\boldsymbol{\pi}(n)$			
n	$j = 1$	$j = 2$	$j = 3$	$\sum_{j=1}^{m} \pi_j^+(n)$	$j = 1$	$j = 2$	$j = 3$	$\sum_{j=1}^{m} \pi_j(n)$
0	0.00004868	0.00006013	0.00001016	0.00011897	0.00038899	0.00047898	0.00008155	0.00094952
1	0.00013608	0.00015931	0.00003125	0.00032664	0.00078038	0.00091050	0.00018037	0.00187125
2	0.00029826	0.00033668	0.00006983	0.00070477	0.00135693	0.00151952	0.00032871	0.00320516
3	0.00058021	0.00062865	0.00014311	0.00135197	0.00215227	0.00231709	0.00054621	0.00501557
4	0.00101867	0.00106110	0.00026290	0.00234266	0.00318945	0.00330533	0.00084375	0.00733853
5	0.00164239	0.00165396	0.00043778	0.00373413	0.00449190	0.00448073	0.00123240	0.01020503
6	0.00250359	0.00243222	0.00068979	0.00562560	0.00603202	0.00580511	0.00170795	0.01354508
7	0.00359175	0.00339029	0.00102261	0.00800466	0.00786081	0.00729403	0.00228097	0.01743582
8	0.00495846	0.00453993	0.00145140	0.01094979	0.00982882	0.00884085	0.00292434	0.02159401
9	0.00657802	0.00585746	0.00197587	0.01441135	0.01192347	0.01043486	0.00363478	0.02599311
10	0.00842707	0.00731680	0.00259386	0.01833774	0.01405411	0.01199447	0.00438565	0.03043423
11	0.01046443	0.00887976	0.00329684	0.02264103	0.01614974	0.01347838	0.00515273	0.03478084
12	0.01263479	0.01050121	0.00406946	0.02720546	0.01816347	0.01485219	0.00592368	0.03893933
13	0.01487420	0.01213177	0.00489299	0.03189895	0.01998245	0.01604699	0.00665164	0.04268108
14	0.01710980	0.01371856	0.00574244	0.03657081	0.02160399	0.01707011	0.00733164	0.04600574
15	0.01926790	0.01521007	0.00659157	0.04106954	0.02291669	0.01784194	0.00792259	0.04868123

(continued)

Table 1 (continued)

	$\tilde{\boldsymbol{\pi}}^{+}(n)$				$\boldsymbol{\pi}(n)$			
16	0.02127666	0.01655768	0.00741266	0.04524700	0.02392453	0.01837740	0.00841785	0.05071978
17	0.02307095	0.01771890	0.00817917	0.04896902	0.02458984	0.01865511	0.00880066	0.05204561
18	0.02459654	0.01866013	0.00886731	0.05212398	0.02490957	0.01868216	0.00906087	0.05265259
19	0.02581198	0.01935705	0.00945738	0.05462640	0.02490958	0.01848029	0.00920483	0.05259470
20	0.02668835	0.01979435	0.00993414	0.05641684	0.02458135	0.01804905	0.00922353	0.05185393
21	0.02721281	0.01996779	0.01028854	0.05746915	0.02398589	0.01743455	0.00913549	0.05055594
22	0.02738405	0.01988054	0.01051661	0.05778120	0.02312599	0.01664187	0.00894240	0.04871026
23	0.02721577	0.01954559	0.01062059	0.05738195	0.02206311	0.01571670	0.00866390	0.04644370
24	0.02673127	0.01898137	0.01060753	0.05632017	0.02083268	0.01468190	0.00831452	0.04382911
25	0.02596313	0.01821199	0.01048900	0.05466412	0.01947692	0.01356849	0.00790920	0.04095461
26	0.02495051	0.01726482	0.01028035	0.05249568	0.01804730	0.01241036	0.00746958	0.03792724
27	0.02373547	0.01616822	0.00999917	0.04990286	0.01656731	0.01122302	0.00700618	0.03479650
28	0.02236327	0.01495164	0.00966537	0.04698029	0.01508710	0.01003939	0.00654006	0.03166656
29	0.02087740	0.01364223	0.00929926	0.04381889	0.01362261	0.00886730	0.00608096	0.02857086
30	0.01932076	0.01226591	0.00892172	0.04050839	0.01311208	0.00833111	0.00601537	0.02745855
Total	0.47133324	0.35084609	0.17782066	1.00000000	0.46971135	0.35782977	0.17191435	0.99945547

$\boldsymbol{p} = [0.00021638\ 0.00028686\ 0.00004129]$, $\boldsymbol{pe} = 0.00054453$, $L_q = 18.21935011$, $W_q = 3.84083056$, $PBL_F = 0.02775198$, $PBL_A = 0.10810429$, $PBL_L = 0.16922842$

Table 2 Queue length distribution at departure epoch and arbitrary epoch for $BMAP/PH_n/1/20$ queue with $m = 4$, $\lambda^* = 9.5645$ and $\lambda_g = 2.2221$

	$\tilde{\boldsymbol{\pi}}^+(n)$					$\boldsymbol{\pi}(n)$				
n	$j = 1$	$j = 2$	$j = 3$	$j = 4$	$\sum_{j=1}^{m} \pi_j^+(n)$	$j = 1$	$j = 2$	$j = 3$	$j = 4$	$\sum_{j=1}^{m} \pi_j(n)$
0	0.000026	0.000069	0.000024	0.000070	0.000190	0.000024	0.000062	0.000022	0.000062	0.000170
1	0.000033	0.000083	0.000030	0.000093	0.000240	0.000031	0.000078	0.000029	0.000087	0.000225
2	0.000050	0.000121	0.000045	0.000134	0.000350	0.000048	0.000117	0.000044	0.000130	0.000339
3	0.000078	0.000180	0.000070	0.000196	0.000524	0.000077	0.000177	0.000069	0.000193	0.000517
4	0.000105	0.000253	0.000096	0.000279	0.000733	0.000104	0.000252	0.000095	0.000278	0.000729
5	0.000154	0.000368	0.000140	0.000405	0.001067	0.000154	0.000370	0.000141	0.000407	0.001073
6	0.000228	0.000541	0.000209	0.000590	0.001569	0.000232	0.000548	0.000212	0.000598	0.001590
7	0.000324	0.000775	0.000295	0.000845	0.002239	0.000329	0.000788	0.000299	0.000859	0.002275
8	0.000470	0.001125	0.000427	0.001222	0.003244	0.000480	0.001150	0.000435	0.001249	0.003315
9	0.000687	0.001644	0.000622	0.001773	0.004726	0.000704	0.001687	0.000638	0.001820	0.004849
10	0.001007	0.002405	0.000909	0.002572	0.006893	0.001036	0.002477	0.000935	0.002648	0.007096
11	0.001478	0.003522	0.001328	0.003726	0.010054	0.001526	0.003638	0.001371	0.003845	0.010380
12	0.002182	0.005169	0.001951	0.005393	0.014695	0.002260	0.005353	0.002020	0.005577	0.015209
13	0.003238	0.007597	0.002878	0.007789	0.021502	0.003363	0.007883	0.002987	0.008066	0.022299
14	0.004837	0.011174	0.004269	0.011205	0.031484	0.005037	0.011615	0.004442	0.011619	0.032712
15	0.007292	0.016442	0.006387	0.016040	0.046162	0.007615	0.017117	0.006664	0.016645	0.048041
16	0.011130	0.024182	0.009671	0.022791	0.067774	0.011655	0.025206	0.010120	0.023659	0.070640
17	0.017256	0.035493	0.014883	0.032035	0.099668	0.018129	0.037027	0.015625	0.033247	0.104029
18	0.027304	0.051865	0.023406	0.044327	0.146902	0.028790	0.054130	0.024665	0.045955	0.153541
19	0.044311	0.075173	0.037844	0.059902	0.217230	0.046909	0.078439	0.040055	0.061960	0.227365
20	0.074106	0.107387	0.063280	0.077983	0.322756	0.067782	0.097349	0.057881	0.070298	0.293310
Total	0.196294	0.345570	0.168766	0.289370	1.000000	0.196286	0.345465	0.168749	0.289203	0.999703

$\boldsymbol{p} = [0.000037\ 0.000099\ 0.000035\ 0.000126]$, $\boldsymbol{pe} = 0.000297$, $L_q = 17.79105941$, $W_q = 1.860114703$, $PBL_F = 0.297813723$, $PBL_A = 0.643984446$, $PBL_L = 0.693968795$

The *BMAP* representation is taken as

$$D_0 = \begin{bmatrix} -3.7893 & 0.456 & 0.284 & 0.423 \\ 0.329 & -2.703 & 0.063 & 0.488 \\ 0.271 & 0.480 & -3.638 & 0.236 \\ 0.071 & 0.0 & 0.231 & -2.4764 \end{bmatrix}, \quad D_1 = \begin{bmatrix} 0.303 & 0.417 & 0.0 & 0.045 \\ 0.087 & 0.452 & 0.141 & 0.156 \\ 0.252 & 0.0 & 0.053 & 0.198 \\ 0.059 & 0.275 & 0.148 & 0.051 \end{bmatrix},$$

$$D_4 = \begin{bmatrix} 0.220 & 0.440 & 0.0423 & 0.0 \\ 0.049 & 0.088 & 0.0 & 0.001 \\ 0.219 & 0.186 & 0.380 & 0.410 \\ 0.343 & 0.166 & 0.272 & 0.094 \end{bmatrix} \text{ and } D_7 = \begin{bmatrix} 0.472 & 0.333 & 0.246 & 0.108 \\ 0.135 & 0.349 & 0.046 & 0.319 \\ 0.097 & 0.086 & 0.291 & 0.479 \\ 0.217 & 0.170 & 0.362 & 0.0174 \end{bmatrix}.$$

For service time, *PH* distribution is taken as $\boldsymbol{\beta}_n = \begin{pmatrix} 1.0 & 0.0 \end{pmatrix}$ and $\mathbf{S}_n = \begin{pmatrix} -2\mu_n & \mu_n \\ 0 & -2\mu_n \end{pmatrix}$ for $1 \le n \le N$, where $\mu_n = 2.5 + \frac{1}{1+n}$ for $0 \le n \le N$.

In Figs. 1, 2 and 3, we carry out a comparative study of our model, where service time changes depending on the queue length at service initiation epoch, with *BMAP*/*G*/1/*N* queue, where service time remains constant for fixed *N*. This comparison clearly demonstrates that the model under consideration may help in optimizing the system performance, in terms of minimizing average queue length, average waiting time of a customer in the queue and loss probability, by controlling the service rate depending on the queue length.

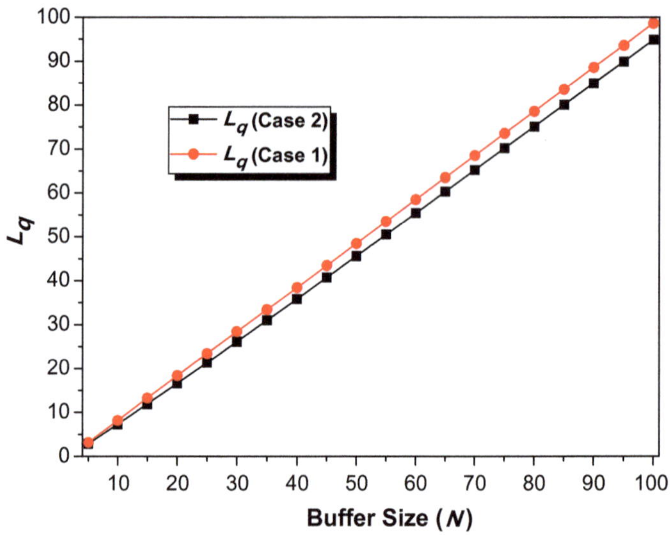

Fig. 1 Effect of *N* on average queue length

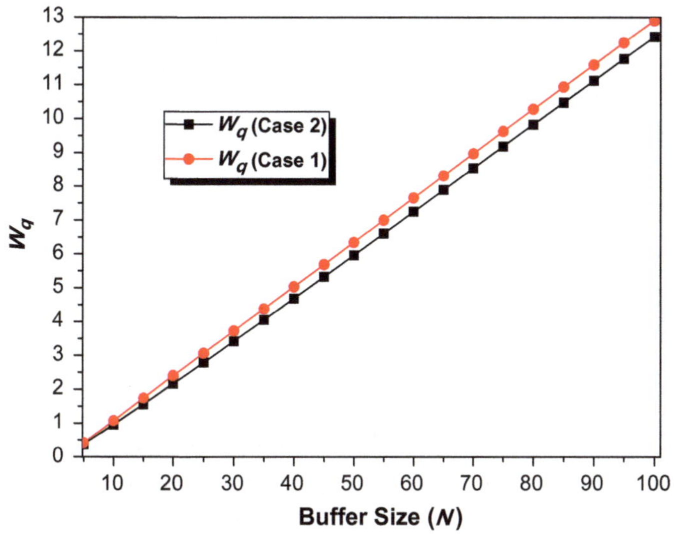

Fig. 2 Effect of N on average waiting time of a customer in the queue

In Figs. 1, 2 and 3, the effect of N (buffer size) on L_q, W_q and loss probability (PBL_F, PBL_A, PBL_L), respectively, has been studied for $BMAP/PH_n/1/N$ (N varies from 5 to 100) queue with the following input parameters: The $BMAP$ representation is taken as $D_0 = \begin{bmatrix} -2.823 & 0.0 \\ 0.188 & -3.121 \end{bmatrix}$, $D_1 = \begin{bmatrix} 0.139 & 0.193 \\ 0.176 & 0.104 \end{bmatrix}$, $D_2 = \begin{bmatrix} 0.182 & 1.228 \\ 1.050 & 0.513 \end{bmatrix}$ and $D_4 = \begin{bmatrix} 0.258 & 0.823 \\ 0.367 & 0.723 \end{bmatrix}$. For service time, PH distribution is taken as $\boldsymbol{\beta}_n = \left(0.4\ 0.6\right)$, $\mathbf{S}_n = \begin{pmatrix} -2\mu_n & \mu_n \\ 0 & -2\mu_n \end{pmatrix}$, for $1 \le n \le N$.

Here we consider the following two cases where Case 1 represents the constant service time distribution whereas Case 2 represents the queue length dependent service process.
Case 1: $\mu_n = \frac{(N+2)\mu}{2N}$ for $0 \le n \le N$ and $\mu = 3.5$,

Case 2: $\mu_n = \frac{(n+1)\mu}{N}$ for $0 \le n \le N$ and $\mu = 3.5$.

It is clear from Figs. 1 and 2 that as N increases L_q and W_q increases for both the cases, and the value of L_q and W_q is little high for Case 1 in comparison to Case 2 for fixed value of N. Therefore, in this example although there is no considerable differences in the values of L_q and W_q are observed for Case 1 and Case 2, however, it can be concluded that queue length dependent service policy minimizes L_q and W_q. Now from Fig. 3, it can be observed that as N increases loss/blocking probability (PBL_F, PBL_A, PBL_L) decreases for Case 2 whereas it increases for Case 1.

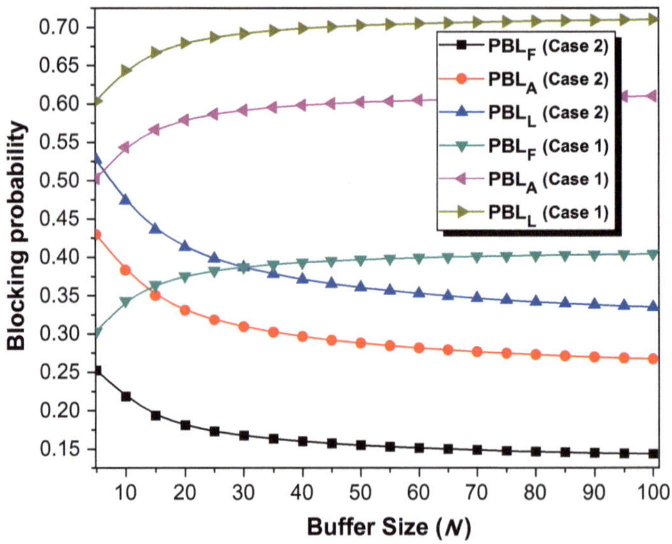

Fig. 3 Effect of N on Blocking Probability

Also it is observed that for fixed N, the values of PBL_F, PBL_A, PBL_L for Case 1 are very high in comparison with the corresponding values of PBL_F, PBL_A, PBL_L for Case 2. With this observation we can conclude that the queue length dependent service policy decreases loss probability very significantly. Therefore, in total we can come to an conclusion that the queue length dependent service mechanism which is studied in this paper with $BMAP/G/1/N$ queue may help in reducing congestion as this is reducing its system performance measures as discussed above.

6 Conclusion

In this paper, we have considered finite buffer queue where customers arrive according to *BMAP*. We analysed the model by considering general service time distribution where service rates are changing dynamically depending on the number of customers waiting for service in the queue at service initiation epoch. Using the embedded Markov chain technique, we obtained the departure epoch probabilities. Then using the supplementary variable technique, we derive a relation with departure epoch probabilities to obtain arbitrary epoch probabilities. Finally, we give an computational procedure by considering phase-type distribution. In future, it would be interesting to study the model with vacation or working vacation.

Acknowledgements The authors thank the anonymous referee for their valuable comments. The second author acknowledges the Department of Science and Technology (DST), Govt. of India, for the partial financial support under the project grant $SB/FTP/MS-048/2013$.

References

1. Lucantoni, D.M., Meier-Hellstern, K.S., Neuts, M.F.: A single-server queue with server vacations and a class of non-renewal arrival processes. Adv. Appl. Probab. **22**(3), 676–705 (1990)
2. Lucantoni, D.M.: New results on the single server queue with a batch Markovian arrival process. Stoch. Models **7**(1), 1–46 (1991)
3. Dudin, A., Klimenok, V.: Queueing system $BMAP/G/1$ with repeated calls. Math. Comput. Modell. **30**(34), 115–128 (1999)
4. Lee, H.W., Park, N.I., Jeon, J.: A new approach to the queue length and waiting time of $BMAP/G/1$ queues. Comput. Oper. Res. **30**(13), 2021–2045 (2003)
5. Shin, Y.W.: $BMAP/G/1$ queue with correlated arrivals of customers and disasters. Oper. Res. Lett. **32**(4), 364–373 (2004)
6. Banik, A., Gupta, U., Pathak, S.: $BMAP/G/1/N$ queue with vacations and limited service discipline. Appl. Math. Comput. **180**(2), 707–721 (2006)
7. Banik, A.D.: Queueing analysis and optimal control of $BMAP/G^{(a,b)}/1/N$ and $BMAP/MSP^{(a,b)}/1/N$ systems. Comput. Ind. Eng. **57**(3), 748–761 (2009)
8. Saffer, Z., Telek, M.: Analysis of *BMAP* vacation queue and its application to IEEE 802.16e sleep mode. J. Ind. Manag. Optim. **6**(3), 661–690 (2010)
9. Baek, J.W., Lee, H.W., Lee, S.W., Ahn, S.: A workload factorization for $BMAP/G/1$ vacation queues under variable service speed. Oper. Res. Lett. **42**(1), 58–63 (2014)
10. Sikdar, K., Samanta, S.K.: Analysis of a finite buffer variable batch service queue with batch markovian arrival process and server's vacation. Opsearch **53**(3), 553–583 (2016)
11. Choi, B.D., Choi, D.I.: Queueing system with queue length dependent service times and its application to cell discarding in ATM networks. J. Appl. Math. Stoch. Anal. **12**(1), 35–62 (1999)
12. Choi, D.I., Knessl, C., Tier, C.: A queueing system with queue length dependent service times, with applications to cell discarding in ATM networks. J. Appl. Math. Stoch. Anal. **12**(1), 35–62 (1999)
13. Banerjee, A.: Analysis of finite buffer queue with state dependent service and correlated customer arrivals. J. Egypt. Math. Soc. **24**, 295–302 (2015)
14. Neuts, M.F., Li, J.-M.: An algorithm for the $P(n, t)$ matrices of a continuous BMAP. In: Chakravarthy, S.R., Alfa, A.S. (eds.) Matrix-Analytic Methods in Stochastic Models. Marcel Dekker (1996)
15. Neuts, M.F.: Matrix-Geometric Solutions in Stochastic Models: An Algorithmic Approach. Johns Hopkins University Press, Baltimore (1981)
16. Banerjee, A., Gupta, U.C., Chakravarty, S.R.: Analysis of a finite-buffer bulk-service queue under Markovian arrival process with batch-size-dependent service. Comput. Oper. Res. **60**, 138–149 (2015)

Computational Analysis of a Single Server Queue with Batch Markovian Arrival and Exponential Single Working Vacation

A. D. Banik, Souvik Ghosh and Debasis Basu

Abstract In this paper, an infinite buffer queue with a single server and non-renewal batch arrival is studied. The service discipline is considered as exhaustive type under single exponential working vacation policy. Further, both the service times during the working vacation and normal busy period are assumed to be generally distributed random variables. It is also assumed that the service times and the arrival process are independent of each others. Moreover, it is accepted that at the end of an exponentially distributed working vacation, the first customer in the front of the queue is likely to receive service rate as per normal busy period service rate irrespective of received service in the working vacation period as the server shifts from working vacation mode to normal period mode. The system-length distributions at different epochs, such as post-departure and arbitrary epoch are obtained. The *RG*-factorization technique is applied to obtain the distribution of the system length at post-departure epoch. Henceforth, the system-length distribution at arbitrary epoch is determined by supplementary variable technique along with some simple algebraic manipulations. Some useful performance measures to be particular the mean system length of the model and the mean waiting time of an arbitrary customer in the system is discussed in the numerical section. Finally, some numerical results are presented for the model, in the form of the table and graphs. A possible application of the model in communication network is outlined in the paper.

Keywords Infinite buffer queue · Batch Markovian arrival process · Single working vacation · Single server · *RG* factorization

---

A. D. Banik (✉) · S. Ghosh
School of Basic Sciences, Indian Institute of Technology Bhubaneswar, Bhubaneswar, India
e-mail: adattabanik@iitbbs.ac.in; banikad@gmail.com

S. Ghosh
e-mail: sg19@iitbbs.ac.in; souvikghosh589@gmail.com

D. Basu
School of Infrastructure, Indian Institute of Technology Bhubaneswar, Bhubaneswar, India
e-mail: dbasu@iitbbs.ac.in; basudebasis2k@gmail.com

S. Kar et al. (eds.), *Operations Research and Optimization*, Springer Proceedings in Mathematics & Statistics 225, https://doi.org/10.1007/978-981-10-7814-9_5

1 Introduction

Queueing theory was originated through the early works of Erlang (1915) while designing telephone network and call availability during the peak hours of service. During the early days of queueing analysis, arrival process was realized as Poisson process. But the correlation in arrival processes in present days computer or communication networks is not described suitably by the so-called Poisson process. The correlation among the inter-arrival times can be fairly explained by Markovian arrival process (*MAP*), introduced by Lucantoni et al. [1]. Further, *MAP* was extended to batch Markovian arrival process (*BMAP*) by Lucantoni [2] to illustrate the correlation among the batch arrivals of variable capacity, which is also able to describe the versatile Markovian point process (*VMPP*), see Neuts [3] and Ramaswami [4].

Vacation queueing systems are considered to be a handy tool to model and analyze todays complex networks arising in communication, computer, and many other engineering systems. A vacation queuing model presumes that the server remains unproductive while it goes for a vacation. However, service requirements in modern fast and explosive networks demand an active server with reduced service rate during vacation. This kind of vacations are termed as working vacations (*WV*) and was studied by Servi and Finn [5]. They studied an *M*/*M*/1 queue with multiple working vacations (*MWV*), which means that after completing a working vacation if the server does not find any customer waiting in the queue then he again starts a working vacation. Besides this, if the server is unable to spot a customer on completing a *WV* and then remains idle, then it is called a single working vacation (*SWV*) rule. A discrete-time $Geo^{[X]}/G/1$ queue under *MWV* and *SWV* policies was examined by Li et al. [6] and Gao and Liu [7], respectively. Chae et al. [8] investigated both the continuous-time *GI*/*M*/1 queue and the discrete-time *GI*/*Geo*/1 queue with *SWV*. Considering both finite and infinite system capacity, Banik [9] studied single working vacation *GI*/*M*/1 queue. *GI*/*M*/1 queue with *SWV* is also studied by Li and Tian [10]. Recently, assuming rational-type *MWV* policy, Banik [11] investigated $BMAP/R/1/\infty$ queue using matrix analytic procedure.

In this paper, a $BMAP/G/1/\infty/SWV$ queue is investigated, where G indicates that the service times are generally distributed random variables for both the server's working vacation and busy mode. The working vacation is exponentially distributed, and the customers can receive service in a reduced rate during the working vacation. The computational analysis is based on the calculation of post-departure epoch probabilities using *RG*-factorization technique, see Li [12]. Henceforth, the supplementary variable method is used to get the arbitrary epoch probabilities using the relation between arbitrary and post-departure epoch probabilities. The queueing system studied in this paper can be modeled in a computer grid.

2 Application of the Model

A grid is an architecture of geographically distributed computer resources, which is used to solve a large-scale complex problem. The architecture builds with heterogeneous resources, i.e., each resource in the grid is dedicated to do different jobs. End users submitted their jobs to the grid and receive the executed results. However, resource management and scheduling is one of the key interests to meet the quality of service (QoS) requirements of the job such as the time bound to finish the job, the number of resources used. It may happen in the grid environment that some of the resources are overburdened while at the same time other resources are underutilized. This kind of unequal job distribution to the resources causes load imbalance in the grid. Analyzing the log file of the Large Hadron Collider Computing Grid (LCG), Yin et al. [13] have shown that there is an activity peak in daily cycle. They also observed that during the peak hours, there are more than 10% resources stands unproductive which cause the load imbalance in the grid. Therefore, it is needed to transfer the job requests from the overloaded resources to the light-loaded or idle resources to utilize the underused or unused resources and reduce the average job response time. We propose the $M/G/1-SWV$ model in the grid environment to achieve better result in terms of server utilization and QoS (job response time).

The vacation period of a resource may be utilized to reduce the workload of other resources. The consignment during vacation period is considered as secondary tasks to the resources. Resources may do the secondary tasks in a reduced speed as after a busy period a resource needs some maintenance. After the maintenance, the resources are again eligible to give service in a normal speed. We can model a grid environment in such a working vacation scenario to reduce the workload of the highly loaded resources as well as the mean job response time. There are three primary components of a grid environment, namely grid portal, scheduler, and grid resources. End users submit their request to the grid through the grid portal. Then, the job requests are delivered to the scheduler. The scheduler matches the user requests and available resources which is efficient to perform those kind of requests. If the scheduler finds that all the pertinent resources are busy while some other resources are idle or underutilized, then the jobs are transferred to those resources. A grid environment has been shown in Fig. 1.

3 Description and Analysis of the Queueing Model

In this section, the analysis of the $BMAP/G/1/\infty$ queue under *SWV* policy is carried out. In *SWV* strategy, the server is allowed to take a conditional vacation when he finds that the queue is empty upon completing a service in a normal busy period. The condition imposed on the sever is that he should provide service during vacation in a reduced rate while taking vacation, and after finishing a *WV* if the server do not find any waiting customer in the queue, then he should remain idle with normal service

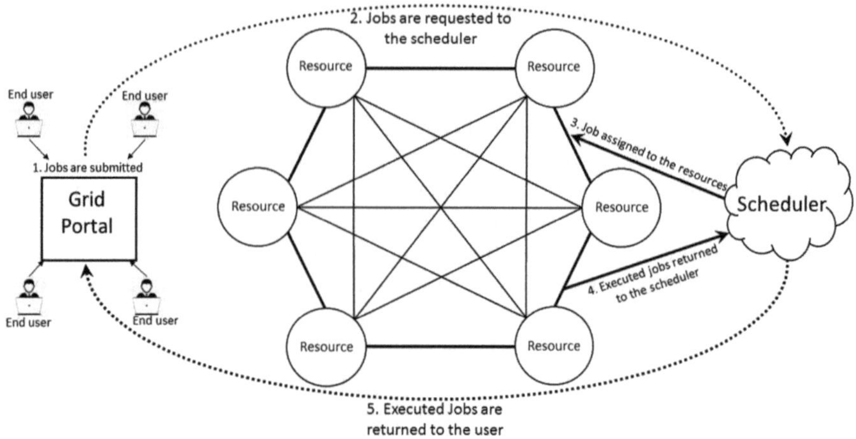

Fig. 1 Architecture of a grid environment

rate rather than taking another *WV*. On the other hand, upon completing a *WV*, if the server finds a customer in the queue then the server should immediately starts a normal busy period. That is, while the server is on *WV* mode, then the first customer on the queue will get service at a reduced rate and can receive service at normal rate if the *WV* terminates. It should be mentioned here that the change of service rate for the leading customer of the waiting queue is independent of the service received during server's *WV*. Let V, S_1, and S_2 are the random variables of the duration of a *WV*, service during normal busy period and service during *WV* period, respectively. Further, it is assumed that V, S_1, S_2, and the arrival process are independent of each others. Let the probability distribution function (DF) and the probability density function (pdf) of a random variable X are denoted by $F_X(x)$ and $f_X(x)$, respectively. Furthermore, $f_X^*(s)$ is symbolized as the Laplace–Stieltjes transform (LST) of $f_X(x)$. Hence, for the independent and identically distributed (iid.) random variable (r.v.) S_1 (S_2) the DF, pdf, and LST are expressed by $F_{S_1}(x)$ ($F_{S_2}(x)$), $f_{S_1}(x)$ ($f_{S_2}(x)$) and $f_{S_1}^*(s)$ ($f_{S_2}^*(s)$). Also, the expected service time during a normal busy period and vacation period are pointed by $1/\mu_1$ and $1/\mu_2$, respectively. Similarly, for exponential vacation times, the LST, pdf, and DF are noted by $f_V^*(s) = \frac{\gamma^*}{\gamma^*+s}$, $f_V(x) = \gamma^* e^{-\gamma^* x}$ and $F_V(x) = 1 - e^{-\gamma^* x}$, respectively, where γ^* (> 0) is assumed as the expected number of vacations per unit of time. One may note that due to lack of memory property, the residual vacation time $\widehat{V}$ is also exponential with parameter γ^*. Therefore, $f_{\widehat{V}}(x) = \gamma^* e^{-\gamma^* x}$ and $f_{\widehat{V}}^*(s) = \frac{\gamma^*}{\gamma^*+s}$.

For the model, it is assumed that the arrival is governed by an m-state batch Markovian arrival process (*BMAP*). The (*BMAP*) is represented by the matrices $\mathbf{D}_k$ ($k = 0, 1, 2, \ldots$), where $\mathbf{D}_0$ is the state transition rate matrix for no arrival and $\mathbf{D}_k$ is the state transition rate matrix for an arrival of batch size k; *BMAP* is discussed in details by Lucantoni [2]. Let the irreducible infinitesimal generator of the underlying Markov chain is denoted by $\boldsymbol{D}$, i.e., $\boldsymbol{D} = \sum_{k=0}^{\infty} \boldsymbol{D}_k$ with $\boldsymbol{De} = \mathbf{0}$, where $\mathbf{e}$ is a column vector with all elements equal to one and appropriate dimension. Then the stationary probability vector of the process, i.e., $\overline{\boldsymbol{\pi}}$, satisfies $\overline{\boldsymbol{\pi}}\mathbf{D} = \mathbf{0}$ and $\overline{\boldsymbol{\pi}}\mathbf{e} = 1$. The mean arrival rate and mean batch arrival rate of the stationary *BMAP* are given by $\lambda^* = \overline{\boldsymbol{\pi}} \sum_{k=1}^{\infty} k\mathbf{D}_k\mathbf{e}$ and $\lambda_g = \overline{\boldsymbol{\pi}} \sum_{k=1}^{\infty} \mathbf{D}_k\mathbf{e} = \overline{\boldsymbol{\pi}}\mathbf{D}_1^{'}\mathbf{e}$, respectively, where $\mathbf{D}_n^{'} = \sum_{i=n}^{\infty} \mathbf{D}_i$ $(n = 1, 2, 3, \ldots)$. Let at time t (≥ 0), $\mathbf{P}(n, t)$ $(n = 0, 1, 2, \ldots)$ denotes an $m \times m$ matrix whose (i,j)-th element is the conditional probability defined as $P_{ij}(n, t) = Pr\{N(t) = n,\ J(t) = j|N(0) = 0, J(0) = i\}$, where n, i and j are non-negative integers with $1 \leq i,j \leq m$. Hence, the matrices $\mathbf{P}(n, t)$ satisfy the following system of difference-differential equations:

$$\mathbf{P}^{(1)}(n, t) = \sum_{k=0}^{n} \mathbf{P}(k, t)\mathbf{D}_{n-k}, \quad n = 0, 1, 2, \ldots, \tag{1}$$

with $\mathbf{P}(0, 0) = \mathbf{I}_m$, where $\mathbf{I}_m$ is an identity matrix of order m and $\mathbf{P}^{(1)}(n, t) = \frac{d}{dt}\mathbf{P}(n, t)$. In the rest of the paper, an identity matrix of proper dimension is denoted by $\mathbf{I}$ and the subscript is mentioned wherever needed. For the continuous variable $|z| \leq 1$, let us define matrix generating function $\mathbf{P}^*(z, t) = \sum_{n=0}^{\infty} \mathbf{P}(n, t)z^n$. Now multiplying Eq. (1) by z^n and then summing over n, one can get $\frac{d}{dt}\mathbf{P}^*(z, t) = \mathbf{P}^*(z, t)\mathbf{D}(z)$, where $\mathbf{D}(z) = \sum_{k=0}^{\infty} \mathbf{D}_k z^k$. Hence, solving the matrix differential equation for the continuous variables $|z| \leq 1$ and $t \geq 0$, it can be derived $\mathbf{P}^*(z, t) = e^{\mathbf{D}(z)t}$. Let us define the $m \times m$ matrices of mass functions as

$$\mathbf{A}_n = \int_0^{\infty} \mathbf{P}(n, t)dF_{S_1}(t),\ \ \mathbf{V}_n = \int_0^{\infty} \mathbf{P}(n, t)dF_{S_2}(t), \quad n = 0, 1, 2, \ldots. \tag{2}$$

The numerical computation of A_n and V_n for *PH*-type and non-*PH*-type distributions are discussed by Neuts [14], Lucantoni [2] and Banik [11].

If $\widehat{V}$ stands for the remaining vacation time, then the distribution of $\widehat{V}$ is exactly same as V. The probability that the remaining vacation time exceeds the service time during a *WV* is given by $\varepsilon = \int_0^{\infty}(1 - F_{\widehat{V}}(x))f_{S_2}(x)\,dx = f_{S_2}^*(\gamma^*)$. Similarly, the probability that remaining vacation time exceeds remaining service time in a working vacation is formulated by $\tau = \int_0^{\infty}(1 - F_{\widehat{V}}(x))f_{\widehat{S}_2}(x)\,dx = f_{\widehat{S}_2}^*(\gamma^*) = \mu_2(1 - f_{S_2}^*(\gamma^*)/\gamma^*$. From these formulae, ε and τ are determined if the corresponding distribution functions, i.e., $F_{S_2}(x)$ and $F_{\widehat{S}_2}(x)$, are given. The service time distribution during a normal busy period and a working vacation can be framed by $F_{S_1}(x) = 1 - \exp\{-\int_0^x u(t)dt\}$ and $F_{S_2}(x) = 1 - \exp\{-\int_0^x v(t)dt\}$, respectively, where $u(t)$ and $v(t)$ are the hazard rates of the distributions S_1 and S_2, respectively. Similarly, if $y(t)$ be the hazard rate of the distribution of $\widehat{V} + S_1$, then $F_{\widehat{V}+S_1}(x) = 1 - \exp\{-\int_0^x y(t)dt\}$. For stability of the model, it is assumed that the traffic intensity $\rho = \lambda^*/\mu_1$ is less than one.

3.1 Stationary Distribution at Post-departure Epoch

Let t_i $(i = 0, 1, 2, \ldots)$ denotes the service completion epochs, while t_i^+ $(i = 0, 1, 2, \ldots)$ stands for the time epoch just after completion of a service. If at any t_i^+ $(i = 0, 1, 2, \ldots)$, the system length, the state of the batch arrival process, and the server's state are symbolized by $N_q(t_i^+)$, $J(t_i^+)$, and $\xi(t_i^+)$, then the state of the system at that epoch can be defined as $\zeta_i = \{N_q(t_i^+),\ J(t_i^+),\ \xi(t_i^+)\}$. Moreover, $\xi(t_i^+)$ can take values 0 or 1, if the server is on *WV* or in a normal busy period, respectively, and for $\rho < 1$, the embedded Markov chain ζ_i $(i = 0, 1, 2, \ldots)$ is irreducible, aperiodic, and ergodic. Hence for $\rho < 1$, the steady-state solution of the Markov chain exists. For $n = 0, 1, 2, \ldots,$ let the limiting probabilities are defined as

$$p_{0,j}^+(n) = \lim_{i\to\infty} P\{N_q(t_i^+) = n, J(t_i^+) = j,\ \xi(t_i^+) = 0\}, \quad j = 1, 2, \ldots, m,$$
$$p_{1,j}^+(n) = \lim_{i\to\infty} P\{N_q(t_i^+) = n, J(t_i^+) = j,\ \xi(t_i^+) = 1\}, \quad j = 1, 2, \ldots, m,$$

with the row vectors $\mathbf{p}_0^+(n) = [p_{0,1}^+(n), p_{0,2}^+(n), \ldots p_{0,m}^+(n)]$ and $\mathbf{p}_1^+(n) = [p_{1,1}^+(n), p_{1,2}^+(n), \ldots p_{1,m}^+(n)]$. Now the transition probability matrix (TPM) which describes the probability of transitions among the states ζ_i $(i = 0, 1, 2, \ldots)$ can be obtained by relating two consecutive embedded post-departure epochs and can be given by

$$\mathcal{P} = \begin{pmatrix} \overline{\mathbf{B}}_{0,0} & \overline{\mathbf{B}}_{0,1} & \overline{\mathbf{B}}_{0,2} & \overline{\mathbf{B}}_{0,3} & \overline{\mathbf{B}}_{0,4} & \cdots & \cdots \\ \overline{\mathbf{B}}_{1,0} & \overline{\mathbf{A}}_1 & \overline{\mathbf{A}}_2 & \overline{\mathbf{A}}_3 & \overline{\mathbf{A}}_4 & \cdots & \cdots \\ \mathbf{0} & \overline{\mathbf{A}}_0 & \overline{\mathbf{A}}_1 & \overline{\mathbf{A}}_2 & \overline{\mathbf{A}}_3 & \cdots & \cdots \\ \mathbf{0} & \mathbf{0} & \overline{\mathbf{A}}_0 & \overline{\mathbf{A}}_1 & \overline{\mathbf{A}}_2 & \cdots & \cdots \\ \vdots & \vdots & \vdots & \vdots & \vdots & \vdots & \ddots \end{pmatrix} \tag{3}$$

where $\overline{\mathbf{A}}_k = \begin{bmatrix} \varepsilon\mathbf{V}_k & (1-\varepsilon)\mathbf{B}_k \\ \mathbf{0} & \mathbf{A}_k \end{bmatrix}$, $k = 0, 1, 2, \ldots,$ $\overline{\mathbf{B}}_{0,0} = \begin{bmatrix} \varepsilon\overline{\mathbf{D}}_1\mathbf{V}_0 & (1-\varepsilon)\overline{\mathbf{D}}_1\mathbf{B}_0 \\ \overline{\mathbf{D}}_1\mathbf{A}_0 & \mathbf{0} \end{bmatrix}$,

$\overline{\mathbf{B}}_{1,0} = \begin{bmatrix} \varepsilon\mathbf{V}_0 & (1-\varepsilon)\mathbf{B}_0 \\ \mathbf{A}_0 & \mathbf{0} \end{bmatrix}$, $\overline{\mathbf{B}}_{0,k} = \begin{bmatrix} \varepsilon \sum_{i=1}^{k+1} \overline{\mathbf{D}}_i\mathbf{V}_{k-i+1} & (1-\varepsilon) \sum_{i=1}^{k+1} \overline{\mathbf{D}}_i\mathbf{B}_{k-i+1} \\ \mathbf{0} & \sum_{i=1}^{k+1} \overline{\mathbf{D}}_i\mathbf{A}_{k-i+1} \end{bmatrix}$, $k = 1, 2, 3, \ldots,$ and $\overline{\mathbf{D}}_k = (-\mathbf{D}_0)^{-1}\mathbf{D}_k$ $(k = 1, 2, 3, \ldots)$ represents the phase transition matrix during an inter-batch arrival time of customers which is accepted in the system. For $n \geq 0$, the stationary probabilities $\mathbf{p}_0^+(n)$ and $\mathbf{p}_1^+(n)$ are determined by applying *RG* factorization to the TPM $\mathcal{P}$ as follows. The state space of the Markov chain $\mathcal{P}$ may be rewritten as $\Omega = \{(n, r) : n = 0, 1, 2, \ldots$ and $r = 1, 2, \ldots, 2m\}$, where n be the level variable and r be the phase variable of the Markov chain. For $k = 0, 1, 2, \ldots,$ let us denote $L_{\leq k} = \bigcup_{i=0}^{k} L_i$, i.e., $L_{\leq k}$ be the set of all the states in the levels up to k. Now if $L_{\leq n}$ $(n = 0, 1, 2, \ldots)$ is set as the censored set, then the block-partitioned censored TPM can be written by

$$\mathcal{P}^{\leq n} = \begin{pmatrix} \boldsymbol{\phi}_{0,0}^{(n)} & \boldsymbol{\phi}_{0,1}^{(n)} & \cdots & \boldsymbol{\phi}_{0,n}^{(n)} \\ \boldsymbol{\phi}_{1,0}^{(n)} & \boldsymbol{\phi}_{1,1}^{(n)} & \cdots & \boldsymbol{\phi}_{1,n}^{(n)} \\ \vdots & \vdots & & \vdots \\ \boldsymbol{\phi}_{n,0}^{(n)} & \boldsymbol{\phi}_{n,2}^{(n)} & \cdots & \boldsymbol{\phi}_{n,n}^{(n)} \end{pmatrix}. \tag{4}$$

For non-negative integers n, i, j with $0 \leq i, j \leq n$, the expression of $\boldsymbol{\phi}_{i,j}^{(n)}$ was derived by Li [12, Lemma 2.4] and can be given by $\boldsymbol{\phi}_{i,j}^{(n)} = \mathcal{P}_{i,j} + \sum_{k=n+1}^{\infty} \boldsymbol{\phi}_{i,k}^{(k)} \sum_{l=0}^{\infty} [\boldsymbol{\phi}_{k,k}^{(k)}]^l \boldsymbol{\phi}_{k,j}^{(k)}$. For $j < i-1$, it should be noted that $\boldsymbol{\phi}_{i,j}^{(n)} = \mathbf{0}$. Then the U-, R-, and G-measures as described by Li [12] can be given by

$$\boldsymbol{\Psi}_n = \boldsymbol{\phi}_{n,n}^{(n)}, \quad n = 0, 1, 2, \ldots, \tag{5}$$

$$\boldsymbol{R}_{i,j} = \boldsymbol{\phi}_{i,j}^{(j)} \left(\boldsymbol{I} - \boldsymbol{\Psi}_j\right)^{-1}, \quad i = 0, 1, 2, \ldots, j, \tag{6}$$

$$\boldsymbol{G}_n = \left(\boldsymbol{I} - \boldsymbol{\Psi}_n\right)^{-1} \boldsymbol{\phi}_{n,n-1}^{(i)}, \quad n = 1, 2, 3, \ldots. \tag{7}$$

Now from Theorem 2.5 of [12], one may write

$$\boldsymbol{I} - \mathcal{P} = (\boldsymbol{I} - \boldsymbol{R}_U)(\boldsymbol{I} - \boldsymbol{\Psi}_D)(\boldsymbol{I} - \boldsymbol{G}_L), \tag{8}$$

where $\boldsymbol{R}_U$, $\boldsymbol{U}_D$ and $\boldsymbol{G}_L$ are strictly upper diagonal, diagonal, and strictly lower diagonal matrices, respectively, and are illustrated by Li [12]. Let $\boldsymbol{\pi} = [\boldsymbol{\pi}_0, \boldsymbol{\pi}_1, \boldsymbol{\pi}_2, \ldots]$ be the stationary probability vector of the block-structured Markov chain $\mathcal{P}$, then $\boldsymbol{\pi}\mathcal{P} = \boldsymbol{\pi}$ and $\boldsymbol{\pi e} = 1$. Now if z_0 is denoted as the stationary probability vector of $\boldsymbol{\Psi}_0$ to level 0, then the stationary probability vector of the Markov chain $\mathcal{P}$ is expressed by

$$\boldsymbol{\pi}_k = \begin{cases} \eta z_0, & \text{if } k = 0, \\ \sum\limits_{i=0}^{k-1} \boldsymbol{\pi}_i \boldsymbol{R}_{i,k}, & \text{if } k = 1, 2, 3, \ldots, \end{cases} \tag{9}$$

where η is a constant and is uniquely determined by $\sum_{k=0}^{\infty} \boldsymbol{\pi}_k \boldsymbol{e} = 1$, see [12, Theorem 2.9]. Note that $\boldsymbol{\pi}_k$ $(k = 0, 1, 2, \ldots)$ is a vector of dimension $1 \times 2m$, where first m components give the post-departure probabilities at server's ideal state, i.e., $\mathbf{p}_0^+(k)$ and rest m components give the post-departure probabilities when the server is busy, i.e., $\mathbf{p}_1^+(k)$. One may note that the normalization condition is $\sum_{n=0}^{\infty} \left(\mathbf{p}_0^+(n) + \mathbf{p}_1^+(n)\right)$ $\mathbf{e} = 1$.

Lemma 1 *The expected time gap between two consecutive service completion epochs is given by*

$$T_{mean} = \left[\varepsilon E(S_2) + (1-\varepsilon)E(\widehat{V} + S_1)\right] \sum_{n=0}^{\infty} \mathbf{p}_0^+(n)\mathbf{e} + E(S_1) \sum_{n=0}^{\infty} \mathbf{p}_1^+(n)\mathbf{e}$$
$$+(\mathbf{p}_0^+(0) + \mathbf{p}_1^+(0))(-\mathbf{D}_0)^{-1}\mathbf{e}. \tag{10}$$

Proof T_{mean} is given by Eq. (10) where $(\mathbf{p}_0^+(0) + \mathbf{p}_1^+(0))(-\mathbf{D}_0)^{-1}\mathbf{e}$ is the term due to mean inter-batch arrival time of customers accepted in the system while the server is idle during a working vacation or during normal busy period. This can be verified as shown below. Let $\mathbf{Y}$ denote the $m \times m$ matrix whose ijth element Y_{ij} is the mean sojourn time of the system in an idle period with phase j, provided at the initial instant of the idle period phase was i. With the help of Eq. (1), it can be written that $\mathbf{Y} = \int_0^{\infty} \mathbf{P}(0,t)dt = \int_0^{\infty} e^{\mathbf{D}_0 t}dt = (-\mathbf{D}_0)^{-1}$.

3.2 Stationary Distribution at Arbitrary Epoch

System-length distribution at arbitrary epochs are obtained by supplementary variable method using post-departure epochs probabilities, and the solution procedure is given by Banik [11]. Let the states of the server at time t is marked as $\xi(t)$ which can take values 0, 1, and 2 for the server being busy in normal busy period, working vacation period, and after finishing a *WV*. The system length including the customer who is in service and the state of the batch arrival is expressed as $N_s(t)$ and $J(t)$, respectively. Further, the elapsed service time of a customer in a normal busy (working vacation) period is symbolized as $\widetilde{S}_1(t)$ $(\widetilde{S}_2(t))$. Similarly, if the random variable S_3 represents the sum of the two random variables, remaining vacation time $(\widehat{V})$, and a service time during the normal busy period (S_1), then the server's elapsed remaining vacation time plus a customer's service time during the next normal busy period is given by $\widetilde{S}_3(t)$. Hence, for positive integers i and j with $1 \le i \le m$, $n \ge 1$, and the continuous variable x (≥ 0), the joint probabilities of $N_s(t)$, $\xi(t)$, $\widetilde{S}_1$, $\widetilde{S}_2$, and $\widetilde{S}_3$ are, respectively, defined by

$$\pi_i(n,x;t)\Delta x = P\{N_s(t) = n,\ J(t) = i,\ x < \widetilde{S}_1(t) < x + \Delta x,\ \xi(t) = 1\}, \tag{11}$$
$$\omega_i(n,x;t)\Delta x = P\{N_s(t) = n, J(t) = i, x < \widetilde{S}_2(t) < x + \Delta x, \xi(t) = 0\}, \tag{12}$$
$$\psi_i(n,x;t)\Delta x = P\{N_s(t) = n, J(t) = i, x < \widetilde{S}_3(t) < x + \Delta x, \xi(t) = 2\}, \tag{13}$$
$$\nu_i(0;t) = P\{N_s(t) = 0, J(t) = i,\ \xi(t) = 0\}, \tag{14}$$
$$\nu_i(1;t) = P\{N_s(t) = 0, J(t) = i,\ \xi(t) = 1\}. \tag{15}$$

As the model is investigated in steady state, i.e., when $t \to \infty$, the above probabilities will be denoted by $\pi_i(n,x)$, $\omega_i(n,x)$, $\psi_i(n,x)$, $\nu_i(0)$ and $\nu_i(1)$. Let us further define the row vectors of order m as $\boldsymbol{\pi}(n,x) = [\pi_i(n,x)]_{1\times m}$, $\boldsymbol{\omega}(n,x) = [\omega_i(n,x)]_{1\times m}$, $\boldsymbol{\psi}(n,x) = [\psi_i(n,x)]_{1\times m}$ and $\boldsymbol{\nu}(0) = [\nu_i(0)]_{1\times m}$, $\boldsymbol{\nu}(1) = [\nu_i(1)]_{1\times m}$, $1 \le i \le m$. Now using the fact that

$$\boldsymbol{\pi}(n) = \int_0^\infty \boldsymbol{\pi}(n,x)dx, \quad \boldsymbol{\omega}(n) = \int_0^\infty \boldsymbol{\omega}(n,x)dx, \text{ and } \boldsymbol{\psi}(n) = \int_0^\infty \boldsymbol{\psi}(n,x)dx, \tag{16}$$

for $n = 1, 2, 3, \ldots$, the following relations can be obtained after using supplementary variable technique and a few algebraic manipulations.

$$\boldsymbol{\pi}(n) = E(S_1)\sum_{i=1}^{n} \boldsymbol{\nu}(1)\mathbf{D}_i\widehat{\mathbf{A}}_{n-i} + \frac{E(S_1)}{T_{mean}}\sum_{i=1}^{n} \mathbf{p}_1^+(i)\widehat{\mathbf{A}}_{n-i}, \tag{17}$$

$$\boldsymbol{\omega}(n) = \tau E(S_2)\sum_{i=1}^{n} \boldsymbol{\nu}(0)\mathbf{D}_i\widehat{\mathbf{V}}_{n-i} + \frac{\tau E(S_2)}{T_{mean}}\sum_{i=1}^{n} \mathbf{p}_0^+(i)\widehat{\mathbf{V}}_{n-i}, \tag{18}$$

$$\boldsymbol{\psi}(n) = (1-\tau)E(S_3)\sum_{i=1}^{n} \boldsymbol{\nu}(0)\mathbf{D}_i\widehat{\mathbf{B}}_{n-i} + \frac{(1-\tau)E(S_3)}{T_{mean}}\sum_{i=1}^{n} \mathbf{p}_0^+(i)\widehat{\mathbf{B}}_{n-i}, \tag{19}$$

since $S_3 \equiv \widehat{V} + S_1$, i.e., S_3 is the sum of r.vs. $\widehat{V}$ and S_1, and we assume

$$\widehat{\mathbf{A}}_n = \frac{1}{E(S_1)}\int_0^\infty \mathbf{P}(n,x)(1-F_{S_1}(x))dx, \quad n = 0, 1, 2, \ldots, \tag{20}$$

$$\widehat{\mathbf{V}}_n = \frac{1}{E(S_2)}\int_0^\infty \mathbf{P}(n,x)(1-F_{S_2}(x))dx, \quad n = 0, 1, 2, \ldots, \tag{21}$$

$$\widehat{\mathbf{B}}_n = \frac{1}{E(S_3)}\int_0^\infty \mathbf{P}(n,x)(1-F_{S_3}(x))dx, \quad n = 0, 1, 2, \ldots, \tag{22}$$

the unknown vectors $\boldsymbol{\nu}(0)$ and $\boldsymbol{\nu}(1)$ satisfy the following equation:

$$\boldsymbol{\nu}(0) + \boldsymbol{\nu}(1) = \frac{1}{T_{mean}}\Big(\mathbf{p}_0^+(0) + \mathbf{p}_1^+(0)\Big)(-\mathbf{D}_0)^{-1}. \tag{23}$$

It should be noted here that one may not able to find the vectors $\boldsymbol{\nu}(0)$ and $\boldsymbol{\nu}(1)$ explicitly. Hence, the approximation for these vectors can be taken as $\boldsymbol{\nu}(0) \approx \frac{1}{T_{mean}}\mathbf{p}_0^+(0)(-\mathbf{D}_0)^{-1}$ and $\boldsymbol{\nu}(1) \approx \frac{1}{T_{mean}}\mathbf{p}_1^+(0)(-\mathbf{D}_0)^{-1}$. We checked that these approximations give better results for other vector values $\boldsymbol{\pi}(n)$, $\boldsymbol{\omega}(n)$ and $\boldsymbol{\psi}(n)$ for Eqs. (17)–(19). This approximation is presented in the numerical results.

Let $\mathbf{p}(n)$ denotes the row vector whose ith component is the probability of n ($n = 0, 1, 2, \ldots$) customers in the system at an arbitrary epoch and state of the arrival process is i ($i = 1, 2, \ldots, m$). Then, we can write $\mathbf{p}(0) = \boldsymbol{\nu}(0) + \boldsymbol{\nu}(1)$, and for $n \geq 1$, $\mathbf{p}(n) = \boldsymbol{\pi}(n) + \boldsymbol{\omega}(n) + \boldsymbol{\psi}(n)$.

Remark 1 It should be noted here that $\frac{1}{T_{mean}} = \lambda^*$, which can be verified with the Eq. (10). Similarly, it can be checked that $\sum_{n=0}^{\infty} \mathbf{p}(n) = \overline{\boldsymbol{\pi}}$. These facts can be used as the correctness of the numerical results.

4 Performance Measures and Numerical Illustrations

Some useful performance measures of the discussed queueing system can be determined using the steady-state probabilities and are formulated in this section. The mean number of customers in the system at a random epoch is evaluated by $L_s = \sum_{i=1}^{\infty} i\mathbf{p}(i)\mathbf{e}$. Similarly, the expected system lengths of the system when the server serves the customers in a normal busy period and in a working vacation period are expressed by $L_{s1} = \sum_{i=0}^{\infty} i\boldsymbol{\pi}(i)\mathbf{e}$ and $L_{s2} = \sum_{i=1}^{\infty} i\boldsymbol{\omega}(i)\mathbf{e}$, respectively. Finally, the mean system length of the system when the server is busy, i.e., the server serves either in a reduced rate or in a normal rate is computed as $L_{s3} = \sum_{i=1}^{\infty} i\boldsymbol{\psi}(i)\mathbf{e}$. If the expected waiting time of an arbitrary customer in the system is denoted by W_s then from Little's law, one can determine $W_s = L_s/\lambda^*$. All the computations are carried out by MAPLE software.

The system-length distributions at post-departure and arbitrary epoch for $BMAP/PH/1/\infty/SWV$ and $BMAP/D/1/\infty/SWV$ queueing system is presented in Table 1 with the following parameter. The 2-state *BMAP* representation is taken as $\boldsymbol{D}_0 = \begin{bmatrix} -5.0 & 1.5 \\ 1.0 & -4.0 \end{bmatrix}, \boldsymbol{D}_1 = \begin{bmatrix} 0.98 & 0.42 \\ 0.84 & 0.36 \end{bmatrix}, \boldsymbol{D}_2 = \begin{bmatrix} 0.73 & 0.31 \\ 0.63 & 0.27 \end{bmatrix}, \boldsymbol{D}_3 = \begin{bmatrix} 0.245 & 0.105 \\ 0.21 & 0.09 \end{bmatrix}, \boldsymbol{D}_4 = \begin{bmatrix} 0.245 & 0.105 \\ 0.28 & 0.12 \end{bmatrix}$ and $\boldsymbol{D}_k$ $(k \geq 5)$ is null matrix of order 2. For this representation of *BMAP*, $\overline{\boldsymbol{\pi}} = [0.548673\ 0.451327]$, $\lambda^* = 6.548673$, and $\lambda_g = 3.274336$. The shape parameter and the scale parameter of the Weibull distribution for the service time during working vacation is taken as 1.2 and 0.2, respectively, and the pdf of the distribution is given by $f_{S_2}(x) = 8.278378x^{0.2}e^{-6.898648x^{1.2}}$. The LST of the Weibull distribution is approximated by Pade approximation [4/5] using first 20 moments calculated from the above pdf of the Weibull distribution, where [4/5] indicates that the numerator polynomial is of degree 4 and that of the denominator is five. One may note that $\epsilon = f^*_{S_2}(\gamma^*)$, $\tau = f^*_{\hat{S}_2}(\gamma^*)$, $f^*_{\hat{S}_2}(s) = \mu_2(1 - f^*_{S_2}(s))/s$. It is assumed that the vacation time is exponential with parameter $\gamma^* = 1.5$. The parameters of *PH* type and deterministic distribution for the service time during normal busy period is so chosen that $E(S_1) = 0.075$. During normal busy period, the Laplace transform of the deterministic service time is taken as $f^*_{S_1}(s) = e^{(-s/\mu_1)}$; hence, $f^*_{\hat{V}+S_1}(s) = \gamma^*/\gamma^* + s$ e^{-s/μ_1}, where $\mu_1 = 1/E(S_1)$. The *PH*-type representation is taken as $\boldsymbol{\beta}_1 = [0.5\ 0.5]$, $\boldsymbol{U}_1 = \begin{bmatrix} -60 & 60 \\ 0 & -15 \end{bmatrix}$ with $\rho = \lambda^* E(S_1) = 0.491150$.

It can be seen in the above numerical example that $\frac{1}{T_{mean}} = 6.547141$ which is approximately equal to λ^*. Also, $\sum_{n=0}^{\infty} \boldsymbol{p}(n)$ is quite close to $\overline{\boldsymbol{\pi}}$. In Figs. 2 and 3, we have plotted the mean system length and mean waiting time of an arbitrary customer in above-described *BMAP/PH/*1 and *BMAP/D/*1 *SWV* systems for different ρ (generated by varying the parameters of *PH* type and deterministic distribution). It can be seen from the figures that as ρ increases, the mean waiting time and the mean system lengths significantly differ for the cases of deterministic and *PH*-type service times.

Table 1 System-length distribution at post-departure and arbitrary epoch

BMAP/PH/1/∞/SWV

n	$p^+_{0,1}(n)$	$p^+_{0,2}(n)$	$\boldsymbol{p}^+_0(n)\boldsymbol{e}$	$p^+_{1,1}(n)$	$p^+_{1,2}(n)$	$\boldsymbol{p}^+_1(n)\boldsymbol{e}$	$p_1(n)$	$p_2(n)$	$\boldsymbol{p}(n)\boldsymbol{e}$
0	0.040543	0.036221	0.076764	0.001683	0.001578	0.003261	0.073150	0.089299	0.162450
1	0.015237	0.011188	0.026425	0.032465	0.029056	0.061521	0.055296	0.043199	0.098495
2	0.014153	0.010117	0.024269	0.035748	0.030239	0.065987	0.056847	0.043105	0.099952
3	0.011877	0.008327	0.020203	0.036912	0.030037	0.066950	0.054187	0.047603	0.094340
4	0.009023	0.006321	0.015345	0.036232	0.028746	0.064978	0.047603	0.034862	0.082466
5	0.007377	0.005135	0.012512	0.034341	0.026803	0.061145	0.039070	0.028781	0.067850
⋮	⋮	⋮	⋮	⋮	⋮	⋮	⋮	⋮	⋮
sum	0.125539	0.096201	0.221739	0.437188	0.341073	0.778260	0.544988	0.438867	0.983855

BMAP/D/1/∞/SWV

n	$p^+_{0,1}(n)$	$p^+_{0,2}(n)$	$\boldsymbol{p}^+_0(n)\boldsymbol{e}$	$p^+_{1,1}(n)$	$p^+_{1,2}(n)$	$\boldsymbol{p}^+_1(n)\boldsymbol{e}$	$p_1(n)$	$p_2(n)$	$\boldsymbol{p}(n)\boldsymbol{e}$
0	0.040048	0.036721	0.076769	0.001625	0.001565	0.003190	0.072542	0.089868	0.162411
1	0.015238	0.011189	0.026426	0.032936	0.029999	0.062935	0.056794	0.044476	0.101270
2	0.014154	0.010117	0.024271	0.036906	0.031383	0.068289	0.058434	0.044290	0.102724
3	0.011877	0.008327	0.020205	0.038354	0.031141	0.069495	0.055620	0.041109	0.096729
4	0.009024	0.006322	0.015346	0.037578	0.029631	0.067209	0.048694	0.035510	0.084205
5	0.007378	0.005135	0.012513	0.035335	0.027388	0.062723	0.039741	0.029119	0.068860
⋮	⋮	⋮	⋮	⋮	⋮	⋮	⋮	⋮	⋮
sum	0.125049	0.096704	0.221753	0.437274	0.340972	0.778247	0.544942	0.438898	0.983840

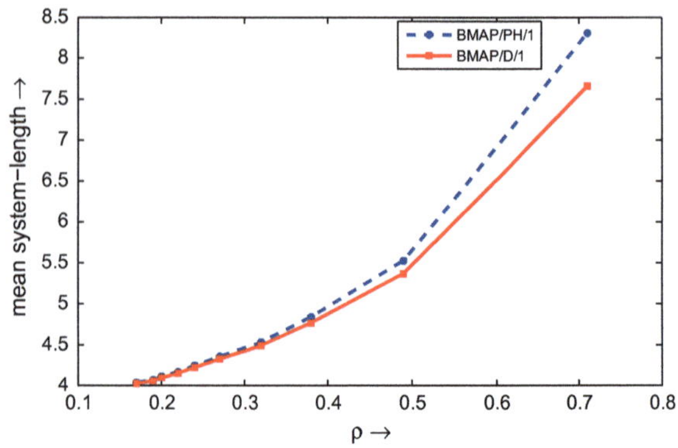

Fig. 2 ρ versus L_s

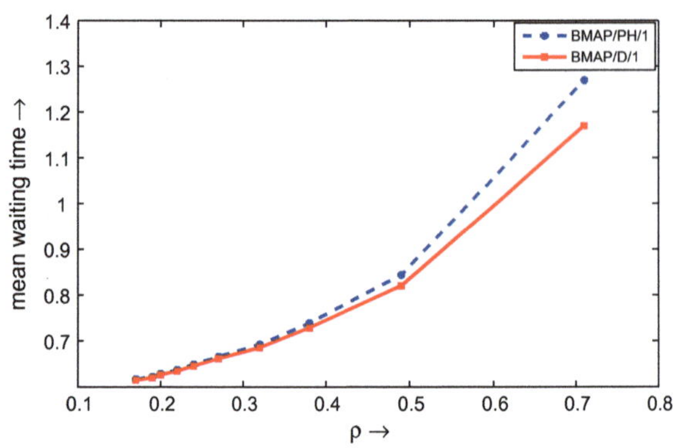

Fig. 3 ρ versus W_s

5 Conclusions and Future Scope

A $BMAP/G/1/\infty$ queue under single working vacation scheme is investigated in this paper. An approach to determine the steady-state distributions of the number of customers in the system at different epochs is outlined here. However, it is interesting to study an equivalent queueing system with correlated service process; for example non-renewal type service time distribution can be considered for the same model. One may also be interested to study the analogous discrete-time queueing model which is effective on digital telecommunication networks. These problems can be investigated in further.

References

1. Lucantoni, D.M., Meier-Hellstern, K.S., Neuts, M.F.: A single-server queue with server vacations and a class of non-renewal arrival processes. Adv. Appl. Probab. 676–705 (1990)
2. Lucantoni, D.M.: New results on the single server queue with a batch Markovian arrival process. Commun. Stat. Stoch. Models **7**, 1–46 (1991)

3. Neuts, M.F.: A versatile Markovian point process. J. Appl. Probab. 764–779 (1979)
4. Ramaswami, V.: The $N/G/1$ queue and its detailed analysis. Adv. Appl. Probab. 222–261 (1980)
5. Servi, L.D., Finn, S.G.: $M/M/1$ queues with working vacations ($M/M/1/WV$). Perform. Eval. **50**, 41–52 (2002)
6. Li, J.H., Liu, W.Q., Tian, N.S.: Steady-state analysis of a discrete-time batch arrival queue with working vacations. Perform. Eval. **67**, 897–912 (2010)
7. Gao, S., Liu, Z.: Performance analysis of a discrete-time $Geo^X/G/1$ queue with single working vacation. World Acad. Sci. Eng. Technol. **56**, 1162–1170 (2011)
8. Chae, K.C., Lim, D.E., Yang, W.S.: The GI/M/1 queue and the $GI/Geo/1$ queue both with single working vacation. Perform. Eval. **66**, 356–367 (2009)
9. Banik, A.D.: Analysis of single working vacation in $GI/M/1/N$ and $GI/M/1/\infty$ queueing systems. Int. J. Oper. Res. **7**, 314–333 (2010)
10. Li, J., Tian, N.: Performance analysis of a $GI/M/1$ queue with single working vacation. Appl. Math. Comput. **217**, 4960–4971 (2011)
11. Banik, A.D.: Stationary analysis of a $BMAP/R/1$ queue with R-type multiple working vacations. Commun. Stat.-Simul. Comput. **46**, 1035–1061 (2017)
12. Li, Q.L.: Constructive Computation in Stochastic Models with Applications: The *RG*-Factorizations. Springer Science & Business Media (2011)
13. Yin, F., Jiang, C., Deng, R., Yuan, J.: Grid resource management policies for load-balancing and energy-saving by vacation queuing theory. Comput. Electr. Eng. **35**, 966–979 (2009)
14. Neuts, M.F.: Matrix-Geometric Solutions in Stochastic Models: An Algorithmic Approach. Johns Hopkins University, Baltimore (1981)

Computational Analysis of the $GI/G/1$ Risk Process Using Roots

Gopinath Panda, A. D. Banik and M. L. Chaudhry

Abstract In this paper, we analyze an insurance risk model wherein the arrival of claims and their sizes occur as renewal processes. Using the duality relation in queueing theory and roots method, we derive closed-form expressions for the ultimate ruin probability, the distribution of the deficit at the time of ruin, and the expected time to ruin in terms of the roots of the characteristic equation. Finally, some numerical computations are portrayed with the help of tables.

Keywords Risk processes · Ruin probability · Duality · Padé approximation Time to ruin · $GI/G/1$ queue · Deficit at the time of ruin

1 Introduction

In recent years, the Sparre Andersen (renewal risk) model has been studied extensively due to its important actuarial applications. Many risk measures like probability of ultimate ruin, time to ruin, surplus prior to ruin, deficit at the time of ruin, and recovery time after ruin, have received substantial attention [1–3]. In insurance risk theory, the principal problem is to find the infinite-time ruin probability. Also, there has been significant development in the areas such as analysis of the deficit at the time of ruin, the surplus before the time of ruin, and the time to ruin. The identification of the above distributions is cumbersome due to the nonexistence of analytic expressions in most of the cases. Gerber and Shiu [4] introduced a unified approach

G. Panda · A. D. Banik (✉)
School of Basic Sciences, Indian Institute of Technology Bhubaneswar, Bhubaneswar, India
e-mail: banikad@gmail.com; adattabanik@iitbbs.ac.in

G. Panda
e-mail: gopinath.panda@gmail.com

M. L. Chaudhry
Department of Mathematics and Computer Science, Royal Military College of Canada, Kingston, ON K7K 7B4, Canada
e-mail: chaudhry-ml@rmc.ca

S. Kar et al. (eds.), *Operations Research and Optimization*, Springer Proceedings in Mathematics & Statistics 225, https://doi.org/10.1007/978-981-10-7814-9_6

with the discounted penalty function (a function of the ruin time, the surplus prior to ruin, and the deficit at ruin) to get the analytic expressions for the above risk measures. The joint analysis of these random variables using the discounted penalty function facilitated them to provide a refined characterization of the ruin measures in terms of a renewal equation [5]. This was further analyzed for some special classes of the Sparre Andersen model by Dickson and Hipp [6, 7], Landriault and Willmot [8], Gerber and Shiu [9], and Li and Garrido [10]. The Sparre Andersen model with Erlang or generalized Erlang distributed inter-claim times has been studied extensively in [11, 12], whereas phase-type inter-claim times are considered in [13, 14].

In a renewal risk model, the probability of ultimate ruin can be identified with the limiting waiting-time distribution in a single server queue with arrival process being a renewal one and service time being generally distributed. The second best alternative to obtain a closed-form solution for ruin probability is a numerical method which helps to compute the ruin probabilities accurately. Many studies of these models have focused on numerical methods for evaluation of ruin probabilities; see [15] for a description of such methods. Some of the main approaches include matrix analytic methods, Laplace transform inversion, and differential and integral equations. Panda et al. [16] studied the $GI/M/1$ risk process employing the characteristic roots and numerically computed several risk measures. An extensive literature on risk theory can be found in the classic text by Asmussen and Albrecher [17].

Prabhu [18] explored the connection between risk theory and other applied probability areas in a queueing context. Since then, the duality results have been used by several authors [19, 21] to find different measures of renewal risk model. We are interested in the closed-form solution for the risk measures which are, however, not easy to obtain in most of the cases with an exception in case of phase-type claims [20]. We extended the model studied in [21], to include both phase-type and non-phase-type claim size distributions as well as for the distributions with rational and non-rational Laplace-Stieltjes transform (LST). This method not only concentrates on the risk models where claim distributions have rational LST but also can be applied for any arbitrary distributions with a little modification, i.e., for distributions with non-rational LST (inverse Gaussian, US Pareto) and no closed-form LST (Weibull, lognormal), we use the Padé approximation method and continued fraction approximation. We have used the duality results and roots method to derive closed-form expressions for the probability of ultimate ruin, expected time to ruin, and the deficit at the time of ruin. Further, we derived the distribution function of the deficit at the time of ruin, from which moments of all orders can be derived easily. The method of roots applied to find the ruin probability is simpler than the Gerber–Shiu penalty function method as the latter involves tedious integrations. Also, the roots method is easier for numerical computation in comparison with the earlier works. The model can be applied to insurance and finance markets.

2 Mathematical Description of Risk Process

We consider a Sparre Andersen risk process in continuous time, associated with an insurance company. Let $R(t)$ be the amount of capital per portfolio for an insurance company at time t. At the beginning ($t = 0$), we assume the capital of the company to be $u\ (> 0)$, that is, $R(0) = u$. The portfolio receives premiums with constant premium rate $p(> 0)$. Claims arrive randomly to the insurance company and are to be settled immediately. Insurance company assumes average claim size to be finite; otherwise, no insurance company would insure such a risk. In the risk theory, the risk reserve process is

$$R(t) = u + pt - \sum_{k=1}^{N(t)} X_k, \tag{1}$$

where $N(t)$ is the total number of claims arrived before time t and X_k is the size of the kth claim. The claim number process $\{N(t), t \geq 0\}$ is a renewal process and is independent of the claim sizes $\{X_k, k \geq 1\}$. Let $T_i, i \geq 1$ denote the interarrival times of claims, where T_1 is the time of the first claim, and T_i for $i = 2, 3, \ldots$, the time between the $(i-1)$th claim and the ith claim. Now, the sequence of positive random variables $\{T_i\}_{i=1}^{\infty}$ is independent identically distributed (i.i.d.) with common distribution function $A(x)$, density $a(x)$, LST $A^*(s) = \int_0^{\infty} e^{-sx} dA(x)$, and mean $E(T)$, where the random variable T is the generic for inter-claim arrival times. We assume that there are finitely many claims in finite-time intervals, i.e., the number $N(t)$ is finite almost surely and $E(N(t))$ is also finite. Let the claim sizes X_k be positive i.i.d. random variables with common distribution function $B(x)$, density $b(x)$, LST $B^*(s) = \int_0^{\infty} e^{-sx} dB(x)$, and mean (mean of single claim) $E(X)$, where X is the generic of claim sizes. The claim amounts and the claim arrival process are mutually independent, i.e., X_i is independent of T_i. The above risk model is characterized by the property $\lim_{t\to\infty} \sum_{i=1}^{N(t)} X_i/t = \rho$, with ρ being the average claim size per unit time. The safety loading factor, denoted by η is the relative amount by which the premium rate p exceeds ρ, i.e., $\eta = (p - \rho)/\rho$. An insurance company will always try to ensure a positive safety loading $\eta > 0$, i.e., the premiums received per unit time should exceed the expected claim payments per unit time. Another process $\{S(t), t \geq 0\}$ in the theory of risk is known as the claim surplus process, which is defined as

$$S(t) = u - R(t) = \sum_{k=1}^{N(t)} X_k - pt. \tag{2}$$

The claim surplus process is more convenient to work with than the original process $\{R(t), t \geq 0\}$. The most important measure in risk theory is the infinite-time ruin probability (or ultimate ruin probability), which is denoted by $\psi(u)$ and is defined as the probability that the risk reserve becomes negative if it started with initial reserve u, $\psi(u) = P(\inf_{t\geq 0} R(t) < 0|R(0) = u)$. On the other hand, the finite-time ruin probability is the ruin probability before a fixed time say, $\bar{T}$ and is defined by $\psi(u, \bar{T}) = P\left(\inf_{0\leq t\leq \bar{T}} R(t) < 0|R(0) = u\right)$. Let $\tau(u)$ denote the time to ruin, that is the time instant for which the risk reserve becomes negative for the first time, then $\tau(u) = \inf\{t \geq 0 : R(t) < 0\} = \inf\{t \geq 0 : S(t) > u\}$ and $M = \sup_{0\leq t<\infty} S(t)$ is the maximum with infinite-time ruin probability. Thus, the ultimate ruin probability is $\psi(u) = P\{\tau(u) < \infty\} = P\{M > u\}$. The following proposition can be found in [17, p. 3].

Proposition 1 *When safety loading factor is negative ($\eta < 0$), then $M = \infty$ and hence, ruin is certain whatever may be the initial reserve ($\psi(u) = 1,\ \forall\ u$). If safety loading factor is positive, then $M < \infty$ and hence, $\psi(u) < 1$, for all sufficiently large u.*

3 GI/G/1 Risk Process and Duality

Consider a risk process in which the claims arrive to the system following a renewal process, i.e., the arrival epochs $t_1, t_2, \ldots,$ of the risk process form a renewal process with $t_n = T_1 + T_2 + \cdots + T_n$. The inter-claim arrival times $\{T_i, i \geq 1\}$ and claim sizes $\{X_i, i \geq 1\}$ are defined in Sect. 2. We assume the premium rate per unit time to be unity ($p = 1$). The average claim amount per unit time is equal to $E(X)/E(T)$. We denote the above risk process as *GI*/*G*/1 risk model. Let $\mathcal{N}(u)$ be the number of claims up to the time of ruin. Then $\mathcal{N}(u)$ is given by

$$\mathcal{N}(u) = \inf\{n : u + \sum_{i=1}^{n} T_i - \sum_{i=1}^{n} X_i < 0\}.$$

The time to ruin is

$$\tau(u) = \sum_{i=1}^{\mathcal{N}(u)} T_i, \tag{3}$$

and the deficit at the time of ruin is

$$\zeta(u) = \sum_{i=1}^{\mathcal{N}(u)} X_i - \sum_{i=1}^{\mathcal{N}(u)} T_i - u = \sum_{i=1}^{\mathcal{N}(u)} X_i - \tau(u) - u. \tag{4}$$

Let $\zeta(0) = \zeta$. Consider the *GI*/*G*/1 risk process. The corresponding queueing system associated with this risk process is the single server *GI*/*G*/1 queue, with an infinite buffer. Let the customers (individual claims) arrive to the system at time epochs

$0 = t_0, t_1, t_2, \dots, t_n, \dots$. The interarrival and service times are respectively the inter-claim arrival times and claim sizes of the risk process. The customers are served by a single server. The service discipline is the classical first-come first-served (FCFS) one. The traffic intensity of the queueing model, $\rho = E(X)/E(T)$, is assumed to be less than unity, i.e., the underlying queueing model will satisfy the stability condition if $\rho < 1$ which is equivalent to $E(X) < E(T)$.

4 Performance Measures of the Risk Process

An insurance risk model is characterized by the ultimate ruin probability, time to ruin, deficit at the time of ruin, and recovery time after a ruin. We analyze the following performance measures.

4.1 *Probability of Ultimate Ruin*

For the *GI*/*G*/1 risk model described above, we consider two cases for ρ: when ρ greater than unity and ρ smaller than unity.

Case 1. ($\rho > 1$) In this case, ruin is certain. When ρ becomes greater than unity, the safety loading factor for the risk process η becomes negative. Hence by Proposition 1, the result follows.

Case 2. ($\rho < 1$) When $\rho < 1$, we use the classical result that the ruin probabilities for the *GI*/*G*/1 risk model are related to the stationary actual waiting-time (W) of an initially empty *GI*/*G*/1 queue by means of $\psi(u) = P(W > u)$. Consider the *GI*/*G*/1 queue, where the claim arrival process serves as the customer arrivals with the same rate and the claim sizes serves as the service times with distribution function $B(x)$. A single server provides service to the incoming claims following the classical first-come first-served discipline. The stability condition for the underlying queueing system is $\rho < 1$. For the above renewal arrival queue with $\rho < 1$, ruin probability is distributed as the tail of the waiting-time in steady state. Let W_n be the actual waiting-time of nth customer, i.e., the time at which the customer arrives to the queue until he starts service. Let W be the stationary actual waiting-time of W_n with LST $W^*(s)$. The literature on queueing theory indicates that distributions bearing rational LSTs address a wide range of distributions that arise in real-life applications. In this perspective, we consider those distributions having rational LSTs of the form $P(s)/Q(s)$, where the polynomials $Q(s)$ and $P(s)$ are of degree n and at most n, respectively. Also they have no common factors. Assuming $B^*(s) = P(s)/Q(s)$, it can be shown by Rouché's theorem that the characteristic equation (c.e.) of the *GI*/*G*/1 queue,

$$Q(s) - A^*(-s)P(s) = 0$$

has n roots (assumed distinct), say $s_k (k = 1, 2, \dots, n)$ with $Re(s_k) < 0$. From (3.5) of Chaudhry et al. [22, p. 112], we can obtain the actual waiting-time LST,

$$W^*(s) = \frac{\prod_{i=1}^{n} -s_i}{Q(0)} + \sum_{i=1}^{n} \frac{A_i}{s - s_i}, \tag{5}$$

where the unknowns are

$$A_i = -s_i \frac{Q(s_i)}{Q(0)} \prod_{j=1, j \neq i}^{n} \left(\frac{-s_j}{s_i - s_j} \right), \; i = 1, 2, \dots, n.$$

The probability that a customer upon arrival will not wait is $W(0) = \prod_{i=1}^{n} -s_i / Q(0)$. Taking the inverse Laplace transform of (5), the pdf and cdf of actual waiting-time distribution are, respectively,

$$w(t) = \frac{1}{Q(0)} \prod_{i=1}^{n} -s_i \, \delta(t) + \sum_{i=1}^{n} A_i \, e^{s_i t} \quad \text{and} \quad W(t) = 1 + \sum_{i=1}^{n} \frac{A_i}{s_i} \, e^{s_i t}, \tag{6}$$

where $\delta(t)$ is the Dirac δ-function. From [17, p. 162], the fundamental duality relations between the steady-state behavior of the *GI*/*G*/1 queue and the ruin probabilities can be found in the proposition below.

Proposition 2 *If* $\eta > 0$ ($\equiv \rho < 1$), *then there exists a random variable* W *such that for* $n \to \infty$, W_n *converges in distribution to* W, *and the probability of ultimate ruin is given by* $\psi(u) = P(W > u)$.

The survival probability $\phi(u)$ for the above renewal risk model is given by $\phi(u) = W(u)$. The ultimate ruin probability becomes

$$\psi(u) = \sum_{i=1}^{n} \frac{A_i}{-s_i} \, e^{s_i u}. \tag{7}$$

Remark 1 The expression for ruin probability $\psi(u)$ in Case 2 ($E(X) < E(T)$) of Frostig [21, p. 396] exactly matches in numerical value with our model when both the claim arrival and claim sizes are phase-type distributed.

4.2 Time to Ruin and Deficit at the Time of Ruin

As time passes, the risk reserve of an insurance company increases with more premium and decreases when a claim is settled. There is a point in time when the risk reserve becomes negative for the first time. This particular time instant is called the ruin time and is denoted by $\zeta(u)$. Here, we consider two cases: when $\rho > 1$ and $\rho < 1$.
Case 1. ($\rho > 1$) In this case, ruin is certain and the kth-order moment of $\tau(u)$ exist for $k = 1, 2, \dots$. The busy period B_p of the dual queue $GI/G_*/1$ is finite, where the interarrival and service times of *GI*/*G*/1 are exchanged in the dual $GI/G_*/1$. *When*

the initial reserve $u = 0$, from the duality between the *GI/G/*1 risk process and the $GI/G_*/1$ queueing system, we obtain $E(\tau(0)) = E(B_p)$ and $E(\zeta(0)) = E(I)$. We continue the analysis of the dual queue $GI/G_*/1$ to find the idle period distribution. The arrival process of the dual queue has LST $B^*(s)$, mean $E(X)$ and that of the service process is $A^*(s)$ and $E(T)$. Hence, the dual queue is stable as $\rho^* = 1/\rho < 1$. The c.e. of the dual queue is $1 - A^*(s)\, B^*(-s) = 0$. Let $A^*(s) = \widehat{P}(s)/\widehat{Q}(s)$, where the degree of $\widehat{Q}(s)$ is m and degree of $\widehat{P}(s)$ is at most m. Following the similar analysis as in case 2 of Sect. 4.1, the actual waiting-time of the dual queue becomes

$$\widehat{W}^*(s) = \frac{\widehat{Q}(s)}{\widehat{Q}(0)} \prod_{i=1}^{m} \frac{-\widehat{s}_i}{(s - \widehat{s}_i)} = \frac{\prod_{i=1}^{m} -\widehat{s}_i}{\widehat{Q}(0)} + \sum_{i=1}^{m} \frac{\widehat{A}_i}{s - \widehat{s}_i}, \tag{8}$$

where $\widehat{s}_i, i = 1, 2, \ldots, m$ are the roots of the c.e. of the dual queue in $Re(\widehat{s}_i) < 0$. The unknowns can be easily calculated and found to be

$$\widehat{A}_i = -\widehat{s}_i \widehat{Q}(\widehat{s}_i) \prod_{j=1, j \neq i}^{m} \left(\frac{-\widehat{s}_j}{\widehat{s}_i - \widehat{s}_j} \right) / \; \widehat{Q}(0), \; i = 1, 2, \ldots, m.$$

Let I be the r.v which represents the distribution of actual idle period observed by an arriving customer. From Eq. (8.106) of [23], the relation between LSTs of idle period distribution and actual waiting-time distribution is given by

$$I^*(s) = 1 - \widehat{W}^*(-s)[1 - A^*(-s)B^*(s)]/\widehat{W}(0). \tag{9}$$

Substituting the value of $\widehat{W}(s)$ and then simplifying the expression for $I^*(s)$ and the partial fraction decomposition results

$$I^*(s) = \sum_{j=1}^{l} \frac{h_j}{s - \alpha_j}, \tag{10}$$

where α_j for $j = 1, 2, \ldots, l$ are distinct roots of $Q(s)$ with $Re(\alpha_j) < 0$ and $h_j, j = 0, 1, \ldots, l$ are constants to be determined by evaluating left-hand side $I^*(s)$ and right-hand side $\sum_{j=1}^{l} \frac{h_j}{s-\alpha_j}$ of Eq. (10) for any l values of $s(= 1, 2, \ldots, l)$. The coefficients h_j satisfy the relation $\sum_{j=1}^{l} \frac{h_j}{\alpha_j} + 1 = 0$. In case $Q(s)$ has repeated roots, say, α_1 with multiplicity k_1 and α_2 with multiplicity k_2 such that $k_1 + k_2 = l$, the partial fractions have to be modified to

$$\sum_{j=1}^{k_1} \frac{h_j}{(s - \alpha_1)^j} + \sum_{j=1}^{k_2} \frac{\bar{h}_j}{(s - \alpha_2)^j}.$$

From Eq. (10), the pdf $f_I(t)$ and cdf $F_I(t)$ of idle period distribution, respectively, are

$$f_I(t) = \sum_{j=1}^{m} h_j e^{\alpha_j t} \quad \text{and} \quad F_I(t) = 1 + \sum_{j=1}^{m} \frac{h_j}{\alpha_j} e^{\alpha_j t}. \tag{11}$$

The kth-order moment of idle period is given by

$$E^k(I) = \sum_{j=1}^{m} h_j / \alpha_j^{k+1}.$$

Suppose that initially the server is idle and he remains so for an arbitrary time. As soon as a customer arrives, he gets busy and remains busy for a random amount of time. So the cycle of idle periods and busy periods will generate two mutually independent sequences which can be studied as an alternating renewal process. The expected busy period, $E(B_p)$, in the queueing system $GI/G/1$ can be found from the relation $E(B_p)/E(I) = \rho^*/(1-\rho^*)$; see [24]. So $E(B_p)$ is given by

$$E(B_p) = \frac{\rho^*}{1-\rho^*} \sum_{j=1}^{m} \frac{h_j}{\alpha_j^2}.$$

Hence, the expected time to ruin and the expected deficit at the time of ruin for $u = 0$ are

$$E(\tau(0)) = \frac{\rho^*}{1-\rho^*} \sum_{j=1}^{m} \frac{h_j}{\alpha_j^2} \quad \text{and} \quad E(\zeta(0)) = \sum_{j=1}^{m} \frac{h_j}{\alpha_j^2}. \tag{12}$$

When the initial reserve $u > 0$, the expected value of the deficit at the time of ruin $\zeta(u)$ is equivalent to the expected value of the excess life of idle period of the associated dual queue at time u. As it is already explained that the idle period and the busy period form an alternating renewal process, we consider the renewal process formed by the idle periods only. Let $r(t)$ be the time measured from t, the instant at which we start observing the renewal process to the next renewal instant after time t. Now, to find the expected value of the excess life (or residual life) of idle period, we have to first calculate the expected value of the number of renewals in $[0, u]$ for the dual queue $GI/G_*/1$. The expected value of excess life of idle period $r(t)$ at time t can be obtained from the definition of excess life

$$r(t) = W_{N(t)+1} - t, t > 0,$$

where $W_{N(t)+1}$ is the waiting-time until the $(N(t)+1)$th renewal, when idle periods take the place of arrivals, i.e., $W_{N(t)+1} = \sum_{i=1}^{N(t)+1} I_i$, where I_i is the inter-occurrence time between $(i-1)$th and ith idle period. Taking expectation of both sides and then applying Wald's identity, the expected number of renewals is

$$E(r(t)) = E(I)[E(N(t)) + 1] - t. \tag{13}$$

Alternatively, we can also derive the expected number of renewals. The expression for the LST of the expected number of renewals $E(N(t))$ is given by

$$V^*(s) = \frac{I^*(s)}{1 - I^*(s)} = \frac{\sum_{j=1}^{l} h_j \prod_{i=1, i\neq j}^{l}(s - \alpha_i)}{\prod_{j=1}^{l}(s - \alpha_j) - \sum_{j=1}^{l} h_j \prod_{i=1, i\neq j}^{l}(s - \alpha_i)}, \quad Re(s) > 0. \tag{14}$$

Clearly, $V^*(s)$ is a proper rational polynomial. Since $V^*(s)$ is convergent for $Re(s) > 0$, using Rouche's theorem it can be shown that $denom(V^*(s)) = 0$ has l roots (assumed distinct), say $\hat{\alpha}_i$ $(i = 1, 2, \ldots, l)$ for which $Re(s_i) \leq 0$. In case of repeated roots, the partial fractions have to be modified similarly as stated earlier. Clearly, one root is at $s = 0$, say s_l. Making partial fractions, we can write (14) as

$$V^*(s) = \sum_{i=1}^{l-1} \frac{d_i}{s - \hat{\alpha}_i} + \frac{d_l}{s - \hat{\alpha}_l}. \tag{15}$$

The constants $d_i, i = 1, 2, \ldots, l$ can be determined by solving $\frac{\sum_{j=1}^{l} h_j \prod_{i=1, i\neq j}^{l}(s-\alpha_i)}{\prod_{j=1}^{l}(s-\alpha_j) - \sum_{j=1}^{l} h_j \prod_{i=1, i\neq j}^{l}(s-\alpha_i)}$ $= \sum_{i=1}^{l-1} \frac{d_i}{s-s_i} + \frac{d_l}{s}$. Taking the inverse LST of (15), we get the renewal density function $v(t)$, as

$$v(t) = \sum_{i=1}^{l-1} d_i e^{s_i t} + d_l. \tag{16}$$

Integrating (16) from 0 to t, we get the explicit expression for the expected number of renewals $E(N(t))$.

$$V(t) = E(N(t)) = d_l t + \sum_{i=1}^{l-1} \frac{d_i}{s_i} e^{s_i t} - \sum_{i=1}^{l-1} \frac{d_i}{s_i}. \tag{17}$$

One can also directly obtain (17) from [25, p. 382] using the relation $E(N(t)) = \mathscr{L}^{-1}\left\{\frac{I^*(s)}{s(1-I^*(s))}\right\}$. We numerically match $E(N(t))$ obtained by different methods as explained above and find good match. Equation (17) is valid for any claim size distribution, whereas $E(\tilde{N}(u))$, calculated using (7.1) of [21] is valid only for phase-type claim sizes. Let $F_\zeta(t)$, the distribution function of the deficit at the time of ruin $\zeta(u)$, be defined as $F_\zeta(t) = P\{\zeta(u) \leq t\} = P\{r(u) \leq t\}$. The closed-form expression of $F_\zeta(t)$ is given by

$$F_\zeta(t) = F_I(u + t) - \int_0^u [1 - F_I(u + t - x)]v(x)dx. \tag{18}$$

The density function f_ζ, expected value and higher-order moments of the deficit at the time of ruin, can easily be derived from (18). For initial reserve $u = 0$, $\zeta(0)$ is distributed as the idle period $F_I(t)$ of dual queue $GI/G_*/1$. The expected value of the deficit at the time of ruin $\zeta(u)$ can be obtained either from (13) and (17) or (18). From Eqs. (13) and (17), the expected value of deficit at the time of ruin is given as

$$E(\zeta(u)) = E(r(u)) = E(I)\left[1 + \sum_{i=1}^{l-1} \frac{d_i}{s_i} e^{s_i u} - \sum_{i=1}^{l-1} \frac{d_i}{s_i}\right] \tag{19}$$

Now to obtain the expected time to ruin, we use (2) and then (4), to get

$$S(\tau(u)) = \sum_{i=1}^{N(\tau(u))} X_i - \tau(u) = u + \zeta(u). \tag{20}$$

Taking expectation of both sides of (20) and then applying Wald's identity, we obtain

$$E(\tau(u)) = \frac{u + E(\zeta(u))}{\rho - 1}. \tag{21}$$

Using Eq. (19), the expected value of the time to ruin of the $GI/G/1$ risk process is

$$E(\tau(u)) = \frac{u + E(I)\left[1 + \sum_{i=1}^{l-1} \frac{d_i}{s_i} e^{s_i u} - \sum_{i=1}^{l-1} \frac{d_i}{s_i}\right]}{\rho - 1}. \tag{22}$$

Case 2. ($\rho < 1$) In this case, the expected time to ruin is infinite. For the dual queue, $\rho^* = 1/\rho > 1$. The fundamental duality relations fail as the dual queue $GI/G_*/1$ becomes unstable. So, using the change of measure technique via exponential family [17, p. 82], we redefine the parameters of our risk process. Define $\kappa(\theta) = B^*(-\theta)A^*(\theta) - 1$. The Lundberg equation $\kappa(\theta) = 0$ has a unique positive solution γ called the Lundberg coefficient. Consider another renewal risk process where the inter-claim arrival times are distributed as T_γ with distribution $A_\gamma(t)$ and with claim size distributed as X_γ, with distribution $B_\gamma(t)$. The densities of claim arrivals and claim sizes are related to the original risk process in the following way:

$$a_\gamma(t) = e^{-\gamma t}\, a(t)/A^*(\gamma), \quad b_\gamma(t) = e^{\gamma t}\, b(t)/B^*(-\gamma).$$

We call this risk process the γ-risk process. Substituting $\gamma = 0$, we get the original risk process. Let P_γ be the probability measure induced by $\{T_{\gamma j}, X_{\gamma j}\}$ and E_γ be the corresponding expectation operator. As the Lundberg conjugation corresponds to interchanging the rates of the inter-claim arrival and the claim sizes, the $\rho_\gamma = E(X_\gamma)/E(T_\gamma) > 1$. Hence, the dual queue of the γ-risk process is stable. This is the condition we have in case 1. Now consider the dual queue $GI/G_*/1$ where the service times are generally distributed as T_γ with mean $E(T_\gamma)$, and interarrival times are distributed as X_γ with mean $E(X_\gamma)$. Since $\rho^*_\gamma = 1/\rho_\gamma$ of the dual queue is less than 1, the steady-state condition holds, i.e., the steady-state solutions of the dual queue exist. Following similar analysis to the one used in case 1, the performance measures of the γ-risk process can be found. These measures can be considered as the upper bounds for the corresponding measures of the original risk process.

5 Numerical Results and Discussion

In this section, we present numerical examples that show the effect of several parameters on the behavior of the different performance measures of the renewal risk process. In particular, we are concerned about the values of the queueing parameters for which the insurance company's business will be stable. Numerical results demonstrated in this paper are performed using Maple 17 program. We have considered several numerical experiments with different sets of system parameters, and for the sake of completeness, few sets of risk parameters are shown below in the numerical computations. The numerical values are computed up to 30 decimal places in Maple 17 program, but due to lack of space only 6 decimal places are presented with rounding off.

We consider a $PH(\boldsymbol{\alpha},\boldsymbol{T})/PH(\boldsymbol{\beta},\boldsymbol{S})/1$ risk process with parameters,

$$\boldsymbol{\alpha} = (0.1, 0.6, 0, 0.3),\ \boldsymbol{T} = \begin{pmatrix} -3 & 1 & 0 & 1 \\ 1 & -5 & 1 & 0 \\ 0 & 2 & -4 & 2 \\ 1 & 0 & 1 & -4 \end{pmatrix}, \boldsymbol{\beta} = (0.5, 0.2, 0.3),\ \boldsymbol{S} = \begin{pmatrix} -3 & 1 & 0 \\ 0 & -6 & 1 \\ 0 & 2 & -5 \end{pmatrix}.$$

Here $\boldsymbol{e}_m$ is the column vector of one's. The expected values of claim arrival and claim sizes are $E(T) = -\boldsymbol{\alpha}\boldsymbol{T}^{-1}\boldsymbol{e}_4 = 0.565333$ and $E(X) = -\boldsymbol{\beta}\boldsymbol{S}^{-1}\boldsymbol{e}_3 = 0.330952$. So $\rho = E(X)/E(T) = 0.585411 < 1$. The queueing model associated with the above risk model is the stable $PH(\boldsymbol{\alpha},\boldsymbol{T})/PH(\boldsymbol{\beta},\boldsymbol{S})/1$ queue. We obtain the stationary distribution actual waiting-time using the roots of the characteristic equation explained in the Sect. 4. The ultimate ruin probability is $\psi(u) = -0.002865\ e^{-7.101841u} + 0.003354\ e^{-3.867988u} + 0.584922\ e^{-1.140942u}$ with $\psi(0) = 0.585411$. The corresponding expression $\psi(u) = \boldsymbol{\beta}^+\boldsymbol{e}^{(\boldsymbol{S}+\boldsymbol{s}\boldsymbol{\beta}^+)u}\boldsymbol{e}$ in [21, p. 396] exactly matches for all u which is presented in the second column (Frostig) in Tables 1 and 2.

In case of inverse Gaussian claims, the LST is not in the rational form. So, Padé approximation formula is used to rationalize the LST. For this, we first obtained N moments $(m_1, m_2, \ldots, m_N)$ of the IG distribution and then form a function series in terms of s, i.e., $\sum_{i=1}^{N}(-1)^i m_i s^i/i!$. Then we approximated this series with the function Padé [4, 5] in Maple to obtain our required LST of IG distribution. Similar procedure is followed in case of Weibull claims to get the LST required for the characteristic equation. Several risk measures are studied for phase-type claim size distribution

Table 1 Ruin probabilities $\psi(u)$ for different claim arrival distributions

u	PH	Frostig	ME	IG	Wb	Pareto
0	0.626887	0.626887	0.498410	0.782942	0.551246	0.757810
1	0.200255	0.200255	0.231030	0.557299	0.269135	0.514416
2	0.063961	0.063961	0.108224	0.399221	0.134552	0.352925
3	0.020436	0.020436	0.050700	0.285989	0.067277	0.242141
4	0.006530	0.006530	0.023752	0.204873	0.033639	0.166133
5	0.002086	0.002086	0.011127	0.146764	0.016820	0.113984
6	0.000667	0.000667	0.005213	0.105137	0.008410	0.078204
7	0.000213	0.000213	0.002442	0.075317	0.004205	0.053656
8	0.000068	0.000068	0.001144	0.053955	0.002103	0.036813
9	0.000022	0.000022	0.000536	0.038651	0.001051	0.025258
10	0.000007	0.000007	0.000251	0.027689	0.000526	0.017329

Table 2 Ruin probabilities $\psi(u)$ for different claim size distributions

u	PH	Frostig	ME	IG	Wb	Pareto
0	0.428468	0.428468	0.794369	0.353593	0.440906	0.588651
1	0.167076	0.167076	0.643123	0.094832	0.288346	0.376409
2	0.065950	0.065950	0.521321	0.030824	0.213110	0.251653
3	0.026074	0.026074	0.422822	0.010576	0.160285	0.169757
4	0.010311	0.010311	0.343019	0.003724	0.122497	0.114737
5	0.004077	0.004077	0.278309	0.001333	0.095017	0.077583
6	0.001612	0.001612	0.225817	0.000483	0.074693	0.052465
7	0.000638	0.000638	0.183230	0.000176	0.059412	0.035480
8	0.000252	0.000252	0.148676	0.000064	0.047738	0.023994
9	0.000100	0.000100	0.120639	0.000024	0.038689	0.016226
10	0.000039	0.000039	0.097889	0.000009	0.031581	0.010973

with representation $(\boldsymbol{\alpha}, \boldsymbol{T})$ and changing the claim arrival distribution to be matrix exponential (ME), inverse Gaussian (IG), and Weibull (Wb) with densities $(1+\frac{1}{(2\pi)^2})(1-cos(2\pi t))e^{-t}$, $\sqrt{\frac{0.5}{2\pi t^3}}e^{-\lambda(t-0.75)^2/(2(0.75)^2 t)}$ and $\frac{0.5}{10}\left(\frac{t}{10}\right)^{0.5-1} e^{-(t/10)^{0.5}}$, respectively.

Also, we consider the US Pareto [17, p. 10] claim size distribution with pdf $\frac{\alpha\, a^{\alpha}}{(a+x)^{\alpha+1}}$, where scale parameter $a = 1.5$ and shape parameter $\alpha = 2.5$. The LST of the Pareto claims, $\alpha(as)^{\alpha}e^{as}\Gamma(-\alpha, as)$, is not rational because of the presence of incomplete Gamma function. We obtained the rational form representation of the incomplete Gamma function using the Legendre's continued fraction approach. For calculation of the LST of the Pareto in rational form, we found five terms in the continued fraction of incomplete Gamma function. For all these risk processes, we carry out similar analysis as in case of $PH/PH/1$ risk process, and the values of ruin probabilities for different initial reserves are presented in Table 1.

Again, we have fixed the claim arrival distribution to be phase-type with representation $PH(\boldsymbol{\alpha}, \boldsymbol{T})$, where $\boldsymbol{\alpha} = (0.1, 0.6, 0, 0.3)$, $\boldsymbol{T} = \begin{pmatrix} -3 & 1 & 0 & 1 \\ 3 & -5 & 1 & 1 \\ 0 & 2 & -4 & 1 \\ 1 & 0 & 2 & -4 \end{pmatrix}$, $E(T) =$ 1.303093 and considered different types of claim amount distributions as follows: Phase-type claims with representation $(\boldsymbol{\beta}_1, \boldsymbol{S}_1)$, matrix exponential claims with representation $(\boldsymbol{\beta}_2, \boldsymbol{S}_2)$, inverse Gaussian density $\sqrt{\frac{\lambda}{2\pi t^3}} e^{-\lambda(t-\mu)^2/(2\mu^2 t)}$, $\lambda = 0.5$, $\mu = 0.5, E(X) = 0.5$, Weibull claims with density $\frac{\lambda}{\mu}\left(\frac{t}{\mu}\right)^{\lambda-1} e^{-(t/\mu)^\lambda}$, $\lambda =$ 0.5,
$\mu = 0.3$, $E(X) = 0.6$ Pareto claims with density $\frac{\alpha\, a^\alpha}{(a+x)^{\alpha+1}}$, $a = 1.5$, $\alpha = 2.5$, where

$$\boldsymbol{\beta}_1 = (0.5, 0.2, 0.3), \boldsymbol{\beta}_2 = (1, 0, 0), \boldsymbol{S}_1 = \begin{pmatrix} -3 & 1 & 0 \\ 0 & -6 & 5 \\ 0 & 3 & -5 \end{pmatrix}, \boldsymbol{S}_2 = \begin{pmatrix} 0 & -1-4\pi^2 & 1+4\pi^2 \\ 3 & 2 & -6 \\ 2 & 2 & -5 \end{pmatrix}.$$

Similar analysis of the model is carried out using the roots method and duality relations. The ultimate ruin probability is calculated from the tail of the waiting-time distribution of the associated queueing system. In case of PH claims, the values are compared with that of Frostig and exact matching is found. The values of the ultimate ruin probabilities against initial reserves of the insurance company for different claim size distributions are presented in Table 2.

It is also important for the insurance companies to know the time to ruin of their business as well as the deficit amount at the time of ruin. Keeping these in mind, we consider different distributions for claim arrivals as well as claim amounts and calculated the expected time to ruin and expected deficit at the time of ruin. We first consider a $PH(\boldsymbol{\alpha}, \boldsymbol{T})/PH(\boldsymbol{\beta}, \boldsymbol{S})/1$ risk process, where $(\boldsymbol{\alpha}, \boldsymbol{T})$ and $(\boldsymbol{\beta}, \boldsymbol{S})$ are

$$\boldsymbol{\alpha} = (0.5, 0.2, 0.3), \boldsymbol{T} = \begin{pmatrix} -3 & 1 & 0 \\ 0 & -6 & 1 \\ 0 & 2 & -5 \end{pmatrix}, \ \boldsymbol{\beta} = (0.1, 0.6, 0, 0.3), \boldsymbol{S} = \begin{pmatrix} -3 & 1 & 0 & 1 \\ 1 & -5 & 1 & 0 \\ 0 & 2 & -4 & 2 \\ 1 & 0 & 1 & -4 \end{pmatrix}.$$

For this risk process, ρ is $1.708201 > 1$. Hence, ruin is certain. To find the time to ruin and deficit at the time of ruin, we consider the dual queue $(PH(\boldsymbol{\beta}, \boldsymbol{S})/PH(\boldsymbol{\alpha}, \boldsymbol{T})/1)$ of the risk process $(PH(\boldsymbol{\alpha}, \boldsymbol{T})/PH(\boldsymbol{\beta}, \boldsymbol{S})/1)$. The dual queue is stable since $\rho^* = 1/\rho < 1$. Using the analysis as explained in Sect. 4.2, we calculate the waiting-time (actual and virtual) distributions and idle period distribution and the excess life of the idle period distribution. Similarly, the expected values of the deficit and the time to ruin are calculated for other claim size distributions described above. The expected deficit at the time of ruin for different claim distributions is presented in Table 3 and that of the expected time to ruin in Table 4. We have found that inverse Gaussian and matrix exponential distributions are more helpful than others for the insurance company.

Table 3 Expected value of the deficit at the time of ruin for different claim sizes

u	PH	Frostig	ME	IG	Wb	Pareto
0	0.628177	0.628177	0.951825	0.473230	2.687979	0.891496
1	0.659026	0.659026	0.981618	0.540957	4.167769	1.020517
2	0.659124	0.659124	0.989147	0.550614	4.904983	1.042251
3	0.659125	0.659125	0.990817	0.552450	5.290328	1.046379
4	0.659125	0.659125	0.991091	0.552911	5.507631	1.047172
5	0.659125	0.659125	0.991089	0.553040	5.643478	1.047325
6	0.659125	0.659125	0.991060	0.553076	5.738733	1.047355
7	0.659125	0.659125	0.991041	0.553086	5.812763	1.047360
8	0.659125	0.659125	0.991032	0.553089	5.874820	1.047361
9	0.659125	0.659125	0.991028	0.553090	5.929389	1.047362
10	0.659125	0.659125	0.991027	0.553091	5.978708	1.047362

Table 4 Expected value of the time to ruin for different claim sizes

u	PH	Frostig	ME	IG	Wb	Pareto
0	0.887003	0.887003	0.452473	0.926465	0.532994	0.641202
1	2.342591	2.342591	0.942010	3.016804	1.024707	1.453243
2	3.754757	3.754757	1.420963	4.993455	1.369176	2.188119
3	5.166785	5.166785	1.897132	6.954796	1.643874	2.910331
4	6.578813	6.578813	2.372636	8.913446	1.885251	3.630145
5	7.990840	7.990840	2.848009	10.871444	2.110476	4.349498
6	9.402868	9.402868	3.323369	12.829261	2.327652	5.068763
7	10.814896	10.814896	3.798735	14.787028	2.540619	5.788010
8	12.226923	12.226923	4.274105	16.744781	2.751213	6.507254
9	13.638951	13.638951	4.749477	18.702529	2.960321	7.226498
10	15.050978	15.050978	5.224851	20.660276	3.168389	7.945741

6 Conclusion and Future Scope

In this paper, a complete analysis of the *GI*/*G*/1 risk process is described for different claim arrivals and claim sizes with distributions having rational LST. Heavy-tailed distributions and distributions without any LST are also considered. We presented the closed-form distributions for the ruin probability and the deficit at the time of ruin. This model can be extended to discrete time renewal risk model, and its counterpart in the continuous time can be approximated. Another challenging area of investigation can be the cross-correlated risk models with dependency between inter-claim times and claim sizes. A further extension of the model to a renewal risk process incorporating different variants like reinsurance, constant dividend barrier, and force of interest is of much importance, and these studies are left for future investigations.

Acknowledgements The third author's research work was partially supported by NSERC, Canada.

References

1. Dickson, D.C.M.: On a class of renewal risk processes. N. Am. Actuar. J. **2**(3), 60–68 (1998)
2. Dong, H., Liu, Z.: A class of Sparre Andersen risk process. Front. Math. China **5**(3), 517–530 (2010)
3. Dufresne, D.: A general class of risk models. Aust. Actuar. J. **7**(4), 755–791 (2011)
4. Gerber, H.U., Shiu, E.S.: The joint distribution of the time of ruin, the surplus immediately before ruin, and the deficit at ruin. Insur. Math. Econ. **21**(2), 129–137 (1997)
5. Gerber, H.U., Shiu, E.S.: On the time value of ruin. N. Am. Actuar. J. **2**(1), 48–72 (1998)
6. Dickson, D., Hipp, C.: Ruin probabilities for Erlang (2) risk processes. Insur. Math. Econ. **22**(3), 251–262 (1998)
7. Dickson, D.C., Hipp, C.: On the time to ruin for Erlang (2) risk processes. Insur. Math. Econ. **29**(3), 333–344 (2001)
8. Landriault, D., Willmot, G.: On the Gerber-Shiu discounted penalty function in the Sparre Andersen model with an arbitrary interclaim time distribution. Insur. Math. Econ. **42**(2), 600–608 (2008)
9. Gerber, H.U., Shiu, E.S.: The time value of ruin in a Sparre Andersen model. N. Am. Actuar. J. **9**(2), 49–69 (2005)
10. Li, S., Garrido, J., et al.: On a general class of renewal risk process: analysis of the Gerber-Shiu function. Adv. Appl. Probab. **37**(3), 836–856 (2005)
11. Dickson, D., Drekic, S.: The joint distribution of the surplus prior to ruin and the deficit at ruin in some Sparre Andersen models. Insur. Math. Econ. **34**(1), 97–107 (2004)
12. Rodríguez-Martínez, E.V., Cardoso, R.M., Dos Reis, A.D.E.: Some advances on the erlang (n) dual risk model. Astin Bull. **45**(01), 127–150 (2015)
13. Asmussen, S., Rolski, T.: Computational methods in risk theory: a matrix-algorithmic approach. Insur. Math. Econ. **10**(4), 259–274 (1992)
14. Avram, F., Usabel, M.: Ruin probabilities and deficit for the renewal risk model with phase-type interarrival times. Astin Bull. **34**, 315–332 (2004)
15. Thorin, O.: Probabilities of ruin. Scand. Actuar. J. **1982**(2), 65–103 (1982)
16. Panda, G., Banik, A.D., Chaudhry, M.L.: Inverting the transforms arising in the $GI/M/1$ risk process using roots. In: Mathematics and Computing 2013, pp. 297–312. Springer (2014)
17. Asmussen, S., Albrecher, H.: Ruin Probabilities, 2nd edn. World Scientific, Singapore (2010)
18. Prabhu, N.U.: On the ruin problem of collective risk theory. Ann. Math. Statist. **3**, 757–764 (1961)
19. Thampi, K.K., Jacob, M.J.: On a class of renewal queueing and risk processes. J. Risk Financ **11**, 204–220 (2010)
20. Drekic, S., Dickson, D.C., Stanford, D.A., Willmot, G.E.: On the distribution of the deficit at ruin when claims are phase-type. Scand. Actuar. J. **2004**(2), 105–120 (2004)
21. Frostig, E.: Upper bounds on the expected time to ruin and on the expected recovery time. Adv. Appl. Probab. **36**, 377–397 (2004)
22. Chaudhry, M.L., Agarwal, M., Templeton, J.G.: Exact and approximate numerical solutions of steady-state distributions arising in the queue $GI/G/1$. Queueing Syst. **10**(1–2), 105–152 (1992)

23. Kleinrock, L.: Queueing Systems Vol 1: Theory. Wiley-Interscience, New York (1975)
24. Komota, Y., Nogami, S., Hoshiko, Y.: Analysis of the *GI*/*G*/1 queue by the supplementary variables approach. Electron. Commun. Jpn (Part I: Commun.) **66**(5), 10–19 (1983)
25. Chaudhry, M.L., Yang, X., Ong, B.: Computing the distribution function of the number of renewals. Am. J. Oper. Res. **3**, 380–386 (2013)

Score-Based Secretary Problem

Jyotirmoy Sarkar

Abstract In the celebrated "Secretary Problem," involving n candidates who have applied for a single vacant secretarial position, the employer interviews them one by one in random order and learns their relative ranks. As soon as each interview is over, the employer must either hire the candidate (and stop the process) or reject her (never to be recalled). We consider a variation of this problem where the employer also learns the scores of the already interviewed candidates, which are assumed to be independent and drawn from a known continuous probability distribution. Endowed with this additional information, what strategy should the employer follow in order to maximize his chance of hiring the candidate with the highest score among all n candidates? What is the maximum probability of hiring the best candidate?

Keywords Analytical expression · Conditional probability · Iterative computation · Recursive relation · Simulation

1 Background and Statement of the Problem

Recall the celebrated secretary problem: There are n applicants who have applied for the single open position of a secretary. The employer will interview them one by one in random order. At the conclusion of each interview, the employer must decide either to hire the candidate (and stop interviewing), or to let her go for good never to be recalled. Thus, the process concludes as soon as a candidate is hired. At any time the employer has a relative ranking of all candidates interviewed so far, but he does not have their absolute ranking among all n candidates (unless and until he interviews all candidates). His objective is to hire the best candidate. Obviously, he must consider hiring a candidate only if she is the best among all interviewed candidates. How can he maximize his probability of hiring the best candidate among all n candidates?

If the employer hires a candidate after interviewing too few candidates, he runs the risk of missing the best candidate who is yet to be interviewed. On the other

J. Sarkar (✉)
Indiana University-Purdue University Indianapolis, Indianapolis, USA
e-mail: jsarkar@iupui.edu

S. Kar et al. (eds.), *Operations Research and Optimization*, Springer Proceedings in Mathematics & Statistics 225, https://doi.org/10.1007/978-981-10-7814-9_7

hand, if he rejects too many candidates and waits too long to hire, he will have too few candidates left to be interviewed, and hence a small chance that one of them will surpass the leading candidate among those who have been already interviewed (and this leading candidate has not been hired). Assuming that all permutations of candidates by rank are equally likely, the optimal strategy turns out to be: "Interview and let go the first $m-1$ candidates, and thereafter hire the first candidate among $m, m+1, \ldots, n-1$ who is the best among all interviewed candidates, if such a candidate comes along; otherwise, hire the last candidate." The value m, of course, depends on n. In fact, $m = \text{argmax}_k\{p_n(k)\}$ where $p_n(1) = 1/n$ and for $k \geq 2, p_n(k) = \frac{k-1}{n} \sum_{j=k}^{n} \frac{1}{j-1}$. The maximum probability of hiring the best candidate by following the optimal strategy is $p_n(m)$. In particular, asymptotically (as $n \to \infty$), the optimal choice is $m \approx n/e$, and the associated highest probability of hiring the best candidate is $1/e = 0.3678794\ldots$.

The secretary problem is also known as the marriage problem, the sultan's dowry problem, the fussy suitor problem, the game of googol, and the best choice problem. It was apparently introduced in 1949 by Flood [1]. R. Palermo proved that all strategies are dominated by a strategy of the form: "Reject the first $k-1$ unconditionally, then accept the next candidate who is the best." The first publication was apparently by Martin Gardner in Scientific American, February 1960, where he presented the analysis by Leo Moser and J.R. Pounder. It was reprinted with additional comments in [2]. The $1/e$-law of best choice is due to Bruss [3]. We refer the reader to Ferguson [4] for an extensive bibliography.

Here we consider a variation of the secretary problem, in which we know the probability distribution from which the scores of each candidate is drawn. Suppose that the candidates appear in a random order. After interviewing each candidate, the employer not only knows her relative rank among all candidates interviewed so far, but also he knows her absolute score on a known scale. In this modified situation, what strategy should the employer follow in order to maximize his chance of hiring the candidate with the highest score among all n candidates? More precisely, we consider the problem below.

Score-Based Secretary Problem: In the context of the secretary problem, suppose that the employer, after interviewing the candidates, can assign absolute scores $X_1, X_2, \ldots$, which are assumed to be drawn independently from the same known continuous distribution function F. What is the employer's best strategy to maximize the chance of winning (that is, hiring the candidate with the highest score among all n candidates)? What is the maximum probability of winning?

In preparation for solving the score-based secretary problem, let us make a few straight-forward observations:

(1) Without loss of generality, we can assume that the scores are independent and identically distributed (IID) as uniform (0, 1). For otherwise, we will simply replace each score X by $U = F(X)$, which will have uniform (0, 1) distribution. See Exercise 1.2 of [5]. Also, by the continuity of F, ties among scores is ruled

out (for when the scores are displayed to infinitely many decimal places, surely they will differ).

(2) The employer will surely let go any candidate who scores below someone else already interviewed. He should consider hiring a candidate only if her score is a **record high** value in the sense that it is the largest among all scores assigned so far. Clearly, X_1 is a record high score. Thereafter, for $m > 1$, X_m is a record high value if $X_m = \max\{X_1, X_2, \ldots, X_m\}$; or equivalently, if $X_m > X_i$ for $i = 1, 2, \ldots, m-1$.

(3) The employer should consider hiring a candidate with a record high score provided her score is large enough so that there is only a small chance that some other candidate yet to be interviewed will score higher than the current candidate. How large a score is large enough for the employer win the game (or to hire a candidate with a record high score)?

(4) The requisite threshold, above which the employer should hire a candidate with a record high score, depends on how many candidates are yet to be interviewed. Finding the threshold corresponding to each possible remaining number of candidates to be interviewed is the crux of the solution to the problem. We contend that the thresholds can be found inductively as we allow more and more candidates.

Let us describe the employer's best strategy and his overall chance of winning.

Definition 1 If there are n applicants, let θ_n denote the threshold such that the employer maximizes his probability of hiring the best candidate by using the strategy "Hire Candidate 1, who scores X_1, if and only if $X_1 > \theta_n$."

Form of the Best Strategy: "If $X_1 > \theta_n$, then hire Candidate 1. Otherwise, wait until a candidate (say, Candidate m) receives a record high score X_m. If $X_m > \theta_{n+1-m}$, then hire Candidate m. Otherwise, let her go and wait until the next record high score is attained. Etc."

Definition 2 If there are n applicants, let ω_n denote the (maximal) probability that the employer wins the game (that is, hires the best candidate) if he uses his above-stated best strategy.

We will now inductively determine θ_n and ω_n for all $n = 1, 2, 3, \ldots$.

2 Exact Solutions for Small Number of Candidates

Recall that the potential scores of the candidates (in the order interviewed) are $X_1, X_2, \ldots, X_n$, which we assume are IID uniform (0, 1). If there is only $n = 1$ applicant, the solution is trivial: Hire her irrespective of her score; that is, $\theta_1 = 0$. In this case, the employer surely hires the best candidate by default. So, $\omega_1 = 1$. Let us proceed to determine $(\theta_2, \omega_2), (\theta_3, \omega_3), \ldots$.

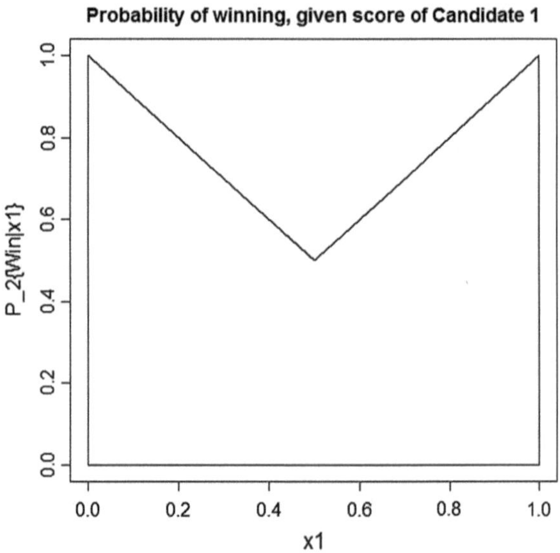

Fig. 1 Conditional probability $f_2(x_1)$ that the employer wins, given $X_1 = x_1$

Suppose that $n = 2$. In this case, after interviewing Candidate 1 who has received score $X_1 = x_1$, there is only one more candidate to be interviewed. If $x_1 > 1/2$, then the employer should hire Candidate 1, because, given $x_1 > 1/2$, the conditional probability that the employer wins (or hires the better candidate) is $P\{X_2 < x_1\} = x_1 > 1/2$. On the other hand, if $x_1 < 1/2$, then the employer should let Candidate 1 go, and hire Candidate 2, because, given $x_1 < 1/2$, the conditional probability that the employer wins is $P\{X_2 > x_1\} = 1 - x_1 > 1/2$. Figure 1 shows the function $f_2(x_1) = \max\{1 - x_1, x_1\}$, the conditional probability of the employer winning, given $X_1 = x_1$.

Since X_1 follows a uniform (0, 1) distribution, the overall probability that the employer wins is the area under the graph of $f_2(x_1)$; that is,

$$\omega_2 = \int_0^1 f_2(x_1)\,dx_1 = \int_0^{1/2} (1 - x_1)\,dx_1 + \int_{1/2}^1 x_1\,dx_1 = \frac{3}{4}\,. \tag{1}$$

We claim that $\theta_2 = 1/2$. The justification is as follows: If the employer adopts a strategy of hiring Candidate 1 if and only if $x_1 > \theta$, with any other threshold value θ, then his overall probability of winning will be, as seen from Fig. 2,

$$\int_0^\theta (1 - x_1)\,dx_1 + \int_\theta^1 x_1\,dx_1 = \theta(1-\theta) + 1/2 = \frac{3}{4} - \left(\theta - \frac{1}{2}\right)^2 \in \left(\frac{1}{2}, \frac{3}{4}\right). \tag{2}$$

Hence, any threshold value θ, other than $\theta_2 = 1/2$, actually lowers the employer's overall probability of winning. Interestingly though, (2) shows that by choosing any sub-optimal threshold $\theta_2 \neq 1/2$, even by choosing $\theta = 0$ (that is, always hire

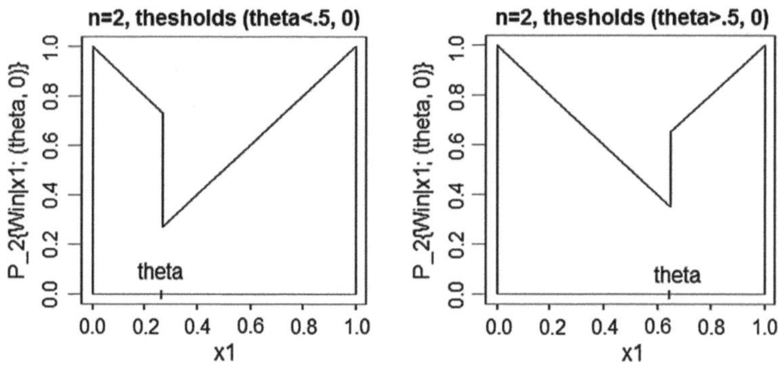

Fig. 2 For $n = 2$, the employer's probability of hiring the better candidate if he uses a suboptimal strategy by choosing thresholds θ and 0 for Candidates 1 and 2 respectively

Candidate 1 irrespective of her score) or $\theta = 1$ (that is, never hire Candidate 1), the employer's chance of winning never falls below 1/2.

Next, suppose that $n = 3$. In this case, after interviewing Candidate 1, who scored $X_1 = x_1$, there are two more candidates to interview. We must determine the value of θ_3 such that the employer's best strategy is to hire Candidate 1 with score $X_1 = x_1$ if and only if $x_1 > \theta_3$. Let us analyze separately what happens if (A) Candidate 1 is hired, and (B) Candidate 1 is let go (or skipped over):

(A) If Candidate 1 is hired, then the interviewer has not seen scores X_2 and X_3, which are independent uniform (0, 1) variables. In this case, the employer's conditional probability of winning is

$$P_3^{\text{hire}}(x_1) = Pr\{X_2 < x_1, X_3 < x_1\} = Pr\{X_2 < x_1\} \cdot Pr\{X_3 < x_1\} = x_1^2 . \qquad (3)$$

Note that $P_3^{\text{hire}}(x_1)$ is an increasing function of x_1 on (0, 1) going from 0 to 1. Thus, whenever it is advantageous for the employer to hire Candidate 1 with a particular score x_1, it must be also advantageous to hire Candidate 1 with any score bigger than x_1.

(B) On the other hand, if the employer lets go Candidate 1 with a score x_1, he gets to interview Candidate 2 and obtain X_2. The problem almost reduces to the case of only two candidates, except that the employer should not apply his best strategy for the two-candidate game unless X_2 is a record high value; that is, unless $X_2 > x_1$. There are three mutually exclusive cases to consider: (1) If $X_2 < x_1$, then the employer should let Candidate 2 go, he should continue to interview Candidate 3 to obtain X_3, and win only if $X_3 > x_1$. (2) If $X_2 > x_1$ but $X_2 < \theta_2$, then (in accordance with the optimal strategy for the two-candidate game) the employer should still let Candidate 2 go, interview Candidate 3 to obtain X_3, and win only if $X_3 > X_2$. (3) If $X_2 > x_1$ and $X_2 > \theta_2$, then the employer should hire Candidate 2, and win if $X_3 < X_2$.

So, given x_1 and the contemplated choice that the employer should let go Candidate 1, the conditional probability of his winning is

$$P_3^{\text{skip}}(x_1) = Pr(\{X_2 < x_1, X_3 > x_1\} \cup \{X_2 > x_1, X_2 < \theta_2, X_3 > X_2\} \cup \{X_2 > x_1, X_2 > \theta_2, X_3 < X_2\})$$

$$= x_1(1 - x_1) + \int_{x_1}^{1} f_2(x)\,dx = \begin{cases} \frac{3}{4} - \frac{x_1^2}{2} & \text{if } x_1 \le 1/2 \\ (1 - x_1)\left(x_1 + \frac{1+x_1}{2}\right) & \text{if } x_1 \ge 1/2 \,. \end{cases}$$

Figure 3a, b shows $P_3^{\text{skip}}(x_1)$ as the total area of the enclosed region for two typical values of x_1—(a) below $1/2$ and (b) above $1/2$.

It can be checked that $P_3^{\text{skip}}(x_1)$ is a decreasing function of x_1 on (0, 1) going from $\omega_2 = 3/4$ to 0, while $P_3^{\text{hire}}(x_1)$ is an increasing function of x_1 on (0, 1) going from 0 to 1. By the intermediate value theorem (see [6], e.g.), these two functions intersect in (0, 1) at a unique point $x_1 = \theta_3 = (1 + \sqrt{6})/5$, which is obtained by solving $(1 - x_1) \cdot (1 + 3x_1)/2 = x_1^2$, or equivalently $5x_1^2 - 2x_1 - 1 = 0$. Therefore, the employer should hire Candidate 1 if $x_1 > \theta_3$, let her go if $x_1 < \theta_3$, and be indifferent if $x_1 = \theta_3$. Hence, given x_1, the conditional probability of the employer winning (if he follows the optimal strategy) is $f_3(x_1) = \max\{P_3^{\text{skip}}(x_1), P_3^{\text{hire}}(x_1)\}$, as shown in Fig. 4.

Finally, as was done in (1), if the employer uses this optimal strategy for $n = 3$, his overall probability of winning is

$$\omega_3 = \int_0^1 f_3(x)\,dx$$

$$= \int_0^{1/2} \left(\frac{3}{4} - \frac{x_1^2}{2}\right) dx_1 + \int_{1/2}^{\theta_3} (1 - x_1)\left(\frac{1 + 3x_1}{2}\right) dx_1 + \int_{\theta_3}^{1} x_1^2\,dx_1$$

$$= \frac{17}{48} + \frac{x_1 + x_1^2 - x_1^3}{2}\Big|_{1/2}^{\theta_3} + \frac{x_1^3}{3}\Big|_{\theta_3}^{1} = \frac{293 + 48\sqrt{6}}{600} = 0.68429251 \,.$$

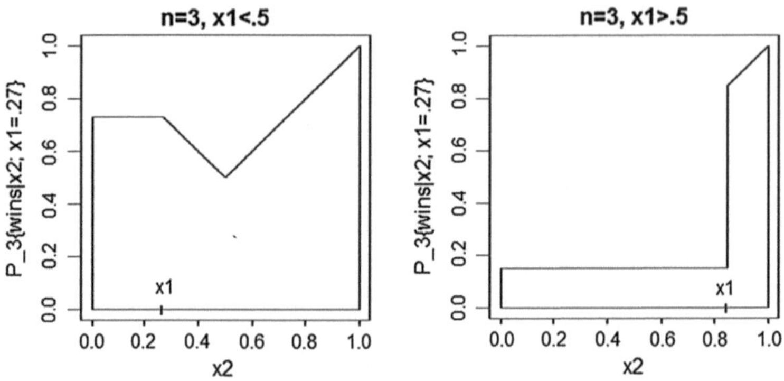

Fig. 3 For **a** $x_1 < 0.5$ and **b** $x_1 > 0.5$, the conditional probability that the employer wins if he lets go Candidate 1, obtains $X_2 = x_2$, and follows his best strategy

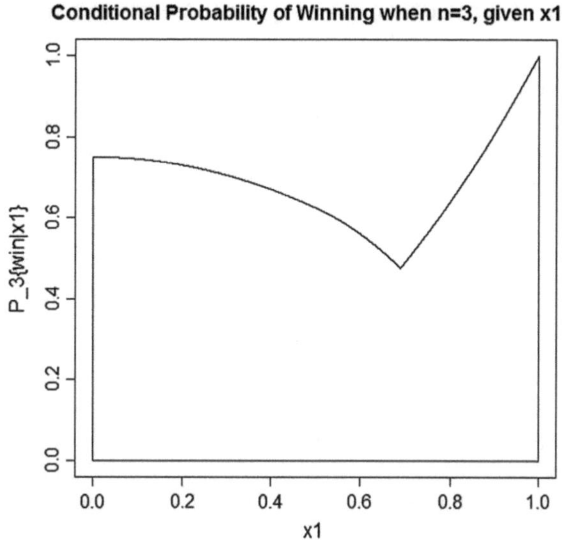

Fig. 4 Conditional probability $f_3(x_1)$ that the employer wins the three-candidate game, given Candidate 1's score x_1, if the employer follows his optimal strategy

For $n \geq 4$ candidates, we proceed inductively. Suppose that the solutions for the n-candidate game—$P_n^{\text{hire}}(x_1), P_n^{\text{skip}}(x_1), f_n(x_1), \theta_n, \omega_n$—are already found. Then we find the solutions for the $(n+1)$-candidate game using Algorithm 1.

Algorithm 1

Step 1 Define $P_{n+1}^{\text{hire}}(x_1) = x_1 \cdot P_n^{\text{hire}}(x_1)$. Starting from (3), by mathematical induction we note that $P_{n+1}^{\text{hire}}(x_1) = x_1^n$, which is an increasing function of x_1 on (0, 1) going from 0 to 1.

Step 2 Define $P_{n+1}^{\text{skip}}(x_1) = x_1 \cdot P_n^{\text{skip}}(x_1) + \int_{x_1}^1 f_n(x_1)\,dx_1$, which is a decreasing function on (0, 1) going from ω_n to 0.

Step 3 The unique solution to $P_{n+1}^{\text{skip}}(x_1) = P_{n+1}^{\text{hire}}(x_1)$ in x_1 yields θ_{n+1}.

Step 4 Define $f_{n+1}(x_1) = \max\{P_{n+1}^{\text{skip}}(x_1), P_{n+1}^{\text{hire}}(x_1)\}$, and obtain the winning probability $\omega_{n+1} = \int_0^1 f_{n+1}(x_1)\,dx_{n+1}$.

Using Algorithm 1, we can obtain $\theta_4 = 0.775845$, by solving $17x_1^3 - 6x_1^2 - 3x_1 - 2 = 0$; and we can evaluate $\omega_4 = \int_0^1 f_4(x_1)\,dx_1 = 0.655397$. In this manner, we can continue to use Algorithm 1 to find $(\theta_5, \omega_5), (\theta_6, \omega_6), \ldots$. But the calculus becomes tedious! In Sect. 3, we describe a computational technique to evaluate (θ_n, ω_n) recursively, and we verify the optimality of the solutions through simulation. In Sect. 4, we develop analytical expressions of (θ_n, ω_n) for all $n \geq 3$.

3 Computing (θ_n, ω_n) and Verifying by Simulation

Nowadays, the ability to compute is at everyone's fingertip—thanks to the invention of personal computer and the advancement of computer languages. For instance, using the programming language (and statistical software) R, which anyone can download for free on to their computer, one can evaluate (θ_n, ω_n).

3.1 Computational Evaluation of (θ_n, ω_n)

First, the integrals in Steps 2 and 4 of Algorithm 1 can be approximated by computing the Riemann sum. To do so, we recommend partitioning the interval (0, 1) into 10^6 equal subintervals of width 10^{-6} each and evaluate the integrand at the middle point of each subinterval. Then the integral is approximated by the mean of the functional values at the 10^6 midpoints. See [6], for example.

Second, because $P_{n+1}^{\text{skip}}(x_1)$ is decreasing from ω_n to 0, while $P_{n+1}^{\text{hire}}(x_1)$ is increasing from 0 to 1 in the interval (0, 1), the n-degree polynomial equation in Step 3 of Algorithm 1 can be solved (approximately) by computing the largest argument at which $P_{n+1}^{\text{skip}}(x_1) > P_{n+1}^{\text{hire}}(x_1)$, plus half of 10^{-6}. The R codes for inductively computing $(\theta_{n+1}, \omega_{n+1})$ are given below.

R Codes to Compute $(\theta_{n+1}, \omega_{n+1})$ Inductively

```
B=10^6;    H=0.5/B
serial=seq(1:B);grid=serial/B-H
theta=c(0); win=c(1) # for n=1
# for n=2 define directly
hire=grid;skip=1-grid;
theta=c(theta, grid[max(which(skip>hire))]+H)
fun=pmax(skip, hire); win=c(win, mean(fun))
# for n>2 define recursively
for (i in 3:100){
hire=hire*grid # Step 1
skip=(cumsum(fun[B:1])[B:1]+(serial-1)*skip)/B # Step 2
theta=c(theta, grid[max(which(skip>hire))]+H) # Step 3
fun=pmax(skip, hire); win=c(win, mean(fun)) # Step 4
}
```

Table 1 gives (θ_n, ω_n) for $n = 1, 2, 3, \ldots, 100$ correct to six decimal places. Figure 5a–d depicts the conditional probability $f_n(x_1)$ that the employer wins given x_1, if he follows his optimal strategy for $n = 5, 10, 20, 100$ candidates.

As seen in Table 1, for the game involving 10 candidates, the employer's overall probability of winning, if he follows his optimal strategy, is about 61%. Likewise,

Table 1 The thresholds of the employer's optimal strategy and his maximal probability of winning corresponding to 1–100 candidates

```
>round(theta, 6)
 [1] 0.000000 0.500000 0.689898 0.775845 0.824590 0.855949 0.877807 0.893910 0.906265 0.916044
[11] 0.923976 0.930539 0.936059 0.940767 0.944829 0.948370 0.951484 0.954243 0.956706 0.958917
[21] 0.960913 0.962724 0.964375 0.965886 0.967274 0.968553 0.969736 0.970834 0.971854 0.972806
[31] 0.973695 0.974528 0.975310 0.976045 0.976738 0.977392 0.978010 0.978595 0.979150 0.979677
[41] 0.980178 0.980654 0.981109 0.981542 0.981956 0.982352 0.982731 0.983094 0.983442 0.983776
[51] 0.984097 0.984405 0.984702 0.984987 0.985262 0.985527 0.985783 0.986030 0.986269 0.986499
[61] 0.986722 0.986938 0.987146 0.987349 0.987544 0.987734 0.987919 0.988097 0.988271 0.988440
[71] 0.988603 0.988763 0.988917 0.989068 0.989215 0.989357 0.989496 0.989632 0.989764 0.989892
[81] 0.990018 0.990140 0.990260 0.990376 0.990490 0.990601 0.990710 0.990816 0.990919 0.991021
[91] 0.991120 0.991217 0.991312 0.991405 0.991495 0.991584 0.991672 0.991757 0.991841 0.991922

>round(win, 6)
 [1] 1.000000 0.750000 0.684293 0.655396 0.639194 0.628784 0.621508 0.616128 0.611986 0.608700
[11] 0.606028 0.603813 0.601948 0.600356 0.598981 0.597781 0.596725 0.595788 0.594952 0.594201
[21] 0.593522 0.592906 0.592344 0.591830 0.591358 0.590922 0.590519 0.590145 0.589797 0.589472
[31] 0.589169 0.588885 0.588618 0.588367 0.588131 0.587908 0.587697 0.587497 0.587307 0.587127
[41] 0.586956 0.586793 0.586638 0.586490 0.586348 0.586213 0.586084 0.585960 0.585841 0.585726
[51] 0.585617 0.585511 0.585410 0.585312 0.585218 0.585127 0.585040 0.584955 0.584874 0.584795
[61] 0.584719 0.584645 0.584573 0.584504 0.584437 0.584372 0.584309 0.584248 0.584189 0.584131
[71] 0.584075 0.584020 0.583967 0.583916 0.583865 0.583817 0.583769 0.583723 0.583678 0.583633
[81] 0.583591 0.583549 0.583503 0.583468 0.583429 0.583391 0.583354 0.583317 0.583282 0.583247
[91] 0.583213 0.583180 0.583147 0.583116 0.583085 0.583054 0.583024 0.582995 0.582966 0.582938
```

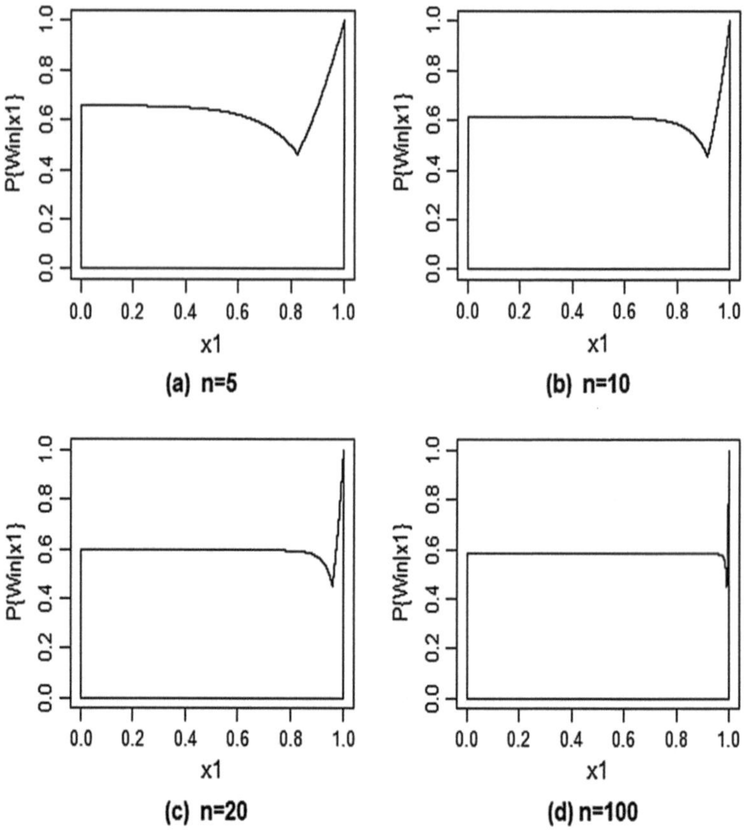

Fig. 5 Conditional probability $f_n(x_1)$ that the employer wins, given the score x_1 of Candidate 1, if he follows his optimal strategy for **a** 5, **b** 10, **c** 20, **d** 100 candidates

for the game involving 100 candidates, the employer's overall probability of winning, if he follows his optimal strategy, is about 58%. However, we also note that his probability of winning is monotonically decreasing as the number of candidates is increasing. Will the game continue to be favorable to him as the number of candidates increase, say to 500 or to 1000? Will there be a limiting value to the employer's probability of winning as the number of candidates increase without bound? Based on the fact that the thresholds and the winning probabilities for 951–1000 candidates have stabilized, as seen from Table 2, we conjecture that the answers are affirmative; but we leave the details to the reader to investigate.

3.2 Verifying Optimality of Computed Values via Simulation

In solving the score-based secretary problem, we made claims of a probabilistic nature. For example, when $n = 5$, we claimed: "The employer's maximal probability of winning a five-candidate game is $\omega_5 = 0.639194$, and he can achieve this probability by adhering to an optimal strategy with $\theta_1 = 0, \theta_2 = 0.5, \theta_3 = 0.689898, \theta_4 = 0.775845, \theta_5 = 0.824590$." The correctness of this claim can be verified by conducting a simulation, which involves imitating a random phenomenon by generating an appropriate sequence of random numbers.

One can simulate the five-candidate game a large number of times to estimate the employer's probability of winning using the R codes below. The win5 function, used in the codes, indicates whether the employer wins in any one particular play. The estimated probability of the employer's winning, $\hat{p}$, is given by the mean of the indicators of win in a series of $M = 10^6$ plays. In fact, in view of the central limit theorem, there is about 95% chance that the estimate $\hat{p}$ will be within $1.96\sqrt{p(1-p)/M} < 1/\sqrt{M}$ of the true probability p. See [7], for example. Therefore, our estimate $\hat{p}$ based on $M = 10^6$ repetitions will be correct to three decimal places with about 95% probability.

With the optimal choice of the threshold values, the estimate is found to be 0.638607, whereas with one particular alternative choice, the estimate turns out to be 0.636515, and with another alternative choice to be 0.592601. Since the estimates are correct to three decimal places, as reasoned above, this simulation provides sufficient justification for the optimal choice of thresholds and declares the other choices of thresholds as suboptimal. We encourage the readers to estimate the probability that the employer wins when other choices of the threshold values $\theta_3, \theta_4, \theta_5$ are made and verify that none will do better than the optimal choices of thresholds our Algorithm 1 has produced. In Sect. 4, we will compute the employer's exact probability of winning, for each of these strategies.

R Codes to Simulate Winning Probability for any Set of Thresholds

```
win5 <- function(theta3,theta4,theta5){
coin=runif(5, 0, 1); top=max(coin)
if (coin[1]>=theta5){ind=(coin[1]==top)}
 else{if (coin[2]>=max(coin[1],theta4)){ind=(coin[2]==top)}
  else{if (coin[3]>=max(coin[1:2],theta3)){ind=(coin[3]==top)}
   else{if (coin[4]>=max(coin[1:3],1/2)){ind=(coin[4]==top)}
    else{ind=(coin[5]==top)}
} } }
ind
}
mean(replicate(10^6, win5((1+sqrt(6))/5,.775845,.824590))) #opt
mean(replicate(10^6, win5(2/3, 3/4, 4/5)))      # alternative 1
mean(replicate(10^6, win5(1/2,.5^(1/5),.5^(1/5),.5^(1/5)))) #2
```

Table 2 The thresholds of the employer's optimal strategy and his maximal probability of winning corresponding to 951–1000 candidates

```
>round(theta[951:1000], 6)
[951] 0.999154 0.999155 0.999156 0.999157 0.999158 0.999158 0.999159 0.999160 0.999161 0.999162
[961] 0.999163 0.999164 0.999165 0.999165 0.999166 0.999167 0.999168 0.999169 0.999170 0.999171
[971] 0.999171 0.999172 0.999173 0.999174 0.999175 0.999176 0.999177 0.999177 0.999178 0.999179
[981] 0.999180 0.999181 0.999182 0.999182 0.999183 0.999184 0.999185 0.999186 0.999187 0.999187
[991] 0.999188 0.999189 0.999190 0.999191 0.999191 0.999192 0.999193 0.999194 0.999195 0.999195

> round(win[951:1000], 6)
[951] 0.580477 0.580477 0.580477 0.580476 0.580476 0.580476 0.580475 0.580475 0.580475 0.580475
[961] 0.580474 0.580474 0.580474 0.580473 0.580473 0.580473 0.580473 0.580472 0.580472 0.580472
[971] 0.580472 0.580471 0.580471 0.580471 0.580470 0.580470 0.580470 0.580470 0.580469 0.580469
[981] 0.580469 0.580469 0.580468 0.580468 0.580468 0.580468 0.580467 0.580467 0.580467 0.580467
[991] 0.580466 0.580466 0.580466 0.580466 0.580465 0.580465 0.580465 0.580464 0.580464 0.580464
```

4 Analytical Expressions of (θ_n, ω_n)

A careful study of Algorithm 1 leads to an analytical expression for θ_n, given in Theorem 1, and thereafter an analytical expression of ω_n, given in Theorem 2. Toward these results, we note the following:

(a) From Step 1, we have $P_{n+1}^{\text{hire}}(x) = x^n$ for all $x \in (0, 1)$.
(b) From Step 4, we have $f_n(x) = \max\{P_n^{\text{skip}}(x), P_n^{\text{hire}}(x)\}$ for all $x \in (0, 1)$. In particular, $f_n(x) = P_n^{\text{hire}}(x)$ for all $x \in (\theta_n, 1)$.
(c) From Steps 2 and 1 and item (b) above, for all $x \in (\theta_n, 1)$, we have

$$P_{n+1}^{\text{skip}}(x) = x\,P_n^{\text{skip}}(x) + \int_x^1 P_n^{\text{hire}}(u)\,du = x\,P_n^{\text{skip}}(x) + \frac{1-x^n}{n} \tag{4}$$

In fact, solving (4), we get Lemma 1 (to be proved momentarily):

Lemma 1 *For all $n \geq 1$, we have*

$$P_{n+1}^{\text{skip}}(x) = \left[\sum_{k=1}^{n} \frac{x^{-k}-1}{k}\right] x^n . \tag{5}$$

(d) From Step 3, we have θ_{n+1} is the solution to $P_{n+1}^{\text{skip}}(x) = P_{n+1}^{\text{hire}}(x)$; or equivalently, in view of items (a) and (c) above, θ_{n+1} is the solution to

$$\sum_{k=1}^{n} \frac{x^{-k}-1}{k} = 1 . \tag{6}$$

In other words, we obtain Theorem 1 (to be proved shortly):

Theorem 1 $\theta_{n+1} = (1+\epsilon_{n+1})^{-1}$, *where ϵ_{n+1} is the unique solution to*

$$\sum_{i=1}^{n} \binom{n}{i} \frac{\epsilon^i}{i} = 1 . \tag{7}$$

Proof of Lemma 1. The proof is by mathematical induction on n. We know that $P_1^{\text{skip}}(x) = 0$ and $P_1^{\text{hire}}(x) = 1$ for all $x \in (0, 1)$. Next, we know that $P_2^{\text{skip}}(x) = 1 - x$ for all $x \in (0, 1)$. Also, putting $n = 1$ in (4), for $x \in (0, 1)$, we have

$$P_2^{\text{skip}}(x) = x\,P_1^{\text{skip}}(x) + \int_x^1 u^0\,du = x \cdot 0 + (1-x) = (x^{-1}-1)x$$

Hence, (6) holds for $n = 1$.

Now suppose that (6) holds for an arbitrary $n \geq 1$. Then from (4), we have

$$P_{n+1}^{\text{skip}}(x) = x \cdot \left[\sum_{k=1}^{n-1} \frac{x^{-k}-1}{k} \right] x^{n-1} + \frac{1-x^n}{n}$$
$$= \left[\sum_{k=1}^{n-1} \frac{x^{-k}-1}{k} \right] x^{n} + \frac{1-x^n}{n} = \left[\sum_{k=1}^{n} \frac{x^{-k}-1}{k} \right] x^{n}.$$

Hence, (6) holds for $n+1$. This completes the proof of Lemma 1. Q.E.D.

Proof of Theorem 1. Write $x = (1+\epsilon)^{-1}$. Then starting from (6), applying the binomial theorem, and simplifying, we have

$$1 = \sum_{k=1}^{n} \frac{x^{-k}-1}{k} = \sum_{k=1}^{n} \frac{(1+\epsilon)^k - 1}{k}$$
$$= \sum_{k=1}^{n} \frac{1}{k} \sum_{i=1}^{k} \binom{k}{i} \epsilon^i = \sum_{i=1}^{n} \sum_{k=i}^{n} \frac{1}{i} \binom{k-1}{i-1} \epsilon^i$$
$$= \sum_{i=1}^{n} \left\{ \sum_{k=i}^{n} \binom{k-1}{i-1} \right\} \frac{\epsilon^i}{i} = \sum_{i=1}^{n} \binom{n}{i} \frac{\epsilon^i}{i}.$$

This completes the proof. Q.E.D.

We already know that $\theta_1 = 0, \theta_2 = 1/2$. Putting $n = 2$ into Theorem 1, we solve $2\epsilon + \epsilon^2/2 = 1$ to obtain $\epsilon_3 = -2 + \sqrt{6}$; whence $\theta_3 = (1+\epsilon_3)^{-1} = (-1+\sqrt{6})^{-1} = (1+\sqrt{6})/5$. Next, putting $n = 3$ into Theorem 1, we solve $3\epsilon + 3\epsilon^2/2 + \epsilon^3/3 = 1$ to obtain $\epsilon_4 = 0.288917$; and hence $\theta_4 = 1/1.288917 = 0.775845$. The advantage of using Theorem 1 over the computational technique of Sect. 3 is that we can now compute θ_n directly without having to compute first $\theta_3, \theta_4, \ldots, \theta_{n-1}$.

Remark 1 Since the left-hand side expression in (7) is increasing in n at each fixed $\epsilon \in (0, 1)$, the solution $\{\epsilon_{n+1}\}$ is a decreasing sequence. Hence, $\{\theta_n\}$ is an increasing sequence.

Remark 2 For $n \geq 100$, we can approximate θ_n as $0.44765^{1/n}$, or more simply approximate it as $1 - 0.8035/n$.

The optimal thresholds $\theta_1 = 0, \theta_2 = 1/2, \theta_3 = (1+\sqrt{6})/5, \theta_4, \ldots, \theta_n$ are used to find an analytical expression for ω_n as given in Theorem 2. In fact, this theorem holds for any non-decreasing sequence of thresholds $\xi_1, \xi_2, \ldots, \xi_n$, which need not be the optimal thresholds. The proof is delegated to the Appendix.

Theorem 2 *For any non-decreasing sequence of thresholds* $\{\xi_n\}$ *with* $\xi_n \in (0, 1)$, *used by the employer to hire a candidate whose score is a running maximum and* *exceeds the corresponding threshold, the probability that the employer wins is given by*

$$
\begin{aligned}
&P_n(\mathrm{Win}; \{\xi_n\}) \\
&= \frac{1}{n} + \left[\frac{\xi_2^{n-1}}{(n-1)\cdot 1} - \frac{\bar{H}_1\, \xi_2^n}{n}\right] + \left[\frac{\xi_3^{n-2}}{(n-2)\cdot 2} + \frac{\xi_3^{n-1}}{(n-1)\cdot 1} - \frac{\bar{H}_2\, \xi_3^n}{n}\right] \\
&\quad + \cdots + \left[\frac{\xi_n}{1\cdot(n-1)} + \frac{\xi_n^2}{2\cdot(n-2)} + \cdots + \frac{\xi_n^{n-1}}{(n-1)\cdot 1} - \frac{\bar{H}_{n-1}\, \xi_n^n}{n}\right].
\end{aligned}
$$

where $\bar{H}_n = 1 + H_n$ *and* $H_n = 1 + 1/2 + \cdots + 1/n$ *is the n-th harmonic number.*

Using Theorem 2, we can find the employer's probability of winning if he uses any set of non-decreasing thresholds. In particular, in the simulation study of Sect. 3, for $n = 5$, we estimated $P\{\mathrm{Win}\}$ based on 10^6 plays of the game for the optimal strategy as well as for two alternative strategies. We can now compute the exact value of $P\{\mathrm{Win}\}$ for each strategy: It is 0.6391947 for optimal strategy, 0.6370689 for alternative 1, and 0.5923543 for alternative 2. Thus, we exhibit that indeed the alternative strategies are sub-optimal.

5 Conclusion

We reiterate that there is no need to restrict the candidates' scores to have the uniform (0, 1) distribution. The results of this paper will continue to hold if the scores are drawn from any known continuous distribution function F with only one small change: simply replace the thresholds of the employer's optimal strategy by the corresponding percentiles of F. The employer's maximal probability of winning will remain unaltered.

Also, we can solve the problem of hiring the candidate with the minimum score (say, on some undesirable negative trait) among all n candidates) by simply defining $Y = 1 - X$ and hiring the candidate with the maximum Y-score. Likewise, we can solve the problem of hiring the candidate with a score closest to any specific percentile of F,

Whereas in the usual secretary problem (with only relative ranks available) the limiting (as $n \to \infty$) probability of winning is $1/e \approx 0.368$, in our score-based secretary problem, we conjecture that the limiting probability of winning is about 0.580. The details are left to the reader.

Acknowledgements The author thanks the Indian Statistical Institute Kolkata and Calcutta University Department of Statistics for hosting his sabbatical leave visit, while this research was conducted. The research is partially supported by the Purdue Research Foundation.

Appendix

Proof of Theorem 2. Suppose that the employer hires Candidate k, where $k = 1, 2, \ldots, n$. Recall that the employer hires Candidate 1 iff $X_1 > \xi_n$. Otherwise, he hires Candidate $k \geq 2$ provided he has not already hired anyone earlier and $X_k = s$ exceeds both the previous maximum $r = \max\{X_1, \ldots, X_{k-1}\}$ and the threshold value ξ_{n+1-k} for Candidate k. In order to express the probability density function of r, subject to the fact that no candidate among Candidates $1, 2, \ldots, k-1$ has been hired, we must account for which candidate possibly achieved the record score r among Candidates $1, 2, \ldots, k-1$. Therefore, we split the range $(0, \xi_n)$ of r as

$$(\xi_{n-1}, \xi_n), (\xi_{n-2}, \xi_{n-1}), \ldots, (\xi_{n+1-k}, \xi_{n+2-k}), (0, \xi_{n+1-k}) \ ;$$

and note that the number of possible candidates who might have achieved r is $1, 2, \ldots, k-1, k-1$, respectively, in these intervals. See Fig. 6.

Having hired Candidate k, the employer wins (or hires the best candidate) only if all future Candidates also score below Candidate k. Adding up the joint probability that the employer hires Candidate k and he wins, we obtain

$$P_n(\text{Win}) = \sum_{k=1}^{n} P_n(\text{Hire } k \text{ and Win})$$

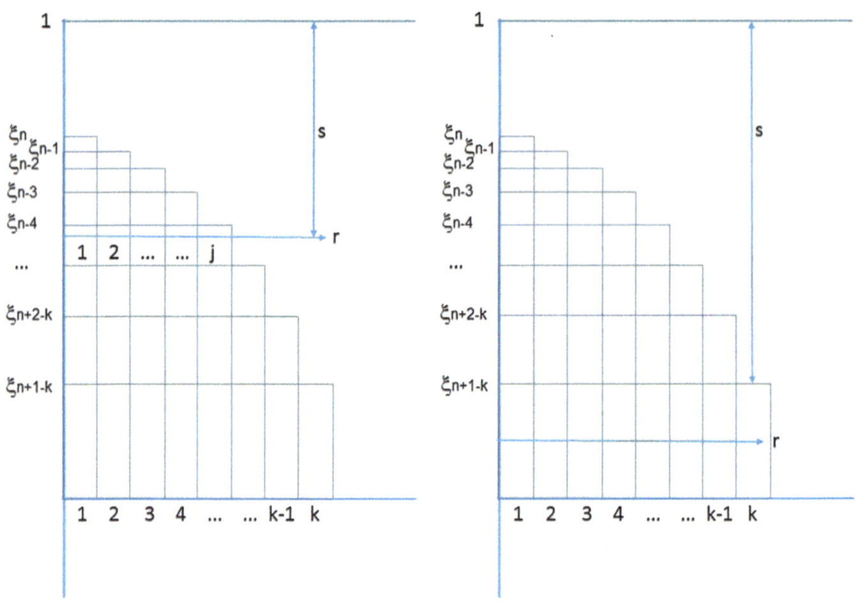

Fig. 6 If Candidate $k \geq 2$ with score $X_k = s$ is hired, then what is the likely score $r = \max\{X_1, \ldots, X_{k-1}\}$ of the previous leading candidate, and which candidate is she?

$$= \int_{\xi_n}^{1} s^{n-1}\, ds + \sum_{k=2}^{n-1} \left\{ \sum_{j=1}^{k-1} \int_{\xi_{n-j}}^{\xi_{n+1-j}} j\, r^{k-2} \int_{r}^{1} s^{n-k}\, ds\, dr \right.$$

$$\left. + \int_{0}^{\xi_{n+1-k}} (k-1)\, r^{k-2} \int_{\xi_{n+1-k}}^{1} s^{n-k}\, ds\, dr \right\}$$

$$+ \sum_{j=1}^{n-1} \int_{\xi_{n-j}}^{\xi_{n+1-j}} j\, r^{n-2} \int_{r}^{1} s^{0}\, ds\, dr$$

$$= \frac{1-\xi_n^n}{n} + \sum_{k=2}^{n-1} \left\{ \sum_{j=1}^{k-1} \int_{\xi_{n-j}}^{\xi_{n+1-j}} j\, \frac{r^{k-2} - r^{n-1}}{n+1-k}\, dr \right.$$

$$\left. + \int_{0}^{\xi_{n+1-k}} (k-1)\, r^{k-2} \frac{1-\xi_{n+1-k}^{n+1-k}}{n+1-k}\, dr \right\}$$

$$+ \sum_{j=1}^{n-1} \int_{\xi_{n-j}}^{\xi_{n+1-j}} j\, (r^{n-2} - r^{n-1})\, dr$$

$$= \frac{1-\xi_n^n}{n} + \sum_{k=2}^{n} \frac{1}{n+1-k} \{ \sum_{j=1}^{k-1} j \left(\frac{\xi_{n-j}^{k-1} - \xi_{n+1-j}^{k-1}}{k-1} - \frac{\xi_{n-j}^{n} - \xi_{n+1-j}^{n}}{n} \right)$$

$$+ \left(\xi_{n+1-k}^{k-1} - \xi_{n+1-k}^{n} \right) \}$$

$$= \frac{1}{n} [1 - \xi_n^n - \xi_{n-1}^n - \cdots - \xi_2^n - \xi_1^n]$$

$$+ \frac{1}{n-1} \left(\xi_n - \frac{\xi_n^n}{n} \right) + \frac{1}{n-2} \left[\left(\frac{\xi_n^2}{2} - \frac{\xi_n^n}{n} \right) + \left(\frac{\xi_{n-1}^2}{2} - \frac{\xi_{n-1}^n}{n} \right) \right]$$

$$+ \frac{1}{n-3} \left[\left(\frac{\xi_n^3}{3} - \frac{\xi_n^n}{n} \right) + \left(\frac{\xi_{n-1}^3}{3} - \frac{\xi_{n-1}^n}{n} \right) + \left(\frac{\xi_{n-2}^3}{3} - \frac{\xi_{n-2}^n}{n} \right) \right]$$

$$+ \cdots$$

$$+ \frac{1}{1} \left[\left(\frac{\xi_n^{n-1}}{n-1} - \frac{\xi_n^n}{n} \right) + \left(\frac{\xi_{n-1}^{n-1}}{n-1} - \frac{\xi_{n-1}^n}{n} \right) + \cdots + \left(\frac{\xi_2^{n-1}}{n-1} - \frac{\xi_2^n}{n} \right) \right]$$

$$= \frac{1}{n} [1 - \xi_n^n - \xi_{n-1}^n - \cdots - \xi_2^n]$$

$$+ \left[\frac{\xi_n}{1 \cdot (n-1)} + \frac{\xi_n^2 + \xi_{n-1}^2}{2 \cdot (n-2)} + \cdots + \frac{\xi_n^{n-1} + \xi_{n-1}^{n-1} + \cdots + \xi_2^{n-1}}{(n-1) \cdot 1)} \right]$$

$$- \frac{\xi_n^n}{n} H_{n-1} - \frac{\xi_{n-1}^n}{n} H_{n-2} - \cdots - \frac{\xi_2^n}{n} H_1 \,,$$

whence the theorem follows. Q.E.D.

References

1. Flood, M.R.: A letter written in 1958. In: The Martin Gardner Papers at Stanford University Archives, series 1, box 5, folder 19
2. Gardner, M.: New Mathematical Diversions from Scientific American: Chapter 3, Problem 3. New York, NY, Simon and Schuster (1966)
3. Bruss, F.T.: A unified approach to a class of best choice problems with an unknown number of options. Ann. Probab. **12**(3), 882–891 (1984)
4. Ferguson, T.S.: Who solved the secretary problem? Stat. Sci. **4**(3), 282–296 (1989)
5. Ross, S.M.: Stochastic Processes, 2nd edn. Wiley, New York, NY (1996)
6. Bittinger, M.L., Ellenbogenn, D.J., Surgent, S.J.: Calculus and Its Applications, 11th edn. Upper Saddle River, NJ, Pearson (2016)
7. Hogg, R.V., McKean, J., Craig, A.T.: Introduction to Mathematical Statistics, 7th edn. Upper Saddle River, NJ, Pearson (2013)

The General Solutions to Some Systems of Adjointable Operator Equations

Nan-Bin Cao and Yu-Ping Zhang

Abstract We consider two systems of adjointable operator equations $A_1X = C_1$, $XB_2 = C_2, A_3XB_3^* - B_3X^*A_3^* = C_3$ and $A_1X = C_1, A_2X = C_2, A_3XB_3^* - B_3X^*A_3^* = C_3$ over the Hilbert C^*-modules. Necessary and sufficient conditions for the existence and the expressions of the general solutions to those systems are established.

Keywords Hilbert C^*-modules · System of operator equations · General solution · Inner inverse of a operator · Moore–Penrose inverse of a operator

2000 AMS subject classifications: 15A09 · 15A24 · 46L08 · 47A48 · 47A62

1 Introduction

It is a very active research topic to investigate the solutions to operator equations, etc., (e.g., [1–4]). For instance, in 2005, Wang [1] studied the systems of matrix equation

$$A_1X = C_1, XB_2 = C_2, A_3XB_3^* = C_3, A_4XB_4 = C_4,$$

$$A_1X = C_1, A_2X = C_2, A_3XB_3^* = C_3, A_4XB_4 = C_4,$$

Xu [3] studied the operator equation

$$AXB^* - BX^*A^* = C$$

in general setting of the adjointable operators between Hilbert C^*-modules. We in this paper consider two systems of adjointable operator equations

N.-B. Cao · Y.-P. Zhang (✉)
School of Mathematics and Science, Hebei GEO University, Shijiazhuang 050031, People's Republic of China
e-mail: yuping.zh@163.com

N.-B. Cao
e-mail: nanbincao@sina.com

S. Kar et al. (eds.), *Operations Research and Optimization*, Springer Proceedings in Mathematics & Statistics 225, https://doi.org/10.1007/978-981-10-7814-9_8

$$A_1X = C_1, XB_2 = C_2, A_3XB_3^* - B_3X^*A_3^* = C_3 \tag{1.1}$$

and

$$A_1X = C_1, A_2X = C_2, A_3XB_3^* - B_3X^*A_3^* = C_3 \tag{1.2}$$

over the Hilbert C^*-modules.

In Sect. 2, we will give some knowledge about the Hilbert C^*-modules. In Sect. 3, we will study the general solution to the system of adjointable operator equations (1.1). In Sect. 4, we will study the general solution to the system of adjointable operator equations (1.2).

2 Preliminaries

Throughout this paper, $\mathscr{U}$ is a C^*-algebra. By a projection, we mean an idempotent and self-adjoint element in a certain C^*-algebra. Let H and K be two Hilbert C^*-modules, denote by $\mathscr{L}(H,K)$ the set of the adjointable operators from H to K. In case $H = K$, $\mathscr{L}(H,H)$ which we abbreviate to $\mathscr{L}(H)$, is a C^*-algebra, whose unit is denoted by I_H. For any $A \in \mathscr{L}(H,K)$, the range and the null space of A are denoted by $R(A)$ and $N(A)$, respectively.

Throughout the rest of this section, H and K be two Hilbert C^*-modules, and A is an element of $\mathscr{L}(H,K)$. By Lance [4], [Theorem 3.2, Remark 1.1] we have the following lemma:

Lemma 2.1 *The closeness of any one of the following sets implies the closeness of the remaining three sets:*

$$R(A), R(A^*), R(AA^*), R(A^*A).$$

If $R(A)$ is closed, then $R(A) = R(AA^*)$, $R(A^*) = R(A^*A)$, and the following orthogonal decompositions hold:

$$H = N(A) \oplus R(A^*), K = R(A) \oplus N(A^*)$$

Definition 2.1 Any element A^- of $A\{1\} = \{X \in \mathscr{L}(K,H) \mid AXA = A\}$ will be called a $\{1\}$-inverse of A. Note that A has a $\{1\}$-inverse if and only if $R(A)$ is closed.

Definition 2.2 The Moore–Penrose inverse of A (if it exists) is defined as the operator $A^\dagger \in \mathscr{L}(K,H)$ satisfying the Penrose equations,

$$AA^\dagger A = A, A^\dagger AA^\dagger = A^\dagger, (A^\dagger A)^* = A^\dagger A, (AA^\dagger)^* = AA^\dagger.$$

As in the Hilbert space case, $A^{\dagger}$ exists if and only if $R(A)$ is closed. In this case, $(A^{\dagger})^{*} = ((A)^{*})^{\dagger}, R(A^{\dagger}) = R(A^{*})$ and $A^{\dagger} |_{R(A)^{\perp}} = 0$, $A^{\dagger} |_{R(A)^{\perp}}$ is the restriction of $A^{\dagger}$ to the orthogonal complement of $R(A)$.

In this paper, the notations $A^{-}, A^{\dagger}, F_A$ and E_A are reserved to denote any $\{1\}$-inverse of A, the Moore–Penrose inverse of A, the projections $I_H - A^{\dagger}A$ and $I_K - AA^{\dagger}$, respectively. If in addition $H = K$, then we define

$$H^{(+,*)} = A + A^{*}, H^{(-,*)} = A - A^{*}.$$

Throughout the rest of this paper, $H_i (i = 1, 2, \dots, 5)$ are Hilbert C^{*}-modules.

3 The General Solution to the System of Adjointable Operator Equations (1.1)

We begin with the following lemma which proof just like one over the complex field.

Lemma 3.1 (See Lemma 2.2 in [2]). *Let $A_1 \in \mathscr{L}(H_1, H_2), B_2 \in \mathscr{L}(H_4, H_3)$ have closed range, and let $C_1 \in \mathscr{L}(H_3, H_2), C_2 \in \mathscr{L}(H_4, H_1)$. Then, the system of adjointable operator equations*

$$A_1 X = C_1, XB_2 = C_2 \tag{3.1}$$

is consistent if and only if

$$E_{A_1} C_1 = 0, C_2 F_{B_2} = 0, A_1 C_2 = C_1 B_2. \tag{3.2}$$

In that case, the general solution of (3.1) is

$$X = A_1^{\dagger} C_1 + F_{A_1} C_2 B_2^{\dagger} + F_{A_1} Y E_{B_2}, \tag{3.3}$$

where $Y \in \mathscr{L}(H_3, H_1)$ is arbitrary.

From Xu [3], we have the following lemma.

Lemma 3.2 *Let $A \in \mathscr{L}(H_3, H_2), B \in \mathscr{L}(H_1, H_2)$ have closed ranges such that $R(B) \subseteq R(A)$. Let $C \in \mathscr{L}(H_2)$ and $D = E_B A$ such that $R(D)$ is also closed. Then, the equation*

$$AXB^{*} - BXA^{*} = C, X \in \mathscr{L}(H_1, H_3) \tag{3.4}$$

has a solution if and only if

$$C^{*} = -C \quad \text{and} \quad H^{(-,*)}((AA^{\dagger} + DD^{\dagger})CBB^{\dagger}) = 2C. \tag{3.5}$$

In this case, the general solution X to Eq. (3.4) is of the form

$$X = X_0 + V - \frac{1}{2}A^\dagger AVB^\dagger B + \frac{1}{2}A^\dagger BV^*A^*(B^\dagger)^* - \frac{1}{2}A^\dagger BV^*(B^\dagger AD^\dagger A)^* - \frac{1}{2}D^\dagger AVB^\dagger B, \tag{3.6}$$

where $V \in \mathscr{L}(H_1, H_3)$ *is arbitrary, and* X_0 *is a particular solution to Eq. (3.4) defined by*

$$X_0 = \frac{1}{2}A^\dagger C(B^\dagger)^* - \frac{1}{2}A^\dagger BB^\dagger C(B^\dagger AD^\dagger)^* + \frac{1}{2}D^\dagger C(B^\dagger)^*. \tag{3.7}$$

In the next theorem, for simply express, we define

$$A_1^\dagger C_1 + F_{A_1}C_2B_2^\dagger = T, A_3F_{A_1} = M, E_{B_2}B_3^* = N^*, E_{B_3}A_3 = D_1,$$
$$E_NM = D_2, C_3 + B_3T^*A_3^* - A_3TB_3^* = S.$$

Theorem 3.3 *Let* $A_1 \in \mathscr{L}(H_1, H_2), A_3 \in \mathscr{L}(H_1, H_5), B_2 \in \mathscr{L}(H_4, H_3), B_3 \in \mathscr{L}(H_3, H_5), C_1 \in \mathscr{L}(H_3, H_2), C_2 \in \mathscr{L}(H_4, H_1), C_3 \in \mathscr{L}(H_5)$, *and let* $A_1, B_2, A_3, B_3, M, N, D_1, D_2$ *have closed range such that* $R(B_3) \subseteq R(A_3), R(N) \subseteq R(M)$. *Then, the system of adjointable operator equations (1.1) is consistent if and only if*

$$E_{A_1}C_1 = 0, C_2F_{B_2} = 0, A_1C_2 = C_1B_2, \tag{3.8}$$

and

$$C_3^* = -C_3 \quad \text{and} \quad H^{(-,*)}((A_3A_3^\dagger + D_1D_1^\dagger)C_3B_3B_3^\dagger) = 2C_3, \tag{3.9}$$

and

$$H^{(-,*)}((MM^\dagger + D_2D_2^\dagger)SNN^\dagger) = 2S. \tag{3.10}$$

In that case, the general solution of (1.1) is

$$\begin{aligned} X = T + F_{A_1}(X_0 + V - \tfrac{1}{2}M^\dagger MVN^\dagger N + \tfrac{1}{2}M^\dagger NV^*M^*(N^\dagger)^* \\ - \tfrac{1}{2}M^\dagger NV^*(N^\dagger MD_2^\dagger M)^* - \tfrac{1}{2}D_2^\dagger MVN^\dagger N)E_{B_2}, \end{aligned} \tag{3.11}$$

where $V \in \mathscr{L}(H_3, H_1)$ *is arbitrary, and* X_0 *is defined by*

$$X_0 = \frac{1}{2}M^\dagger S(N^\dagger)^* - \frac{1}{2}M^\dagger NN^\dagger S(N^\dagger MD_2^\dagger)^* + \frac{1}{2}D_2^\dagger S(N^\dagger)^*. \tag{3.12}$$

Proof For the necessary part. Suppose that the system of adjoint operator equations (1.1) has a common solution X, then X is a common solution to the system of adjoint operator equations (3.1), and X is a solution of operator equation

$$A_3XB_3^* - B_3X^*A_3^* = C_3, \tag{3.13}$$

from Lemmas 3.1 and 3.2, we can get (3.8) and (3.9). Because X is a common solution to the system of adjoint operator equations (3.1), then X can be expressed as

$$X = A_1^{\dagger}C_1 + F_{A_1}C_2B_2^{\dagger} + F_{A_1}YE_{B_2}, \tag{3.14}$$

where $Y \in \mathscr{L}(H_3, H_1)$ is arbitrary. Taking (3.14) into (3.13), we can get

$$MYN^* - NY^*M^* = S, \tag{3.15}$$

then (3.15) must has a solution, from Lemma 3.2, we can get (3.10) and

$$S^* = -S, \tag{3.16}$$

that is

$$(C_3 + B_3T^*A_3^* - A_3TB_3^*)^* = -(C_3 + B_3T^*A_3^* - A_3TB_3^*), \tag{3.17}$$

it means from $C_3^* = -C_3$ we can derive $S^* = -S$, so (3.16) can be omit.

We now proceed to prove the sufficient part of the theorem. From (3.8) and (3.9), we know the system of operator equations (3.1) has a common solution and the operator equation (3.13) has a solution, the general common solution of (3.1) is given by (3.14), taking (3.14) into (3.13) we can get (3.15), from (3.10) we know (3.15) is consistent, which means the system of adjoint operator equations (1.1) has a common solution.

In the rest of this theorem, we will give the proof that for any solution to the system of adjointable operator equations (1.1) can be expressed as (3.11). Suppose that X is a common solution to the system of adjoint operator equations (1.1), then

$$\begin{aligned} F_{A_1}X &= X - A_1^{\dagger}A_1X \\ &= X - A_1^{\dagger}C_1. \end{aligned}$$

So,

$$\begin{aligned} F_{A_1}XE_{B_2} &= X - XB_2B_2^{\dagger} - A_1^{\dagger}C_1 + A_1^{\dagger}C_1B_2B_2^{\dagger} \\ &= X - C_2B_2^{\dagger} - A_1^{\dagger}C_1 + A_1^{\dagger}A_1C_2B_2^{\dagger}. \end{aligned}$$

Thereby,

$$X = A_1^{\dagger}C_1 + C_2B_2^{\dagger} - A_1^{\dagger}A_1C_2B_2^{\dagger} + F_{A_1}XE_{B_2},$$

then,

$$A_3XB_3^* - B_3X^*A_3^* = MXN^* - NX^*M^* + A_3TB_3^* - B_3T^*A_3^* = C_3,$$

it means

$$MXN^* - NX^*M^* = S. \tag{3.18}$$

For any $V \in \mathscr{L}(H_3, H_1)$, let

$$\begin{aligned}\phi(V) = V - \tfrac{1}{2}M^\dagger MVN^\dagger N + \tfrac{1}{2}M^\dagger NV^*M^*(N^\dagger)^* \\ - \tfrac{1}{2}M^\dagger NV^*(N^\dagger MD_2^\dagger M)^* - \tfrac{1}{2}D_2^\dagger MVN^\dagger N.\end{aligned}$$

We have

$$\begin{aligned}M\phi(V)N^* &= \tfrac{1}{2}(MVN^* + NV^*M^*NN^\dagger - NV^*(NN^\dagger MD_2^\dagger M)^* - MD_2^\dagger MVN^*) \\ &= \tfrac{1}{2}MVN^* + \tfrac{1}{2}NV^*M^*NN^\dagger - \tfrac{1}{2}NV^*((I - E_N)MD_2^\dagger D_2)^* - \tfrac{1}{2}MD_2^\dagger MVN^* \\ &= \tfrac{1}{2}MVN^* + \tfrac{1}{2}NV^*M^*NN^\dagger - \tfrac{1}{2}NV^*D_2^\dagger D_2A^* + \tfrac{1}{2}NV^*D_2^* - \tfrac{1}{2}MD_2^\dagger MVN^* \\ &= \tfrac{1}{2}MVN^* + \tfrac{1}{2}NV^*M^*NN^\dagger + \tfrac{1}{2}NV^*M^*E_N - H^{(+,*)}(\tfrac{1}{2}NV^*D_2^\dagger D_2M^*) \\ &= H^{(+,*)}(\tfrac{1}{2}MVN^*) - H^{(+,*)}(\tfrac{1}{2}NV^*D_2^\dagger D_2M^*).\end{aligned}$$

It follows that $\phi(V)$ is a solution to the following equation:

$$MXN^* - NX^*M^* = 0. \tag{3.19}$$

On the other hand, given any solution X to Eq. (3.19), let $V = X$. Then we have

$$\begin{aligned}\phi(X) &= X - \tfrac{1}{2}M^\dagger MXN^\dagger N + \tfrac{1}{2}M^\dagger(NX^*M^*)(N^\dagger)^* \\ &\quad - \tfrac{1}{2}M^\dagger(NX^*M^*)(N^\dagger MD_2^\dagger)^* - \tfrac{1}{2}D_2^\dagger MXN^\dagger N \\ &= X - \tfrac{1}{2}M^\dagger MXN^\dagger N + \tfrac{1}{2}M^\dagger MXN^*(N^\dagger)^* \\ &\quad - \tfrac{1}{2}M^\dagger MXN^*(N^\dagger MD_2^\dagger)^* - \tfrac{1}{2}D_2^\dagger MXN^*(N^\dagger)^* \\ &= X - \tfrac{1}{2}M^\dagger MX(N^\dagger MD_2^\dagger N)^* - \tfrac{1}{2}D_2^\dagger NX^*M^*(N^\dagger)^* \\ &= X.\end{aligned}$$

We have proved that the general solution to Eq. (3.19) has a form $\phi(V)$ for some $V \in \mathscr{L}(H_3, H_1)$. It is easy to check that X_0 defined by

$$X_0 = \frac{1}{2}M^\dagger S(N^\dagger)^* - \frac{1}{2}M^\dagger NN^\dagger S(N^\dagger MD_2^\dagger)^* + \frac{1}{2}D_2^\dagger S(N^\dagger)^*$$

is a particular solution to Eq. (3.18), we conclude that the general solution X to the system of adjointable operator equations (1.1) has a form of (3.11). □

4 The General Solution to the System of Adjointable Operator Equations (1.2)

We begin with the following lemma which proof just like one over the complex field.

Lemma 4.1 (See Lemma 2.2 in [1]). *Let $A_1 \in \mathscr{L}(H_1, H_2), A_2 \in \mathscr{L}(H_1, H_4)$ have closed range, and let $C_1 \in \mathscr{L}(H_3, H_2), C_2 \in \mathscr{L}(H_3, H_4), W = A_2F_{A_1}, G = E_WA_2$. Then, the system of adjointable operator equations*

$$A_1X = C_1, A_2X = C_2 \tag{4.1}$$

is consistent if and only if

$$E_{A_1}C_1 = 0, E_{A_2}C_2 = 0, G(A_2^{\dagger}C_2 - A_1^{\dagger}C_1) = 0. \tag{4.2}$$

In that case, the general solution of (4.1) is

$$X = A_1^{\dagger}C_1 + F_{A_1}W^{\dagger}A_2(A_2^{\dagger}C_2 - A_1^{\dagger}C_1) + F_{A_1}F_WY, \tag{4.3}$$

where $Y \in \mathscr{L}(H_3, H_1)$ is arbitrary.

In the next theorem, for simply express, we define

$$W = A_2F_{A_1}, G = E_WA_2, A_1^{\dagger}C_1 + F_{A_1}W^{\dagger}A_2(A_2^{\dagger}C_2 - A_1^{\dagger}C_1) = T,$$

$$A_3F_{A_1}F_W = M, E_{B_3}A_3 = D_1, E_{B_3}M = D_2, C_3 + B_3T^*A_3^* - A_3TB_3^* = S.$$

Theorem 4.2 *Let $A_1 \in \mathscr{L}(H_1, H_2), A_2 \in \mathscr{L}(H_1, H_4), A_3 \in \mathscr{L}(H_1, H_5), B_3 \in \mathscr{L}(H_3, H_5), C_1 \in \mathscr{L}(H_3, H_2), C_2 \in \mathscr{L}(H_3, H_4), C_3 \in \mathscr{L}(H_5)$, and let $A_1, A_2, A_3, B_3, M, D_1, D_2$ have closed range such that $R(B_3) \subseteq R(M)$. Then, the system of adjointable operator equations (1.2) is consistent if and only if*

$$E_{A_1}C_1 = 0, E_{A_2}C_2 = 0, G(A_2^{\dagger}C_2 - A_1^{\dagger}C_1) = 0, \tag{4.4}$$

and

$$C_3^* = -C_3 \quad \text{and} \quad H^{(-,*)}((A_3A_3^{\dagger} + D_1D_1^{\dagger})C_3B_3B_3^{\dagger}) = 2C_3, \tag{4.5}$$

and

$$H^{(-,*)}((MM^{\dagger} + D_2D_2^{\dagger})SB_3B_3^{\dagger}) = 2S. \tag{4.6}$$

In that case, the general solution of (1.2) is

$$\begin{aligned} X = T + F_{A_1}F_W(X_0 + V - \tfrac{1}{2}M^{\dagger}MVB_3^{\dagger}B_3 + \tfrac{1}{2}M^{\dagger}B_3V^*M^*(B_3^{\dagger})^* \\ - \tfrac{1}{2}M^{\dagger}B_3V^*(B_3^{\dagger}MD_2^{\dagger}M)^* - \tfrac{1}{2}D_2^{\dagger}MVB_3^{\dagger}B_3), \end{aligned} \tag{4.7}$$

where $V \in \mathscr{L}(H_3, H_1)$ *is arbitrary, and* X_0 *is defined by*

$$X_0 = \frac{1}{2}M^{\dagger}S(B_3^{\dagger})^* - \frac{1}{2}M^{\dagger}B_3B_3^{\dagger}S(B_3^{\dagger}MD_2^{\dagger})^* + \frac{1}{2}D_2^{\dagger}S(B_3^{\dagger})^*. \tag{4.8}$$

Proof From $R(B_3) \subseteq R(M)$, we can get $R(B_3) \subseteq R(A_3)$.

For the necessary part. Suppose that the system of adjoint operator equations (1.2) has a common solution X, then X is a common solution to the system of adjoint operator equations (4.1) and X is a solution of operator equation

$$A_3XB_3^* - B_3X^*A_3^* = C_3; \tag{4.9}$$

from Lemmas 4.1 and 3.2, we can get (4.4) and (4.5). Because X is a common solution to the system of adjoint operator equations (4.1), X can be expressed as

$$X = A_1^{\dagger}C_1 + F_{A_1}W^{\dagger}A_2(A_2^{\dagger}C_2 - A_1^{\dagger}C_1) + F_{A_1}F_WY, \tag{4.10}$$

where $Y \in \mathscr{L}(H_3, H_1)$ is arbitrary. Taking (4.10) into (4.9), we can get

$$MYB_3^* - B_3Y^*M^* = S, \tag{4.11}$$

then (4.11) must has a solution, from Lemma 3.2, we can get (4.6) and

$$S^* = -S, \tag{4.12}$$

that is

$$(C_3 + B_3T^*A_3^* - A_3TB_3^*)^* = -(C_3 + B_3T^*A_3^* - A_3TB_3^*), \tag{4.13}$$

it means from $C_3^* = -C_3$ we can derive $S^* = -S$, so (4.12) can be omit.

We now proceed to proved the sufficient part of the theorem. From (4.4) and (4.5), we know the system of operator equations (4.1) has a common solution and the operator equation (4.9) has a solution, the general common solution of (4.1) is given by (4.10), taking (4.10) into (4.9) we can get (4.11), from (4.6) we know (4.11) is consistent, which means the system of adjoint operator equations (1.2) has a common solution.

In the rest of this theorem, we will give the proof that for any solution to the system of adjointable operator equations (1.2) can be expressed as (4.7). Suppose that X is a common solution to the system of adjoint operator equations (1.2), then

$$\begin{aligned} F_WX &= X - W^{\dagger}WX \\ &= X - W^{\dagger}A_2(X - A_1^{\dagger}C_1) \\ &= X - W^{\dagger}(C_2 - A_2A_1^{\dagger}C_1) \\ &= X - W^{\dagger}A_2(A_2^{\dagger}C_2 - A_1^{\dagger}C_1). \end{aligned}$$

So,

$$F_{A_1}F_W Y = X - A_1^\dagger C_1 - F_{A_1}W^\dagger A_2(A_2^\dagger C_2 - A_1^\dagger C_1).$$

Thereby,

$$X = A_1^\dagger C_1 + F_{A_1}W^\dagger A_2(A_2^\dagger C_2 - A_1^\dagger C_1) + F_{A_1}F_W Y,$$

then,

$$A_3XB_3^* - B_3X^*A_3^* = MXB_3^* - B_3X^*M^* + A_3TB_3^* - B_3T^*A_3^* = C_3,$$

it means

$$MXB_3^* - B_3X^*M^* = S. \tag{4.14}$$

For any $V \in \mathscr{L}(H_3, H_1)$, let

$$\begin{aligned}\phi(V) = V &- \tfrac{1}{2}M^\dagger MVB_3^\dagger B_3 + \tfrac{1}{2}M^\dagger B_3V^*M^*(B_3^\dagger)^* \\ &- \tfrac{1}{2}M^\dagger B_3V^*(B_3^\dagger MD_2^\dagger M)^* - \tfrac{1}{2}D_2^\dagger MVB_3^\dagger B_3.\end{aligned}$$

We have

$$\begin{aligned}M\phi(V)B_3^* &= \tfrac{1}{2}(MVB_3^* + B_3V^*M^*B_3B_3^\dagger - B_3V^*(B_3B_3^\dagger MD_2^\dagger M)^* - MD_2^\dagger MVB_3^*) \\ &= \tfrac{1}{2}MVB_3^* + \tfrac{1}{2}B_3V^*M^*B_3B_3^\dagger - \tfrac{1}{2}B_3V^*((I - E_{B_3})MD_2^\dagger D_2)^* - \tfrac{1}{2}MD_2^\dagger MVB_3^* \\ &= \tfrac{1}{2}MVB_3^* + \tfrac{1}{2}B_3V^*M^*B_3B_3^\dagger - \tfrac{1}{2}B_3V^*D_2^\dagger D_2A^* + \tfrac{1}{2}B_3V^*D_2^* - \tfrac{1}{2}MD_2^\dagger MVB_3^* \\ &= \tfrac{1}{2}MVB_3^* + \tfrac{1}{2}B_3V^*M^*B_3B_3^\dagger + \tfrac{1}{2}B_3V^*M^*E_{B_3} - H^{(+,*)}(\tfrac{1}{2}B_3V^*D_2^\dagger D_2M^*) \\ &= H^{(+,*)}(\tfrac{1}{2}MVB_3^*) - H^{(+,*)}(\tfrac{1}{2}B_3V^*D_2^\dagger D_2M^*).\end{aligned}$$

It follows that $\phi(V)$ is a solution to the following equation:

$$MXB_3^* - B_3X^*M^* = 0. \tag{4.15}$$

On the other hand, given any solution X to Eq. (4.15), let $V = X$. Then we have

$$\begin{aligned}\phi(X) &= X - \tfrac{1}{2}M^\dagger MXB_3^\dagger B_3 + \tfrac{1}{2}M^\dagger(B_3X^*M^*)(B_3^\dagger)^* \\ &\quad - \tfrac{1}{2}M^\dagger(B_3X^*M^*)(B_3^\dagger MD_2^\dagger)^* - \tfrac{1}{2}D_2^\dagger MXB_3^\dagger B_3 \\ &= X - \tfrac{1}{2}M^\dagger MXB_3^\dagger B_3 + \tfrac{1}{2}M^\dagger MXB_3^*(B_3^\dagger)^* \\ &\quad - \tfrac{1}{2}M^\dagger MXB_3^*(B_3^\dagger MD_2^\dagger)^* - \tfrac{1}{2}D_2^\dagger MXB_3^*(B_3^\dagger)^* \\ &= X - \tfrac{1}{2}M^\dagger MX(B_3^\dagger MD_2^\dagger B_3)^* - \tfrac{1}{2}D_2^\dagger B_3X^*M^*(B_3^\dagger)^* \\ &= X.\end{aligned}$$

We have proved that the general solution to Eq. (4.15) has a form $\phi(V)$ for some $V \in \mathscr{L}(H_3, H_1)$. It is easy to check that X_0 defined by

$$X_0 = \frac{1}{2}M^{\dagger}S(B_3^{\dagger})^* - \frac{1}{2}M^{\dagger}B_3B_3^{\dagger}S(B_3^{\dagger}MD_2^{\dagger})^* + \frac{1}{2}D_2^{\dagger}S(B_3^{\dagger})^*$$

is a particular solution to Eq. (4.14), we conclude that the general solution X to the system of adjointable operator equations (1.2) has a form of (4.7). □

Acknowledgements This research was supported by the grants from the youth funds of Natural Science Foundation of Hebei Province (A2012403013), the Education Department Foundation of Hebei Province (QN2015218), and the Natural Science Foundation of Hebei Province (A2015403050).

References

1. Wang, Q.W.: The general solution to a system of real quaternion matrix equations. Comput. Math. Appl. **49**, 665–675 (2005)
2. Wang, Q.W., Chang, H.X., Ning, Q.: The common solution to six quaternion matrix equations with application. Appl. Math. Comput. **198**, 209–226 (2008)
3. Xu, Q., Sheng, L., Gu, Y.: The solutions to some operator equations. Linear Algebra Appl. **429**, 1997–2024 (2008)
4. Lance, E.C.: Hilbert -modules-A Toolkit for Operator Algebraists. Cambridge University Press (1995)

Global Stability of a Delayed Eco-Epidemiological Model with Holling Type III Functional Response

Hongfang Bai and Rui Xu

Abstract In this paper, we consider an eco-epidemiological model with Holling type III functional response and a time delay representing the gestation period of the predator. In the model, it is assumed that the predator population suffers a transmissible disease. By means of Lyapunov functionals and Laselle's invariance principle, sufficient conditions are obtained for the global stability of the endemic coexistence of the system.

Keywords Eco-epidemiological model · Delay · Laselle's invariance principle
Global stability

1 Introduction

Epidemiological models have received considerate attention in the literature to explain the spread and control of infectious disease [1–4]. Most of these models descend from the pioneering work of Kermack and Mckendrick [5], who proposed the classical SIR model . Seeing that species do not exist alone in the nature world, so it is very important to study the system of two or more interacting species subjected to disease [6].

Recently, great attention has been paid to study the relationships between demographic processes among different populations and diseases (see, e.g., [7–11]). Such as, Zhang et al. [7] studied the following eco-epidemiological model with Holling type I response function

This work was supported by the National Natural Science Foundation of China (No. 11371368) and the Natural Science Foundation of Hebei Province (No. A2014506015).

H. Bai · R. Xu (✉)
Department of Mathematics, Shijiazhuang Mechanical Engineering College,
Shijiazhuang 050003, People's Republic of China
e-mail: rxu88@yahoo.com.cn

S. Kar et al. (eds.), *Operations Research and Optimization*, Springer Proceedings in Mathematics & Statistics 225, https://doi.org/10.1007/978-981-10-7814-9_9

$$\begin{aligned}\dot{x}(t) &= rx(t) - a_{11}x^2(t) - a_{12}x(t)S(t),\\ \dot{S}(t) &= a_{21}x(t-\tau)S(t-\tau) - r_1S(t) - \beta S(t)I(t),\\ \dot{I}(t) &= \beta S(t)I(t) - r_2I(t),\end{aligned} \tag{1.1}$$

where $x(t), S(t), I(t)$ denote the densities of the prey, the susceptible predator, and the infected predator population, respectively.

In system (1.1), it assumes that the per capita rate of predation depends on the prey numbers only. But Holling found that each predator increased its consumption rate when exposed to a higher prey density, and also predator density increased with increasing prey density [12, 13]. So he suggested the following three kinds of functional responses referring to the number of prey eaten per predator per unit time.

$$(1)\ p_1(x) = ax, \quad (2)\ p_2(x) = \frac{ax}{m+x}, \quad (3)\ p_3(x) = \frac{ax^2}{m+x^2},$$

where x denotes the density of prey, $a > 0$ is the search rate of the predator, $m > 0$ is half-saturation constant, $p_1(x), p_2(x)$, and $p_3(x)$ represent Holling type I, II, and III functional responses, respectively.

Holling type III functional response reveals that the risk of being preyed upon is small at low prey density but increases up to a certain point as prey density increases, which is in accordance with some phenomena of natural world. Also, we know that many factors contribute to a type III functional response such as prey refuge, predator learning, and the presence of alternative prey [14].

Motivated by the works of Holling [14] and Zhang et al. [7], in this paper, we consider a delayed eco-epidemiological model with Holling type III functional response, which suffers a transmissible disease. Thus, we study the following eco-epidemiological model:

$$\begin{aligned}\dot{x}(t) &= rx(t) - a_{11}x^2(t) - \frac{a_{12}x^2(t)S(t)}{1+mx^2(t)} - \frac{a_{13}x^2(t)I(t)}{1+mx^2(t)},\\ \dot{S}(t) &= k\frac{a_{12}x^2(t-\tau)S(t-\tau)}{1+mx^2(t-\tau)} - r_1S(t) - \beta S(t)I(t),\\ \dot{I}(t) &= \beta S(t)I(t) + k\frac{a_{13}x^2(t-\tau)I(t-\tau)}{1+mx^2(t-\tau)} - r_2I(t).\end{aligned} \tag{1.2}$$

where $x(t)$, $S(t)$, and $I(t)$ represent the densities of the prey, the susceptible predator, and the infected predator population, respectively. r is the intrinsic growth rate of prey population without disease, r/a_{11} is the environmental carrying capacity, a_{12} is the capturing rate of the susceptible predators. The infected predator also can catch the prey; here, a_{13} denotes the capturing rate of the infected predator. k is the conversion rate of nutrients into the reproduction of predators by consuming prey, β is the disease transmission coefficient, r_1 is the natural death rate of the susceptible predators, r_2 is the natural and disease-related mortality rate of the infected predator. Here, $r_1 < r_2$. τ is a time delay representing a duration of τ time units elapses when an individual prey is killed and the moment when the corresponding addition is made to the predator population. All the parameters are positive.

The initial conditions for system (1.2) are

$$\begin{aligned} &x(\theta) = \phi_1(\theta), S(\theta) = \phi_2(\theta), I(\theta) = \phi_3(\theta), \theta \in [-\tau, 0], \\ &\phi_i \in C([-\tau, 0], R^3_+), \quad \phi_i > 0, \quad i = 1, 2, 3, \end{aligned} \tag{1.3}$$

where $R^3_+ = (x_1, x_2, x_3) : x_1 \geq 0, x_2 \geq 0, x_3 \geq 0$.

The organization of this paper is as follows. In Sect. 2, the positivity and the equilibria of system (1.2) are presented. In Sect. 3, we consider about the permanence of system (1.2) by using the persistence theory on infinite dimensional systems developed by Hale and Waltman [15]. In Sect. 4, we establish sufficient conditions for the global asymptotic stability of the endemic-coexistence equilibrium of system (1.2) by constructing suitable Lyapunov functionals and adopting Lasalle's invariance principle. Finally, we discuss the biological meaning of the result obtained in this paper.

2 Preliminaries

In this section, we consider the positivity of solutions and the equilibria of system (1.2).

2.1 *Positivity of Solutions*

Theorem 2.1 *Suppose that $(x(t), S(t), I(t))$ is a solution of system (1.2) with initial conditions (1.3). Then, $x(t) \geq 0$, $S(t) \geq 0$, and $I(t) \geq 0$ for all $t \geq 0$.*

Proof From the first equation of system (1.2), we have

$$x(t) = x(0)\exp\left\{\int_0^t \left[r - a_{11}x(u) - a_{12}x(u)S(u)/(1+mx^2(u)) - a_{13}x(u)I(u)/(1+mx^2(u))\right] \mathrm{d}u\right\} > 0.$$

Hence, $x(t)$ is positive.

In order to prove that $S(t)$ is positive on $[0, \infty]$, suppose that there exists $t_1 > 0$ such that $S(t_1) = 0$, and $S(t) > 0$ for $t \in [0, t_1]$. Then, $\dot{S}(t_1) \leq 0$. From the second equation of (1.2), we have

$$\dot{S}(t_1) = k\frac{a_{12}x^2(t_1-\tau)S(t_1-\tau)}{1+mx^2(t_1-\tau)} - r_1S(t_1) - \beta S(t_1)I(t_1)$$
$$= k\frac{a_{12}x^2(t_1-\tau)S(t_1-\tau)}{1+mx^2(t_1-\tau)} > 0,$$

which is a contradiction.

In order to show that $I(t)$ is positive on $[0,\infty]$, suppose that there exists $t_2 > 0$ such that $I(t_2) = 0$, and $I(t) > 0$ for $t \in [0, t_2]$. Then, $\dot{I}(t_2) \le 0$. From the third equation of (1.2), we have

$$\dot{I}(t_2) = \beta S(t_2)I(t_2) + k\frac{a_{13}x^2(t_2-\tau)I(t_2-\tau)}{1+mx^2(t_2-\tau)} - r_2I(t_2)$$
$$= k\frac{a_{13}x^2(t_2-\tau)I(t_2-\tau)}{1+mx^2(t_2-\tau)} > 0,$$

which is a contradiction. □

2.2 Equilibria

System (1.2) possesses the following equilibria in general.

(i) The trivial equilibrium $E_0 = (0, 0, 0)$.
(ii) The predator-extinction equilibrium $E_1 = (r/a_{11}, 0, 0)$.
(iii) The disease-free equilibrium $E_2 = (x_2, S_2, 0)$, where

$$x_2 = \sqrt{\frac{r_1}{ka_{12} - r_1m}},$$
$$S_2 = \frac{k}{\sqrt{r_1(ka_{12} - r_1m)}}\left(r - a_{11}\sqrt{\frac{r_1}{ka_{12} - r_1m}}\right). \tag{2.1}$$

We denote an ecological threshold parameter by $\mathfrak{R}_1 = \frac{k}{r_1}\frac{r^2a_{12}}{a_{11}^2 + mr^2}$. It is easy to show that if $\mathfrak{R}_1 > 1$, then $x_2 > 0$, $I_2 > 0$.

(iv) The planar equilibrium $E_3 = (x_3, 0, I_3)$, where

$$x_3 = \sqrt{\frac{r_2}{ka_{13} - r_2m}},$$
$$I_3 = \frac{k}{\sqrt{r_2(ka_{13} - r_2m)}}\left(r - a_{11}\sqrt{\frac{r_2}{ka_{13} - r_2m}}\right). \tag{2.2}$$

Similar, we denote $\mathfrak{R}_2 = \frac{k}{r_2} \frac{r^2 a_{13}}{a_{11}^2 + mr^2}$. It is easy to show that if $\mathfrak{R}_2 > 1$, then $x_3 > 0, I_3 > 0$.

(v) The endemic-coexistence equilibrium $E^* = (x^*, S^*, I^*)$, where

$$\begin{aligned} I^* &= \frac{ka_{12}x^{*2}}{\beta(1+mx^{*2})} - \frac{r_1}{\beta}, \\ S^* &= \frac{r_2}{\beta} - \frac{ka_{13}x^{*2}}{\beta(1+mx^{*2})}, \end{aligned} \tag{2.3}$$

in which x^* is a positive real root of the following cubic equation:

$$m\beta a_{11}x^3 - mr\beta x^2 + (a_{11}\beta + r_2 a_{12} - a_{13}r_1)x - r\beta = 0. \tag{2.4}$$

It can be seen that if

(H1) $r_2(ka_{12} - r_1 m) > r_1(ka_{13} - r_2 m)$,

then system (1.2) has a endemic-coexistence equilibrium E^*.

3 Permanence

In this section, we study the permanence of system (1.2). Before starting our theorem, we give some basic concepts and corresponding theory.

Definition 3.1 System (1.2) is said to be permanent (uniformly persistent) if there are positive m_i and $M_i (i = 1, 2, 3)$ such that each positive solution $(x(t), S(t), I(t))$ of system (1.2) satisfies

$$\begin{aligned} m_1 &< \liminf_{t\to+\infty} x(t) \le \limsup_{t\to+\infty} x(t) \le M_1, \\ m_2 &\le \liminf_{t\to+\infty} S(t) \le \limsup_{t\to+\infty} S(t) \le M_2, \\ m_3 &\le \liminf_{t\to+\infty} I(t) \le \limsup_{t\to+\infty} I(t) \le M_3. \end{aligned}$$

Definition 3.2 System (1.2) is said to be permanent if there exists a compact region $\Omega_0 \in \mathtt{int}\Omega$ such that every solution of Eqs. (1.2) with initial condition (1.3) will eventually enter and remain in region Ω_0.

It is easy to see that for a dissipative system, uniform persistence is equivalent to permanence. For the sake of convenience, we present the uniform persistence theory for infinite dimensional systems.

Let X be a complete metric space with metric d. Suppose that T is a continuous semiflow on X, that is, a continuous mapping $T : [0, +\infty] \times X \to X$ with the following properties

$$T_t \circ T_s = T_{t+s}, \quad t, s \geq 0, \quad T_0(x) = x, x \in X,$$

where T_t denotes the mapping from X to X given by $T_t(x) = T(t, x)$.

The distance d(x, Y) of a point $x \in X$ from a subset Y of X is defined by

$$\mathrm{d}(x, Y) = \inf_{y \in Y} \mathrm{d}(x, y).$$

Recall that the positive orbit $\gamma^+(x)$ through x is defined as $\gamma^+(x) = \cup_{t \geq 0}\{T(t)x\}$, and its $\omega-$ limit set is $\omega(x) = \cap_{s \geq 0} \overline{\cup_{t \geq s}\{T(t)x\}}$. Define $W^s(A)$ the strong stable set of a compact invariant set A as

$$W^s(A) = \{x : x \in X, \omega(x) \neq \emptyset, \omega(x) \subset A\}.$$

Suppose that X^0 is open and dense in X and $X^0 \cup X_0 = X$, $X^0 \cap X_0 = \emptyset$. Moreover, the C^0-semigroup $T(t)$ on X satisfies

$$T(t) : X^0 \to X^0, T(t) : X_0 \to X_0. \tag{3.1}$$

Let $T_b(t) = T(t) \mid_{X_0}$ and A_b be the global attractor for $T_b(t)$.

Lemma 3.1 (Hale and Waltman [15]) *Suppose that $T(t)$ satisfies (3.1). If the following hold*

(i) there is a $t_0 \geq 0$ such that $T(t)$ is compact for $t > t_0$;
(ii) $T(t)$ is point dissipative in X; and
(iii) $\bar{A}_b = \cup_{x \in A_b} \omega(x)$ is isolated and has an acyclic covering $\hat{M}_t$, where

$$\hat{M}_t = \{\tilde{M}_1, \tilde{M}_2, \ldots, \tilde{M}_n\};$$

(iv) $W^s(\tilde{M}_i) \cap X^0 = \emptyset$ for $i = 1, 2, \ldots, n$.

Then, X_0 is a uniform repeller with respect to X^0; that is, there is an $\varepsilon > 0$ such that for any $x \in X^0$, $\liminf_{t \to +\infty} \mathrm{d}(T(t)x, X_0) \geq \varepsilon$.

We also need the following result to study the permanence of system (1.2).

Lemma 3.2 *There are positive constants M_1 and M_2 such that for any positive solution $(x(t), S(t), I(t))$ of system (1.2) with initial conditions (1.3),*

$$\limsup_{t \to +\infty} x(t) < M_2, \quad \limsup_{t \to +\infty} S(t) < M_1, \quad \limsup_{t \to +\infty} I(t) < M_1. \tag{3.2}$$

Proof Let $(x(t), S(t), I(t))$ be any solution of system (1.2) with initial conditions (1.3). Consider the function

$$V(t) = kx(t) + S(t + \tau) + I(t + \tau).$$

From system (1.2), we get

$$\begin{aligned}\dot{V}(t) =& krx(t) - ka_{11}x^2(t) - r_1S(t+\tau) - r_2I(t+\tau)\\ =& k(r+r_1)x(t) - ka_{11}x^2(t) - r_1V(t) + (r_1 - r_2)I(t+\tau)\\ \le& M_1 - r_1V(t),\end{aligned}$$

where $M_1 = \dfrac{k(r+r_1)^2}{4a_{11}}$. Which yields $\limsup_{t\to+\infty} V(t) \le M_1$. If we choose $M_2 = M_1/k$, then (3.2) follows. This complete the proof. □

In the following, we investigate the permanence of system (1.2).

Theorem 3.1 *If $\beta S_2 > r_2$ holds, then system (1.2) is permanent.*

Proof Let $C^+([-\tau,0],\mathbb{R}^3_+)$ denote the space of continuous functions mapping $[-\tau,0]$ into $\mathbb{R}^3_+$. Define

$$C_1 = \left\{(\phi_1,\phi_2,\phi_3) \in C^+([-\tau,0],\mathbb{R}^3_+) \,:\, \phi_1(\theta) \ne 0, \phi_2(\theta) = \phi_3(\theta) = 0, \theta \in [\tau,0]\right\},$$
$$C_2 = \left\{(\phi_1,\phi_2,\phi_3) \in C^+([-\tau,0],\mathbb{R}^3_+) \,:\, \phi_1(\theta)\phi_2(\theta) \ne 0, \phi_3(\theta) = 0, \theta \in [\tau,0]\right\}.$$

Denote $C_0 = C_1 \cup C_2$, $X = C^+([-\tau,0],\mathbb{R}^3_+)$, and $C^0 = \text{int}C^+([-\tau,0],\mathbb{R}^3_+)$.

We verify below that the conditions in Lemma 3.1 are satisfied. By the definition of C^0 and C_0, it is easy to know that C^0 and C_0 are positively invariant. Moreover, the conditions (i) and (ii) in Lemma 3.1 are clearly satisfied. Thus, we need only to verify that the conditions (iii) and (iv) hold. System (1.2) has two constant solutions in C_0 : $\bar{E}_1 \in C_1, \bar{E}_2 \in C_2$ corresponding, respectively, to $x(t) = r/a_{11}, S(t) = 0, I(t) = 0$ and $x(t) = x_2$, $S(t) = S_2$, $I(t) = 0$.

Firstly, we verify the condition (iii) of Lemma 3.1. If $(x(t), S(t), I(t))$ is a solution of system (1.2) initiating from C_1, then $\dot{x}(t) = rx(t) - a_{11}x^2(t)$, which yields $x(t) \to r/a_{11}$ as $t \to +\infty$. If $(x(t), S(t), I(t))$ is a solution of system (1.2) initiating from C_2 with $\phi_1(\theta) > 0$ and $\phi_2(\theta) > 0$, then we have

$$\begin{aligned}\dot{x}(t) =& rx(t) - a_{11}x^2(t) - \frac{a_{12}x^2(t)S(t)}{1+mx^2(t)},\\ \dot{S}(t) =& ka_{12}\frac{x^2(t-\tau)S(t-\tau)}{1+mx^2(t-\tau)} - r_1S(t).\end{aligned} \tag{3.3}$$

It is obvious that if $\beta S_2/r_2 > 1$, then $\mathfrak{R}_1 > 1$. Using Lemmas 3.1 and 3.2, it is easy to prove that if $\mathfrak{R}_1 > 1$ holds, then system (3.3) is uniformly persistent. Noting that $C_1 \cap C_2 = \emptyset$, this shows that the invariant sets $\bar{E}_1$ and $\bar{E}_2$ are isolated. Hence, $\left\{\bar{E}_1, \bar{E}_2\right\}$ is isolated and is an acyclic covering.

Secondly, we show that $W^s(\tilde{E}_i)\bigcap C^0 = \emptyset (i = 1, 2)$. Here, we restrict out attention to show $W^s(\tilde{E}_2)\bigcap C^0 = \emptyset$ holds because the proof of $W^s(\tilde{E}_1)\bigcap C^0 = \emptyset$ is simple. Assuming the contrary, namely $W^s(\tilde{E}_2)\bigcap C^0 \ne \emptyset$. Then, there exists a positive solution $(x(t), S(t), I(t))$ satisfying $\lim_{t\to+\infty}(x(t), S(t), I(t)) = (x_2, S_2, 0)$.

Since $\beta S_2 > r_2$, we can choose $\varepsilon > 0$ small enough such that

$$\beta(S_2 - \varepsilon) > r_2. \tag{3.4}$$

Noting that $\lim_{t\to+\infty} S(t) = S_2$, for $\varepsilon > 0$ sufficiently small satisfying (3.3), there is a $t_0 > 0$ such that if $t > t_0$, $S_2 - \varepsilon < S(t) < S_2 + \varepsilon$. For $\varepsilon > 0$ sufficiently small satisfying (3.4), it follows from the third equation of system (1.2) that for $t > t_0 + \tau$, $\dot{I}(t) > \beta(S_2 - \varepsilon)I(t) - r_2 I(t)$, which, follows from (3.4), yields $\lim_{t\to+\infty} I(t) = +\infty$. This is contradicts Lemma 3.2. Thus, we have $W^s(\tilde{E}_2) \bigcap C^0 = \emptyset$. By Lemma 3.1, we conclude that C_0 repels positive solutions of system (1.2) uniformly, and therefore, system (1.2) is permanent. The proof is complete. □

4 Global Stability

Theorem 4.1 *If the endemic-coexistence equilibrium E^* of system (1.2) exists, then E^* is globally asymptotically stable provided that*
(H2): $\underline{x} \geq r/(2a_{11})$.
Here, $\underline{x}$ is the persistency constant for x satisfying $\liminf_{t\to+\infty} x \geq \underline{x}$.

Proof Assume that $(x(t), S(t), I(t))$ is any positive solution of system (1.2) with initial conditions (1.3). Denote $\phi(x(t)) = \dfrac{x^2(t)}{1 + mx^2(t)}$. Define

$$\begin{aligned} V_{11}(t) =& k\left(x(t) - x^* - \int_{x^*}^{x} \frac{\phi(x^*)}{\phi(x(u))} du\right) + S(t) - S^* - S^* \ln \frac{S(t)}{S^*} \\ &+ I(t) - I^* - I^* \ln \frac{I(t)}{I^*}. \end{aligned} \tag{4.1}$$

Calculating the derivative of $V_{11}(t)$ along positive solutions of system (1.2), it follows that

$$\begin{aligned} \frac{d}{dt}V_{11}(t) =& k\left(1 - \frac{\phi(x^*)}{\phi(x(t))}\right)\left[rx(t) - a_{11}x^2(t) - a_{12}\phi(x(t))S(t) - a_{13}\phi(x(t))I(t)\right] \\ &+ \left(1 - \frac{S^*}{S(t)}\right)\left(ka_{12}\phi(x(t-\tau))S(t-\tau) - r_1 S(t) - \beta S(t)I(t)\right) \\ &+ \left(1 - \frac{I^*}{I(t)}\right)\left(\beta S(t)I(t) + ka_{13}\phi(x(t-\tau))I(t-\tau) - r_2 I(t)\right). \end{aligned} \tag{4.2}$$

On substituting $rx^* - a_{11}x^{*2} - a_{12}\phi(x^*)S^* - a_{13}\phi(x^*)I^* = 0$, $ka_{12}\phi(x^*)S^* - r_1 S^* - \beta S^* I^* = 0$, and $\beta S^* I^* + ka_{13}\phi(x^*)I^* - r_2 I^* = 0$ into Eq. (4.2), we derive that

$$\begin{aligned}\frac{d}{dt}V_{11}(t) =& k\left(1-\frac{\phi(x^*)}{\phi(x(t))}\right)\left[rx(t)-rx^*-a_{11}(x^2(t)-x^{*2})+a_{12}\phi(x^*)S^*+a_{13}\phi(x^*)I^*\right]\\ &-ka_{12}\phi(x(t))S(t)+ka_{12}\phi(x(t-\tau))S(t-\tau)-ka_{13}\phi(x(t))I(t)+ka_{12}\phi(x(t-\tau))I(t-\tau)\\ &-ka_{12}S^*\phi(x^*)\frac{\phi(x(t-\tau))S(t-\tau)}{S(t)\phi(x^*)}+ka_{12}S^*\phi(x^*)\\ &-ka_{13}I^*\phi(x^*)\frac{\phi(x(t-\tau))I(t-\tau)}{I(t)\phi(x^*)}+ka_{13}I^*\phi(x^*)\\ &+ka_{12}\phi(x^*)S+ka_{13}\phi(x^*)I-r_1S(t)+\beta S^*I-\beta I^*S(t)-r_2I(t).\end{aligned} \tag{4.3}$$

Define

$$\begin{aligned}V_{12}(t) &= ka_{12}\int_{t-\tau}^{t}\left[\phi(x(u))S(u)-\phi(x^*)S^*-\phi(x^*)S^*\ln\frac{\phi(x(u))S(u)}{\phi(x^*)S^*}\right]du,\\ V_{13}(t) &= ka_{13}\int_{t-\tau}^{t}\left[\phi(x(u))I(u)-\phi(x^*)I^*-\phi(x^*)I^*\ln\frac{\phi(x(u))I(u)}{\phi(x^*)I^*}\right]du.\end{aligned} \tag{4.4}$$

Then,

$$\begin{aligned}\frac{d}{dt}V_{12}(t) &= ka_{12}\left(\phi(x(t))S(t)-\phi(x(t-\tau))S(t-\tau)+\phi(x^*)S^*\ln\frac{\phi(x(t-\tau))S(t-\tau)}{\phi(x(t))S(t)}\right),\\ \frac{d}{dt}V_{13}(t) &= ka_{13}\left(\phi(x(t))I(t)-\phi(x(t-\tau))I(t-\tau)+\phi(x^*)I^*\ln\frac{\phi(x(t-\tau))I(t-\tau)}{\phi(x(t))I(t)}\right).\end{aligned} \tag{4.5}$$

Set $V_1(t) = V_{11}(t) + V_{12}(t) + V_{13}(t)$. It follows from (4.1) (4.4), and (4.5) that

$$\begin{aligned}\frac{d}{dt}V_{1}(t) =& k\left(1-\frac{\phi(x^*)}{\phi(x(t))}\right)\left[rx(t)-rx^*-a_{11}(x^2(t)-x^{*2})+a_{12}\phi(x^*)S^*+a_{13}\phi(x^*)I^*\right]\\ &+ka_{12}\phi(x^*)S^*\ln\frac{\phi(x(t-\tau))S(t-\tau)}{\phi(x(t))S(t)}+ka_{13}\phi(x^*)I^*\ln\frac{\phi(x(t-\tau))I(t-\tau)}{\phi(x(t))I(t)}\\ &-ka_{12}S^*\phi(x^*)\frac{\phi(x(t-\tau))S(t-\tau)}{S(t)\phi(x^*)}+ka_{12}S^*\phi(x^*)\\ &-ka_{13}I^*\phi(x^*)\frac{\phi(x(t-\tau))I(t-\tau)}{I(t)\phi(x^*)}+ka_{13}I^*\phi(x^*)\\ &+ka_{12}\phi(x^*)S+ka_{13}\phi(x^*)I-r_1S(t)+\beta S^*I-\beta I^*S(t)-r_2I(t).\end{aligned} \tag{4.6}$$

Noting that

$$\begin{aligned}\ln\frac{\phi(x(t-\tau))S(t-\tau)}{\phi(x(t))S(t)} &= \ln\frac{\phi(x(t-\tau))S(t-\tau)}{S(t)\phi(x^*)}+\ln\frac{\phi(x^*)}{\phi(x(t))},\\ \ln\frac{\phi(x(t-\tau))I(t-\tau)}{\phi(x(t))I(t)} &= \ln\frac{\phi(x(t-\tau))I(t-\tau)}{I(t)\phi(x^*)}+\ln\frac{\phi(x^*)}{\phi(x(t))},\end{aligned} \tag{4.7}$$

we derive from (4.7) that

$$\begin{aligned}\frac{d}{dt}V_1(t) =&k\left(1-\frac{\phi(x^*)}{\phi(x(t))}\right)\left[rx(t)-rx^*-a_{11}(x^2(t)-x^{*2})\right]\\&-ka_{12}\phi(x^*)S^*\left[\frac{\phi(x^*)}{\phi(x(t))}-1-\ln\frac{\phi(x^*)}{\phi(x(t))}\right]-ka_{13}\phi(x^*)I^*\left[\frac{\phi(x^*)}{\phi(x(t))}-1-\ln\frac{\phi(x^*)}{\phi(x(t))}\right]\\&-ka_{12}S^*\phi(x^*)\left[\frac{\phi(x(t-\tau))S(t-\tau)}{S(t)\phi(x^*)}-1-\ln\frac{\phi(x(t-\tau))S(t-\tau)}{S(t)\phi(x^*)}\right]\\&-ka_{13}I^*\phi(x^*)\left[\frac{\phi(x(t-\tau))I(t-\tau)}{I(t)\phi(x^*)}-1-\ln\frac{\phi(x(t-\tau))I(t-\tau)}{I(t)\phi(x^*)}\right]\\&+ka_{12}\phi(x^*)S+ka_{13}\phi(x^*)I-r_1S(t)+\beta S^*I-\beta I^*S(t)-r_2I(t).\end{aligned}\tag{4.8}$$

On substituting $ka_{12}\phi(x^*) = r_1 + \beta I^*$ and $ka_{13}\phi(x^*) = r_2 - \beta S^*$ into Eq. (4.8), we derive that

$$\begin{aligned}\frac{d}{dt}V_1(t) =&k\left(1-\frac{\phi(x^*)}{\phi(x(t))}\right)\left[rx(t)-rx^*-a_{11}(x^2(t)-x^{*2})\right]\\&-ka_{12}\phi(x^*)S^*\left[\frac{\phi(x^*)}{\phi(x(t))}-1-\ln\frac{\phi(x^*)}{\phi(x(t))}\right]-ka_{13}\phi(x^*)I^*\left[\frac{\phi(x^*)}{\phi(x(t))}-1-\ln\frac{\phi(x^*)}{\phi(x(t))}\right]\\&-ka_{12}S^*\phi(x^*)\left[\frac{\phi(x(t-\tau))S(t-\tau)}{S(t)\phi(x^*)}-1-\ln\frac{\phi(x(t-\tau))S(t-\tau)}{S(t)\phi(x^*)}\right]\\&-ka_{13}I^*\phi(x^*)\left[\frac{\phi(x(t-\tau))I(t-\tau)}{I(t)\phi(x^*)}-1-\ln\frac{\phi(x(t-\tau))I(t-\tau)}{I(t)\phi(x^*)}\right].\end{aligned}\tag{4.9}$$

Noting that $\phi(x^*) = \dfrac{x^{*2}(t)}{1+mx^{*2}(t)}$ and $\phi(x) = \dfrac{x^2(t)}{1+mx^2(t)}$, we derive from (4.9) that

$$\begin{aligned}\frac{d}{dt}V_1(t) =&k\frac{(x+x^*)(x(t)-x^*)^2}{x^2(t)(1+mx^{*2})}\left[r-a_{11}(x(t)+x^*)\right]\\&-ka_{12}\phi(x^*)S^*\left[\frac{\phi(x^*)}{\phi(x(t))}-1-\ln\frac{\phi(x^*)}{\phi(x(t))}\right]-ka_{13}\phi(x^*)I^*\left[\frac{\phi(x^*)}{\phi(x(t))}-1-\ln\frac{\phi(x^*)}{\phi(x(t))}\right]\\&-ka_{12}S^*\phi(x^*)\left[\frac{\phi(x(t-\tau))S(t-\tau)}{S(t)\phi(x^*)}-1-\ln\frac{\phi(x(t-\tau))S(t-\tau)}{S(t)\phi(x^*)}\right]\\&-ka_{13}I^*\phi(x^*)\left[\frac{\phi(x(t-\tau))I(t-\tau)}{I(t)\phi(x^*)}-1-\ln\frac{\phi(x(t-\tau))I(t-\tau)}{I(t)\phi(x^*)}\right].\end{aligned}\tag{4.10}$$

Since (H2) holds, there exists a constant $T > 0$ such that if $t \geq T$, $x(t) > r/(2a_{11})$. In this case, we have that, for $t \geq T$,

$$\frac{(x+x^*)(x(t)-x^*)^2}{x^2(t)(1+mx^{*2})}\left[r-a_{11}(x(t)+x^*)\right] \leq 0, \tag{4.11}$$

with equality if and only if $x = x^*$. Seeing that the function $f(x) = x - 1 - \ln x$ is always nonnegative for any $x > 0$, and $f(x) = 0$ if and only if $x = 1$, therefor, if $t \geq T$, $\dot{V}_1(t) \leq 0$, which equality if and only if $x = x^*$, $S(t) = S(t-\tau)$, $I(t) = I(t-\tau)$. We now look for the invariant subset M within the set

$$M = \{(x, S, I) : x = x^*, S(t) = S(t - \tau), I(t) = I(t - \tau)\}. \tag{4.12}$$

Since $x = x^*$, $S(t) = S(t - \tau)$, $I(t) = I(t - \tau)$ on M, it follows from the system (1.2) that

$$\begin{aligned} 0 = \dot{x}(t) &= rx^* - a_{11}x^{*2} - \frac{a_{12}x^{*2}S(t)}{1 + mx^{*2}} - \frac{a_{13}x^{*2}I(t)}{1 + mx^{*2}}, \\ 0 = \dot{S}(t) &= \left[k\frac{a_{12}x^{*2}}{1 + mx^{*2}} - r_1 - \beta I(t)\right] S(t), \\ 0 = \dot{I}(t) &= \left[\beta S(t) + k\frac{a_{13}x^{*2}}{1 + mx^{*2}} - r_2\right] I(t), \end{aligned} \tag{4.13}$$

which yields $S = S^*$ and $I = I^*$. Hence, the only invariant set in M is $\mathbb{M} = (x^*, S^*, I^*)$. Therefore, the global asymptotic stability of E^* follows from Lasalle's invariance principle for delay differential systems [16]. This completes the proof. □

5 Discussion

In this paper, we have proposed and analyzed an eco-epidemiological system with time delay due to the gestation of the predator. We assumed that a transmissible disease spreading among the predator population, meanwhile, both the susceptible predator and the infected predator can catch the prey. Specially, system (1.2) has no intraspecific competition terms in the second and the third equations. In this case, under what conditions will the global stability of a feasible equilibrium of system (1.2) persists independent of the time delay? We established global asymptotic stability of the endemic-coexistence equilibrium of the system by means of Lyapunov functionals and Laselle's invariance principle. According to Theorem 4.1, we can see that the endemic-coexistence equilibrium of system (1.2) is globally asymptotically stable when the prey population is abundant enough.

References

1. Beretta, E., Hara, T., Ma, W., Takeuchi, Y.: Global asymptotic stability of an SIR epidemic model with distributed time delay. Nonlinear Anal. **47**, 4017–4115 (2001)
2. Gakkhar, S., Negi, K.: Pulse vaccination in SIRS epidemic model with non-monotonic incidence rate. Chaos Solitions Fractals **35**, 626–638 (2008)
3. Xu, R., Ma, Z.E., Wang, Z.P.: Global stability of a delayed SIRS epidemic model with saturation incidence and temporary immunity. Comput. Math. Appl. **59**, 3211–3221 (2010)
4. Xu, R.: Global dynamics of an SEIRI epidemioligical model with time delay. Appl. Math. Comput. **232**, 436–444 (2014)
5. Kermack, W.Q., Mckendrick, A.G.: Contributions to the mathematical theory of epidemics (Part I). Proc. R. Soc. A **115**, 700–721 (1927)

6. Anderson, R.M., May, R.M.: Regulation stability of host-parasite population interactions: I. Regulatory processes. J. Anim. Ecol. **47**, 219–267 (1978)
7. Zhang, J., Li, W., Yan, X.: Hopf bifurcation and stability of periodic solutions in a delayed eco-epidemiological system. Appl. Math. Comput. **198**, 865–876 (2008)
8. Debasis, M.: Hopf bifurcation in an eco-epidemic model. Appl. Math. Comput. **217**, 2118–2124 (2010)
9. Bairagi, N.: Direction and stability of bifurcating periodic solution in a delay-induced eco-epidemiological system. Int. J. Differ. Equ. 1–25 (2011)
10. Xu, R., Tian, X.H.: Global dynamics of a delayed eco-epidemiological model with Holling type-III functional response. Math. Method Appl. Sci. **37**, 2120–2134 (2014)
11. Sahoo, B.: Role of additional food in eco-epidemiological system with disease in the prey. Appl. Math. Comput. **259**, 61–79 (2015)
12. Holling, C.S.: The components of predation as revealed by a study of small mammal predation of the European pine sawfly. Can. Entomol. **91**, 293–320 (1959)
13. Holling, C.S.: Some characteristics of simple types of predation and parasitism. Can. Entomol. **91**, 385–398 (1959)
14. Holling, C.S.: The functional response of predators to prey density and its role in mimicry and population regulation. Mem. Entomol. Soc. Can. **45**, 3–60 (1965)
15. Hale, J., Waltman, P.: Persistence in infinite-dimensional systems. SIAM J. Math. Anal. **20**, 383–395 (1989)
16. Haddock, J.R., Terjéki, J.: Liapunov-Razumikhin functions and an invariance principle for functional-differential equations. J. Differ. equ. **48**, 95–122 (1983)

Part II
Engineering and Bio-Medical Applications

Enhancing Comfort of Occupants in Energy Buildings

Monalisa Pal, Amr Alzouhri Alyafi, Sanghamitra Bandyopadhyay, Stéphane Ploix and Patrick Reignier

Abstract As buildings contribute significantly towards global energy consumption, it is essential that the occupants receive the best comfort without utilizing further energy. This work treats building, environment and the occupants as a system, which presents the context, and the occupants also provide their comfort criteria to a black box for yielding the schedule of actions (opening/closing of doors/windows) for optimal comfort. The physical state of an office, situated in France, is recorded over a span of 100 days. This data is utilized by a physical model of the building to simulate the indoor ambience based on random sets of user actions from which an optimal schedule is obtained, representing equally best trade-off among minimal thermal and CO_2-based air quality dissatisfaction. Results indicate that adopting the proposed schedule of user actions can efficiently enhance the occupant's comfort.

Keywords Differential evolution · Energy management · Multi-objective optimization · Pareto optimality · Smart buildings

M. Pal (✉) · S. Bandyopadhyay
Machine Intelligence Unit, Indian Statistical Institute,
203 Barrackpore Trunk Road, Kolkata 700108, India
e-mail: monalisap90@gmail.com

S. Bandyopadhyay
e-mail: sanghami@gmail.com

A. A. Alyafi · S. Ploix
GSCOP Laboratory, Grenoble Institute of Technology,
46, Avenue Felix Viallet, 38031 Grenoble, France
e-mail: amr.alzouhri-alyafi@imag.fr

S. Ploix
e-mail: stephane.ploix@grenoble-inp.fr

A. A. Alyafi
LIG Laboratory, Grenoble Institute of Technology,
46, Avenue Felix Viallet, 38031 Grenoble, France

P. Reignier
University of Grenoble Alpes, CNRS, INRIA, LIG,
38000 Grenoble, France
e-mail: patrick.reignier@inria.fr

S. Kar et al. (eds.), *Operations Research and Optimization*, Springer Proceedings in Mathematics & Statistics 225, https://doi.org/10.1007/978-981-10-7814-9_10

1 Introduction

Considering the ever-growing energy demand and the depletion of non-renewable energy resources, it is important to limit the energy usage in buildings which constitute roughly about 40% of the global energy supply. Thus, it is imperative to satisfy the demands of the building occupants without increasing the present rate of energy consumption in buildings. Occupant, on the other hand, being an integral part of the building system, can affect the indoor environmental condition through their actions. By intelligently utilizing these actions, positive effects can be brought upon the indoor environment. Thus, guiding the occupant's actions, like opening and closing of doors and windows, for significant period of time, can help achieve better comfort in energy building at the same cost of energy consumption.

Attempts have been considered to meet the comfort demands of the occupants by improving building construction techniques and adding insulation to walls and ceilings. Building regulations also play an important role in the overall (both global and local) energy management. However, obtaining positive impacts of occupant's actions in the energy consumption can help to manage their own comfort. It is therefore important to assist the occupants with an optimal energy plan in order to explain that their expectations on comfort can be attained by themselves to some extent.

The approach used in this work considers an office, situated in Grenoble Institute of Technology, France, fitted with 27 sensors, for collection of data like temperature, solar illuminance, wind speed, humidity, moisture, CO_2 concentration, to construct the physical context. Also a physical model of the office is used which based on the physical context of outside environment and neighbouring corridors, and a random set of actions can simulate the indoor ambience. The indoor temperature and CO_2 concentration are responsible for thermal and air quality dissatisfaction of the occupants. This work uses a multi-objective optimization algorithm, viz. differential evolution, to obtain the schedule of user actions that can lead to minimal thermal and air quality dissatisfaction. On presenting the occupants with this optimal schedule, they can compare it with their previous schedule and adopt the new schedule, after trial, if they find significant improvement of comfort.

Depending on the weather changes, the optimal schedule is variable. However, rarely, there are day-to-day changes in the outside temperatures and CO_2 concentrations. Hence, learning from past day's environmental conditions and occupant's actions, the proposed optimal set of actions can be adopted for next day.

Rest of the paper is organized as follows: Section 2 describes the information flow to yield the schedule of actions corresponding to the trade-offs between minimal thermal and air quality dissatisfactions. Section 3 discusses the results to assess the efficacy of the proposed approach. Section 4 presents the conclusion while directing towards future research.

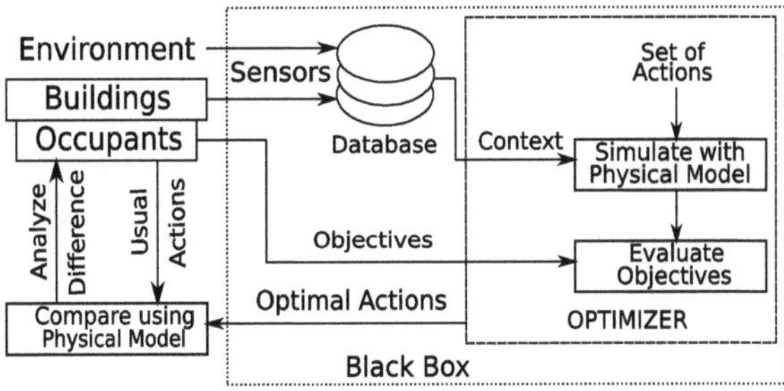

Fig. 1 General schema of energy management in smart buildings

2 Experimental Framework

This work presents optimal schedules of actions (opening/closing of doors/windows) which the occupants can adopt to achieve better thermal and CO_2-based air quality comfort at no extra energy expenditure. As the proposed scheme aims at improving occupant's comfort without increasing energy consumption, this can be considered as an energy management scheme. The general schema of the work is shown in Fig. 1. In this section, the function of each of these modules is explained in detail along with the overall interconnection of the modules to explain the information flow. It is to be mentioned that the environment, buildings and the occupants represent the smart building system where the environment and buildings contribute to the physical context/state, and the occupants, on the other hand, generate actions and provide their comfort criteria to a black box. Finally, the occupants compare their actions with the optimal actions generated by the black box in order to learn the scope of improvement in their comfort. As the internal operation of the optimizer is unknown to the occupant, it has been labelled as a black box.

2.1 Building and Environment

Through an array of sensors, the physical state of the outdoor, indoor and neighbouring zones are recorded. Here, physical state refers to temperature, CO_2 concentration, humidity, etc. The inertia of these physical quantities from the outside environment and neighbouring zones like corridors, staircase, etc., influences the physical quantities inside the room of the occupant. Hourly samples of such quantities are recorded in the database for future reference. Here, as a test bed, an office room at Grenoble Institute of Technology, France, where four researchers work, is fitted with 27 sensors for recording the physical state/context and usual schedule of actions. This acts as the smart building system for the proposed work.

2.2 Occupants

Occupants are the integral part of the entire system. On the one hand, they provide the optimization criteria like comfort, energy consumption, for obtaining an optimal schedule of actions, and on the other hand, their actions influence the system which in turn affects the optimization criteria. In the absence of controllable HVAC (heating, ventilation and air conditioning) system, the only actions of the occupants which can influence the indoor ambient conditions (and hence, their comfort) are opening and closing of doors and windows. The occupants can compare the optimal schedules with their usual schedule and adopt the proposed schedules as per need. For example, if the difference in comfort is negligible, the occupant might not prefer to change their schedule, whereas if the occupant observes significant improvement in comfort, the occupant is expected to adopt the proposed schedules of actions.

After the basic building block, viz. the smart building system, and its purpose, has been explained, the next step is to obtain the optimal schedules of actions. The optimizer module, as explained next, helps in searching for best trade-offs among the objectives provided by occupants.

2.3 Optimizer

The role of the optimizer is to yield a few optimal schedules of actions. The essential specifications for the optimization module are as follows: the representation of the solution vector in order to decode the result, the optimization algorithm, the objectives and their relation to the solution vector, the stopping condition and the algorithmic parameters.

A solution of the optimization problem represents a set of actions. The allowed actions for the occupants are opening and closing of doors and windows. As the data is recorded in an office environment, the actions are noted over 12 working hours i.e. from 8 a.m. to 8 p.m. Hence, the solution is represented by a 24-dimensional binary vector where the first 12 entries imply opening/closing (open = 1, close = 0) of windows and the later 12 entries are for opening/closing of doors for each of the 12 working hours, respectively.

Given the environmental context of the room, the primary objective of the work is to obtain various schedules, such that each of the schedules represent trade-offs among several conflicting objectives like minimizing thermal dissatisfaction, minimizing CO_2-based air quality dissatisfaction, minimizing humidity-based air quality dissatisfaction, minimizing energy consumption, etc. Due to the presence of multiple objectives, a multi-objective version of an optimization algorithm, viz. differential evolution [3, 4], has been employed.

Assuming a physical context (indoor and outdoor environmental variables like CO_2 concentration, temperature, etc.) and a set of actions (opening/closing of doors/windows) as inputs to the smart building system (outdoor environment and building with occupants), it outputs some effects (like thermal and air quality comfort). A physical model [1, 5] representing this system has been used which can simulate the effects corresponding to random sets of actions, in the same context as obtained from the database. These random sets of actions generate various sets of effects. Several best trade-offs among the effects (generating the Pareto-front) are chosen. Then, the occupants can compare the schedule of actions corresponding to these trade-offs with their usual schedules to analyse the difference. Hence, evaluation of objectives is a two-step process. The first step uses the physical model of the office to evaluate indoor environmental variables depending on a true context and a randomly assumed schedule of actions. The model is represented by Eqs. (1) and (2) where the variables are defined in Table 1. Some of these variables represent sensor measurements, whereas the remaining ones are learned by repeated simulation of the physical model to match the office room. The second step evaluates effects (here, thermal and CO_2-based air quality dissatisfaction) from the simulated indoor physical variables. These effects are shown in Eqs. (4) and (5) which represent thermal and air quality dissatisfaction at the ith hour, respectively. The objectives are formulated keeping in mind that the indoor temperature is preferred between 21 and 23 °C and the indoor CO_2 concentration is preferred between 400 and 1500 ppm. The purpose of the optimizer is to optimize (minimize) the effects (dissatisfaction levels) as formulated by Eq. (3).

$$T_{in} = \frac{R}{R_i}\tau + R\left(\frac{1}{R_{out}} + \frac{\zeta_W}{R_W}\right)T_{out} + R\left(\frac{1}{R_n} + \frac{\zeta_D}{R_D}\right)T_n \tag{1}$$

$$\begin{aligned} V\frac{dC_{in}}{dt} = &- \left(Q_0^{out}(t) + Q_0^{cor}(t) + \zeta_W(t)Q_W + \zeta_D(t)Q_D\right)C_{in} \\ &+ \left(Q_0^{out}(t) + \zeta_W(t)Q_W\right)C_{out} + \left(Q_0^{cor}(t) + \zeta_D(t)Q_D\right)C_{cor} \\ &+ S_{CO_2} \times n(t) \end{aligned} \tag{2}$$

$$\text{Minimize: } D(\text{actions}) = \left[d_1(T_{in}), d_2(C_{in})\right] = \left[\frac{\sum_{i=1}^{12} d_1^i}{12}, \frac{\sum_{i=1}^{12} d_2^i}{12}\right] \tag{3}$$

where

$$d_1^i(T_{in}) = \begin{cases} \frac{21-T_{in}}{21-18} & \text{if } T_{in} < 21 \\ 0 & \text{if } 21 \le T_{in} \le 23 \\ \frac{T_{in}-23}{26-23} & \text{if } T_{in} > 23 \end{cases} \tag{4}$$

$$d_2^i(C_{in}) = \begin{cases} 0 & \text{if } C_{in} \le 400 \\ \frac{C_{in}-400}{1500-400} & \text{if } C_{in} > 400 \end{cases} \tag{5}$$

$$\text{actions} = \left[\zeta_W^1, \zeta_W^2, \ldots, \zeta_W^{12}, \zeta_D^1, \zeta_D^2, \ldots, \zeta_D^{12}\right]$$

Table 1 Algorithmic parameters and their values

Module	Parameters	Explanation	Values
Physical model	ζ_W, ζ_D	Status of window (W), door (D)	Open = 1, Close = 0
	T_{in}, T_n, T_{out}	Temperatures of indoor, adjacent corridor, outdoor	From database
	R_D, R_W	Thermal resistances of door (D), window (W)	From database
	R_i, R_n, R_{out}	Resistance of walls, adjacent corridor, outdoor	From database
	R	Equivalent resistance	$R_i \| R_{out} \| R_n \|$ R_W (when $\zeta_W = 1$)$\|$ R_D (when$\zeta_D = 1$)
	τ	Thermal coefficient representing building inertia	From database
	V	Volume of the room (office)	From database
	C_{in}, C_{out}, C_{cor}	CO_2 concentrations indoor, outdoor, in adjacent corridor	From database
	Q^{out}, Q^{cor}	Air speed outdoor, in corridor	From database
	Q_W, Q_D	Air speed through window (W), door (D)	From database
	S_{CO_2}	Breath production of CO_2 per occupant	From database
	$n(t)$	Number of occupants at time t	From database
Differential evolution (optimizer)	NP	Population size	20
	G_{max}	Maximum generations	300
	F	Scale factor	Randomly chosen between 0 and 2
	CR	Crossover rate	0.8
	r	Reference point for ranking and decision making	Ideal point i.e. $(0, 0)$

Like most evolutionary optimization algorithms, the multi-objective version of differential evolution is executed for a predetermined number of generations by which the optimization algorithm is expected to have converged. The description of various parameters and their values for which best results are obtained are noted in Table 1.

At the end of the optimization algorithm, a few optimal schedules of actions are generated. The user can choose among these schedules based on their preference among the multiple objectives of optimality by comparing their usual schedule with the chosen schedule.

2.4 Comparing Schedule of Actions

Based on any schedule of actions, the physical model can generate the corresponding indoor ambience. The indoor ambience based on the usual schedule of the occupant is available in the database. Occupants can compare the simulated ambience (what best could have happened) to their usual ambience (what had actually happened) and understand the difference in effects to gain better comfort.

After the description of the experimental set-up, the next section discusses various results in order to validate the proposed approach.

3 Result and Discussion

This section analyses the performance of the proposed approach using the various results. Dissatisfaction levels as obtained from various simulated schedules are shown in a scatter plot. From these, the set of Pareto-optimal schedules obtained using optimization techniques are marked. Next, the variation of average indoor temperature and CO_2 concentration, obtained from optimal schedules, are compared with usual schedules. Finally, the variation of net dissatisfaction resulting from the proposed optimal schedule is shown to validate the optimization performance. The experimental data is collected for 100 days (1 April 2015 to 9 July 2015), and the analysis is conducted in 10 groups of 10 days each as shown in Table 2.

3.1 Pareto-Front and Optimal Schedules

A set of Pareto-optimal solutions is obtained for every working day in the experimental duration. Depending on the occupant's preference, any one of these schedules can be chosen as the preferred optimal schedules. Considering equal preference for both the objectives, the solution nearest to reference point (at the minima for thermal and air quality dissatisfaction i.e. at $(0, 0)$) is considered as the best schedule. The solution at the end of the Pareto-front is also analysed further for comparison because these represent the best schedules with respect to each objectives (minimal thermal or CO_2-based air quality dissatisfaction), independently. Hence, occupant's usual

Table 2 Groups of experimental data recorded during working hours (8 a.m. to 8 p.m.)

Group	Period	Mean outdoor temperature (°C)	Outdoor CO_2 concentration (ppm)
1	1 to 10 April 2015	10.1528	395
2	11 to 20 April 2015	17.8809	
3	21 to 30 April 2015	17.1548	
4	1 to 10 May 2015	20.7708	
5	11 to 20 May 2015	20.1111	
6	21 to 30 May 2015	15.9167	
7	31 May to 9 June 2015	24.8571	
8	10 to 19 June 2015	21.5119	
9	20 to 29 June 2015	23.5417	
10	30 June to 9 July 2015	28.7812	

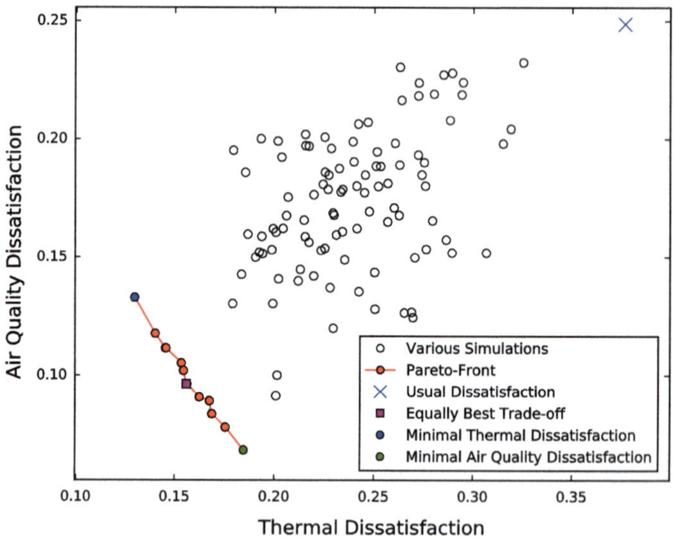

Fig. 2 Pareto-front and schedules of interest

schedule is compared with three other schedules: best schedule, schedule for optimal thermal comfort and schedule for optimal air quality comfort. Using the context from 16 April 2015, the thermal versus air quality dissatisfaction corresponding to various simulated schedules is shown Fig. 2. It also shows the Pareto-front and the three optimal schedules of interest, along with the usual level of dissatisfaction.

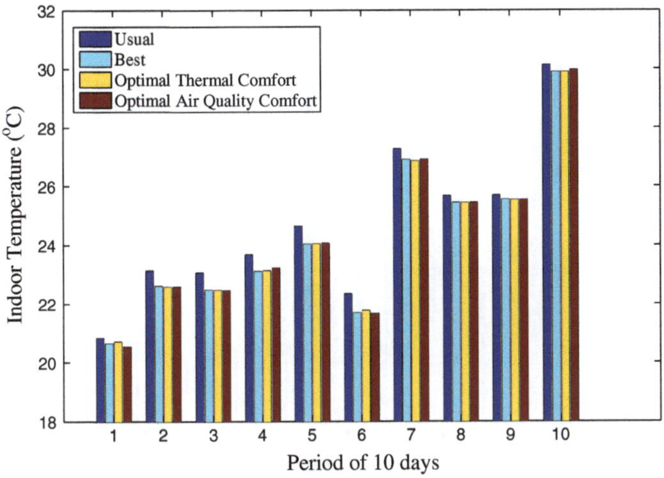

Fig. 3 Variation in indoor temperature resulting from different schedules in the experimental duration

3.2 Comparison of Physical Variables for Different Schedules

As there are very less day-to-day changes in environmental conditions, the average of physical variables of 10 days (working hours only) are considered for comparison of the various schedules. Variations in indoor temperature and CO_2 concentrations due to various schedules are presented in Figs. 3 and 4 for all the 10 groups of 10 days each as mentioned in Table 2. From the figures, it can be noticed that the difference between usual and proposed indoor physical variables (temperatures and CO_2 concentrations) is more during earlier days of experiment. In summer (group 7 to 10), it is difficult to maintain the physical variables in preferred ranges just by varying the schedule of opening/closing of doors/windows. Hence, in extreme cases where outdoor physical variables (Table 2) are higher than preferred ranges, HVAC system is needed to regulate the indoor physical variables. It is also to be noted that the best schedule usually presents higher temperature than the schedules for optimal thermal comfort. Similar observation is also noted for indoor CO_2 concentration. This is due to the fact the best schedule presents a trade-off solution, whereas the other proposed schedules are optimal with respect to one objective at a time.

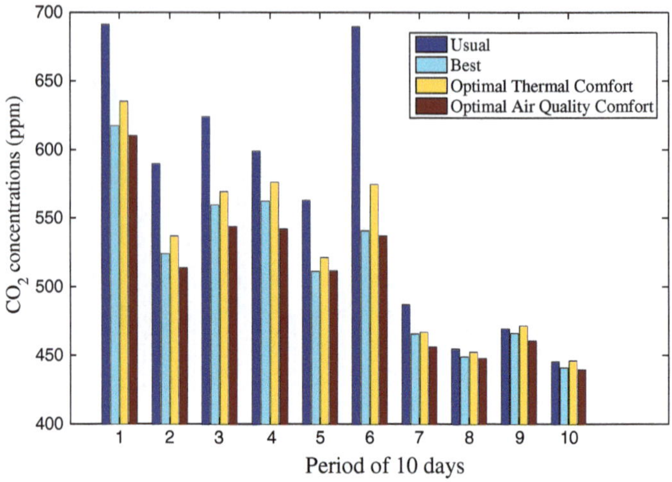

Fig. 4 Variation in indoor CO_2 concentrations resulting from different schedules in the experimental duration

3.3 Performance Analysis of Optimization Algorithm

Optimization algorithm, in this case, presents a set of solutions, called the Pareto-optimal solution. The dissatisfaction values corresponding to these solutions create the Pareto-front. To assess the convergence of the best schedule with respect to the ideal optimal solution, the city-block distance (sum of absolute difference) [2] between the corresponding dissatisfaction levels is measured. This distance represents the net dissatisfaction which is to be minimized. The parameters required for this performance metric, viz. the reference point, are noted in Table 1. Box plots for the distribution of net dissatisfaction for the best schedule of the Pareto-front are noted in Fig. 5 for all the 10 groups which shows that the median is close to 0 and the range of values is very less towards the earlier group and becomes higher for later groups. For a general idea of the range of dissatisfaction values, the theoretical variation of dissatisfaction is plotted against indoor temperature and indoor CO_2 concentration in Fig. 5 which follows from Eqs. (4) to (5). As noted from these plots, a combined dissatisfaction value around 2 or higher indicates that the indoor physical variables are not within preferred ranges. This implies that the proposed optimization approach is reproducible and efficient in yielding optimal schedules for this application during those period when HVAC system is not needed and thus, managing energy efficiently in the studied scenario.

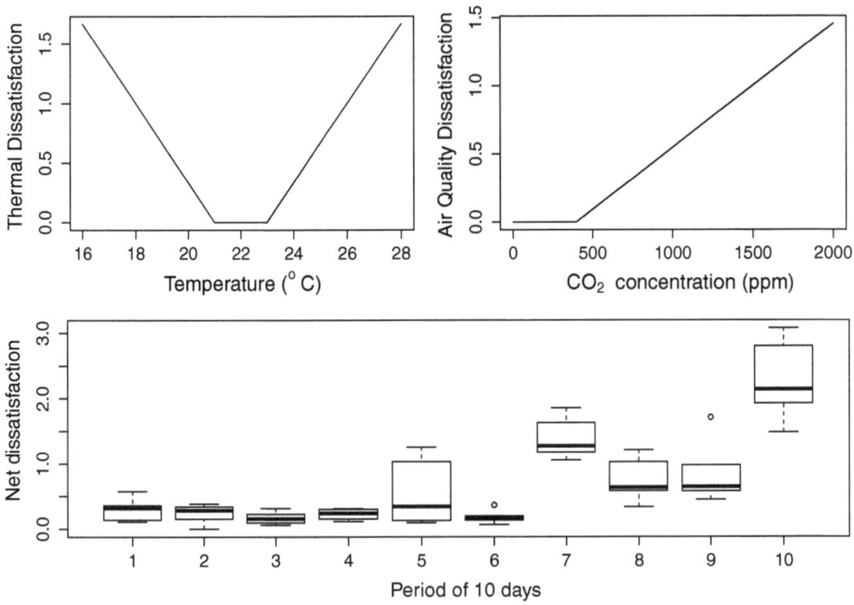

Fig. 5 Variation in net dissatisfaction for the best schedule in the experimental duration

4 Conclusion

The objective of this work is to demonstrate an approach which considers the occupant's schedule of actions as the only controllable parameter of thermal and CO_2-based air quality comfort and yields several schedules of actions which represent the best trade-offs between thermal and air quality dissatisfaction. Among the several schedules, the schedule corresponding to the equally best compromise of air quality and thermal comfort is analysed further. This optimal schedule can lead to efficient energy management. This work also addresses the need of HVAC system which arises when the environmental variables (here, temperature and CO_2 concentration) are too high from the preferred values such that change in schedule of actions can negligibly influence the comfort level of the occupants.

However, just presenting the optimal schedule might not convince the occupants to change their schedule. In order to gain occupant's trust on the system, the effect of small changes in action and the internal working of the black box are to be explained in non-technical terms. This can lead to better comfort without depending on external devices like HVAC system. The authors are working on the explanations and adding more context to improve the parameterization of overall comfort. On the other hand, when several devices including HVAC system (if required) are operational, their energy consumption and environmental impact form other important effects to be considered along with the comfort of the occupants. This forms open area of research along this domain of energy management in smart buildings.

Acknowledgements This study has been supported by the Indian side of the project sanctioned vide DST-INRIA/2015-02/BIDEE/0978 by the Indo-French Centre for the Promotion of Advanced Research (CEFIPRA—IFCPAR).
This work benefits from the support of the INVOLVED ANR-14-CE22-0020-01 project (http://www.agence-nationale-recherche.fr/?Projet=ANR-14-CE22-0020) of the French National Research Agency, ANR: Agence Nationale de la recherche, which aims at implementing new occupant interactive energy services, like MIRROR, WHAT-IF and SUGGEST, into a positive energy building constructed at Strasbourg in France by Elithis.

References

1. Amayri, M., Ploix, S., Bandyopadhyay, S.: Estimating occupancy in an office setting. Sustainable Human Building Ecosystems, pp. 72–80. Carnegie Mellon University, Pittsburgh, USA (2015)
2. De Souza, R.M., De Carvalho, F.D.A.: Clustering of interval data based on city-block distances. Pattern Recogn. Lett. **25**(3), 353–365 (2004)
3. Pal, M., Bandyopadhyay, S.: Many-objective feature selection for motor imagery EEG signals using differential evolution and support vector machine. In: International Conference on Microelectronics, Computing and Communications (MicroCom), pp. 1–6. IEEE (2016)
4. Robič, T., Filipič, B.: Demo: differential evolution for multiobjective optimization. Evolutionary Multi-criterion Optimization, pp. 520–533. Springer (2005)
5. Scanu, L., Bernaud, P.B., Ploix, S., Wurtz, E.: Methodologie pour la comparaison de structures de modeles simplifies. IBPSA, France (2016)

CMA—H^{∞} Hybrid Design of Robust Stable Adaptive Fuzzy Controllers for Non-linear Systems

Kaushik Das Sharma, Amitava Chatterjee, Patrick Siarry and Anjan Rakshit

Abstract The present paper utilizes covariance matrix adaptation (CMA), an evolution strategy, in conjunction with H^{∞}-based robust control law to design a stable adaptive fuzzy controller for a class of non-linear systems. The objective of the design is to develop a self-adaptive optimal/near optimal fuzzy controller, with guaranteed stability and satisfactory robust transient performance. The global search capability of CMA and H^{∞}-based tuning, that provide a fast adaptation utilizing local search method, is employed in tandem with this proposed design methodology. The hybrid control strategy is implemented for benchmark simulation case study, and the results demonstrate the usefulness of the proposed approach.

Keywords Covariance matrix adaptation (CMA) · H^{∞} based robust control
Adaptive fuzzy control · Hybrid fuzzy control · Non-linear systems

1 Introduction

Optimization of design variables is one of the challenges to the engineering community. Many times the fitness functions of such optimizations are mostly non-differentiable, or even if differentiable, their derivatives may not be calculated

K. Das Sharma (✉)
Department of Applied Physics, University of Calcutta, Kolkata, India
e-mail: kdsaphy@caluniv.ac.in

A. Chatterjee · A. Rakshit
Department of Electrical Engineering, Jadavpur University, Kolkata, India
e-mail: achatterjee@ee.jdvu.ac.in

A. Rakshit
e-mail: anjan_rakshit@yahoo.com

P. Siarry
Lab. LiSSi, Université Paris-Est Créteil, 122 Rue Paul Armangot,
94400 Vitry-sur-Seine, France
e-mail: siarry@u-pec.fr

S. Kar et al. (eds.), *Operations Research and Optimization*, Springer Proceedings
in Mathematics & Statistics 225, https://doi.org/10.1007/978-981-10-7814-9_11

explicitly. Furthermore, the problems are non-convex type in general, and so, finding the global optima is quite difficult. To get rid of these kind of problems, Hansen et al. proposed the covariance matrix adaptation evolution strategy (CMA-ES) or simply CMA [1]. It is an evolutionary algorithm creating a number of probable solution points utilizing the normal distribution, and it is a derivative-free optimization. Although the original version is applicable only to the unconstrained problems and can handle small population size during operation, the further improvements show its application to the constrained optimization with large population size [2, 3].

The present paper proposes a systematic design procedure of stable adaptive fuzzy logic controllers (AFLCs) using hybridizations of H^{∞}-based approach (HBA) of robust control [4] and CMA algorithm-based stochastic optimization technique [1]. The hybrid design strategy of fuzzy controller design tries to combine the advantages of both H^{∞} control-based local adaptation and stochastic optimization-based global search method to develop a superior method [5]. The aims of the proposed design scheme are to execute simultaneous adaptations of both the structural features of FLC and its free parameters, such that (i) to guarantee the stability of the closed-loop system and (ii) to accomplish high degrees of automation in the design process, utilizing CMA algorithm, by getting rid of many manually tuned parameters [6]. Hence, the objective of this work is to design stable adaptive fuzzy controllers which can deliver high degree of automation in the design process, guarantee asymptotic stability in the sense of Lyapunov and also achieve adequate transient performance. The proposed scheme is implemented for a benchmark non-linear case study, and the results show the usefulness of the scheme.

The rest of the paper is organized as follows: Sect. 2 presents the H^{∞}-based approach of AFLC design. Section 3 discusses the CMA algorithm and adaptive fuzzy controller design using CMA, and Sect. 4 describes the proposed hybrid controller design technique. Section 5 shows the simulation studies for a non-linear system, and Sect. 6 concludes the paper.

2 H^{∞}-Based Design of Robust Stable Adaptive Fuzzy Controller

Let us consider an nth-order non-linear plant given as [4, 5]:

$$\left.\begin{aligned} x^{(n)} &= f(\underline{x}) + bu + d \\ y &= x \end{aligned}\right\} \tag{1}$$

where $f(\cdot)$ is an unknown continuous function, $u \in R$ and $y \in R$ are the input and output of the plant, b is an unknown positive constant and d is unknown but bounded external disturbances. It is assumed that the state vector is given as $\underline{x} = (x_1, x_2, \ldots, x_n)^T = (x, \dot{x}, \ldots, x^{(n-1)})^T \in R^n$. In order to (1) to be controllable in certain controllability region $U_{\underline{x}} \subset R$, we require that $b \neq 0$. Thus, without loss of generality we can assume that $b > 0$.

The control objective is to track the reference signal $y_m(t)$ and thus, the tracking error is $e = y - y_m$. The objective is to design a stable AFLC for the system described in (1). Here we need to find a feedback control law $u = u_c(\underline{x}|\underline{\theta})$, using fuzzy logic system and an adaptive law for adjusting the parameter vector $\underline{\theta}$ such that the closed-loop system must be globally stable in the sense that all variables must be uniformly bounded, and the $\underline{e} - \underline{\theta}$ space should be stable in the large for the system [7]. Let the error vector be $\underline{e} = (e, \dot{e}, \ldots, e^{(n-1)})^T$ and $\underline{k} = (k_1, k_2, \ldots, k_n)^T \in R^n$ be such that all the roots of the Hurwitz polynomial $s^n + k_n s^{n-1} + \cdots + k_2 s + k_1$ are in the left half of s-plane.

Now the ideal control law for the system in (1) as:

$$u^* = \frac{1}{b}\left[-f(\underline{x}) + y_m^{(n)} + \underline{k}^T \underline{e} - u_r\right] \tag{2}$$

where $y_m^{(n)}$ is the nth derivative of the output of the reference signal, and the control signal u_r is applied to attenuate the external disturbance and the error due to fuzzy approximation of the AFLC [4].

For some specific class of plants the error differential equation as

$$e^{(n)} = -k_1 e - k_2 \dot{e} - \cdots - k_n e^{(n-1)} + u_r - d \tag{3}$$

This definition implies that u^* guarantees perfect tracking, i.e. $y(t) \equiv y_m(t)$ if $\underset{t \to \infty}{Lt}\, e(t) = 0$. As f and b are not known precisely, the ideal u^* of (2) cannot be implemented in practice. The design objective of asymptotically stable tracking may be achieved if the fuzzy approximation error (ε) and the external disturbance d are zero; otherwise, the H$^\infty$ tracking performance will come to the play as [4, 8]:

$$\int_0^T \underline{e}^T Q \underline{e}\, dt \le \underline{e}^T(0) P \underline{e}(0) + \frac{1}{\nu} \underline{\theta}^T(0)\underline{\theta}(0) + \rho^2 \int_0^T \varepsilon^T \varepsilon\, dt \quad \forall T \in [0, \infty) \;\; \varepsilon \in L_2[0.T] \tag{4}$$

for given weighting matrices $Q = Q^T \ge 0$, $P = P^T \ge 0$, an adaptation gain $\nu > 0$ and prescribed attenuation level ρ.

Let us assume that the AFLC is constructed using a zero order Takagi–Sugeno (T-S) fuzzy system. Then $u_c(\underline{x}|\underline{\theta})$ for the AFLC is given in the form [5, 6]:

$$u_c(\underline{x}|\underline{\theta}) = \underline{\theta}^T * \xi(\underline{x}) \tag{5}$$

where $\underline{\theta} = [\theta_1\, \theta_2 \ldots \theta_N]^T$ = the vector of the output singletons, N = the total number of rules and $\underline{\xi}(\underline{x})$ = vector containing normalized firing strength of all fuzzy IF–THEN rules $= (\xi_1(\underline{x}), \xi_2(\underline{x}), \ldots, \xi_N(\underline{x}))^T$, for details see [4, 7].

Now if b is a known constant, then for direct adaptive control, as suggested in [4], the direct AFLC can form as

$$u = u_c(\underline{x}|\underline{\theta}) - \frac{1}{b} u_r(\underline{x}) \tag{6}$$

and

$$u_r(\underline{x}) = -\frac{1}{r} B^T P \underline{e} \tag{7}$$

where $u_r(x)$ is the H^∞ robust term to compensate the fuzzy approximation error and the external disturbance d, r is a positive scalar value and P can be obtained from the solution of Riccati-like equation stated as [4, 8]:

$$A^T P + PA - PB\left(\frac{2}{r} - \frac{1}{\rho^2}\right) B^T P = -Q \tag{8}$$

The parameter update law is chosen as [4, 5]:

$$\underline{\theta} = \nu \underline{e}^T bPB\xi(\underline{x}) \tag{9}$$

where ν is adaptation gain or learning rate.

Now, to ensure stability it is assumed that the control law $u(t)$ is actually the summation of the fuzzy control $u_c(\underline{x}|\underline{\theta})$ and robust control term, given as:

$$u(t) = u_c(\underline{x}|\underline{\theta}) - \frac{1}{b} u_r(\underline{x}) = \underline{\theta}^T * \xi(\underline{x}) - \frac{1}{b} u_r(\underline{x}) \tag{10}$$

Now the closed-loop error equation in (3) becomes

$$\underline{\dot{e}} = \Lambda_c \underline{e} + \underline{b}_c [u_c(\underline{x}|\underline{\theta}) - u^* + \frac{1}{b} u_r] - \frac{\underline{b}_c}{b} d \tag{11}$$

where

$$\Lambda_c = \begin{bmatrix} 0 & 1 & 0 & 0 & \cdots & 0 & 0 \\ 0 & 0 & 1 & 0 & \cdots & 0 & 0 \\ \vdots & \vdots & \vdots & \vdots & \ddots & \vdots & \vdots \\ 0 & 0 & 0 & 0 & \cdots & 0 & 1 \\ -k_1 & -k_2 & \cdots & \cdots & \cdots & -k_{n-1} & -k_n \end{bmatrix} \text{ and } \underline{b}_c = \begin{bmatrix} 0 \\ \vdots \\ 0 \\ b \end{bmatrix} \tag{12}$$

Now defining the Lyapunov function as

$$V = \frac{1}{2}\underline{e}^T P \underline{e} + \frac{1}{2\nu}\underline{\theta}^T P \underline{\theta} \tag{13}$$

where $\underline{\theta}^*$ = the optimal parameter vector and $\underline{\theta} = \underline{\theta} - \underline{\theta}^*$.

After some straightforward calculations and using (9), (10), (11) and (13):

$$\dot{V} \leq -\frac{1}{2}\underline{e}^T Q \underline{e} + \frac{1}{2}\rho^2 \varepsilon^T \varepsilon \tag{14}$$

Using (13) and (14)

$$\frac{1}{2}\int_0^T \underline{e}^T Q \underline{e} dt \leq \frac{1}{2}\underline{e}^T(0) P \underline{e}(0) + \frac{1}{2\nu}\underline{\theta}^T(0) P \underline{\theta}(0) + \frac{1}{2}\rho^2 \int_0^T \varepsilon^T \varepsilon \tag{15}$$

Equation (15) is identical to that of (4), i.e. the condition for the H$^\infty$ tracking control. Thus, using the control law $u_r(\underline{x})$, one can guarantee the boundedness in $\underline{x}$. Hence, the closed-loop stability is guaranteed [4, 5]. The flowchart representation of H$^\infty$-based approach (HBA) of tracking control scheme is shown in Fig. 1.

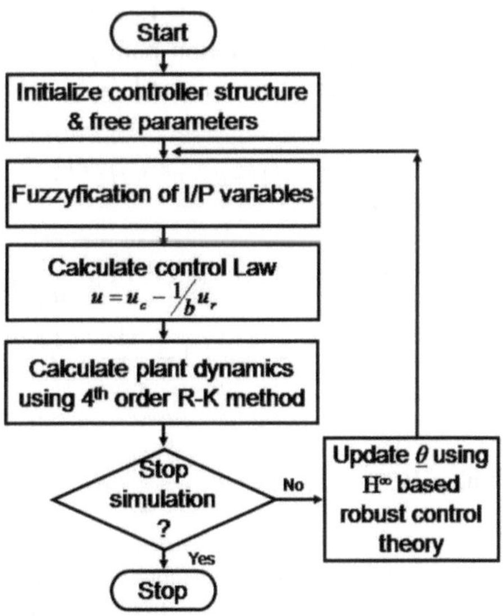

Fig. 1 Flowchart representation of H$^\infty$-based approach (HBA) of tracking control

3 Covariance Matrix Adaptation (CMA)

CMA was introduced by Hansen et al. [1] in 1995, and further improvements were developed by Hansen et al. in 2001 [9] and 2003 [10]. Originally, CMA was designed for small population sizes and was interpreted as a robust local search strategy [1, 9]. The drawbacks of basic CMA were the complexity of the adaptation process and the dependence on the design parameter formulae for which very little theoretical knowledge is available. In [10], the rank-μ-update version of CMA was proposed and it exploits the advantage of large population size without affecting the performance for small population size. In 2005, Hansen et al. proposed restart version [10] of CMA which arguably proved the best performing evolutionary algorithm for continuous optimization for a set of test functions [2]. The present paper utilizes the rank-μ-update and weighted recombination version CMA [9].

Rank-μ-update and weighted recombination version CMA

It is considered that there λ number of potential candidate solutions, and each λ individual is generated at (g + 1)th generation as [9]:

$$z_k^{g+1} = N\left(\langle z\rangle_w^g, \sigma^{g^2} C^g\right) = \langle z\rangle_w^g + \sigma^g B^g D^g N(0, I), \quad k = 1, \ldots, \lambda \tag{16}$$

where $N(m, C)$ = normally distributed random vector with mean m and covariance matrix C. $\langle z\rangle_w^g = \sum_{i=1}^{\mu} \omega_i z_{i:\lambda}^g$ = weighted mean of the selected individuals, $\omega_i > 0$, $\forall i = 1, \ldots, \mu$ and $\sum_{i=1}^{\mu} \omega_i = 1$. The adaptation of the covariance matrix C^g is calculated along the evolution path p_c^{g+1} and by the μ weighted difference vectors between the recent parents and $\langle x\rangle_w^g$ as [1, 9]:

$$p_c^{g+1} = (1 - c_c).p_c^g + H_\sigma^{g+1} \sqrt{c_c(2 - c_c)}.\frac{\sqrt{\mu_{eff}}}{\sigma^g}\left(\langle z\rangle_w^{g+1} - \langle z\rangle_w^g\right) \tag{17}$$

$$\begin{aligned} C^{g+1} = {} & (1 - c_{\text{cov}}).C^g + c_{\text{cov}} \frac{1}{\mu_{\text{cov}}} p_c^{g+1} \left(p_c^{g+1}\right)^T \\ & + c_{\text{cov}}.\left(1 - \frac{1}{\mu_{\text{cov}}}\right) \sum_{i=1}^{\mu} \frac{\omega_i}{\sigma^{g^2}} \left(z_{i:\lambda}^{g+1} - \langle z\rangle_w^g\right)\left(z_{i:\lambda}^{g+1} - \langle z\rangle_w^g\right)^T \end{aligned} \tag{18}$$

where $H_\sigma^{g+1} = 1$, if $\frac{\left\|p_\sigma^{g+1}\right\|}{\sqrt{1 - (1 - c_\sigma)^{2(g+1)}}} < \left(1.5 + \frac{1}{n - 0.5}\right) E(\|N(0, I)\|)$, and 0, otherwise.

$\mu_{eff} = 1/\sum_{i=1}^{\mu} \omega_i^2$ = variance effective selection mass and $\mu_{eff} = \mu$, if $\omega_i = 1/\mu$, i.e. the condition of intermediate recombination [2, 9]. The weights ω_i are a matrix with rank $\min(\mu, n)$. $c_{\text{cov}} \approx \min\left(1, 2\mu_{eff}/n^2\right)$ = learning rate for the covariance matrix C.

The adaptation of the global step size σ^g is calculated from the conjugate evolution path p_σ^{g+1} as [9]:

$$p_\sigma^{g+1} = (1 - c_\sigma).p_\sigma^g + \sqrt{c_\sigma(2 - c_\sigma)}.B^g D^{g^{-1}} B^{g^T} \frac{\sqrt{\mu_{eff}}}{\sigma^g} \left(\langle z\rangle_w^{g+1} - \langle z\rangle_w^g\right) \tag{19}$$

B^g, the orthogonal matrix, and D^g, the diagonal matrix, are both obtained from the principal component analysis of C^g. Thus, the global step size can be calculated as [9]:

$$\sigma^{g+1} = \sigma^g.\exp\left(\frac{c_\sigma}{d_\sigma}\left(\frac{\|p_\sigma^{g+1}\|}{E(\|N(0,I)\|)} - 1\right)\right) \tag{20}$$

where $E(\|N(0,I)\|) = \sqrt{2}\Gamma\left(\frac{n+1}{2}\right)/\Gamma\left(\frac{n}{2}\right) \approx \sqrt{n}\left(1 - \frac{1}{4n} + \frac{1}{21n^2}\right)$ = the expected length of p_σ under random selection.

The initial values are to be selected as $p_\sigma^0 = p_c^0 = 0$ and $C^0 = I$, whereas x^0 and σ^0 are problem dependent. The default strategy parameters are given as [3, 11]:

$$\lambda = 4 + \lfloor 3.\ln(n) \rfloor,\ \mu = \lfloor \lambda/2 \rfloor,\ \omega_{i=1\ldots\mu} = \frac{\ln(\mu+1) - \ln(i)}{\sum_{j=1}^{\mu} \ln(\mu+1) - \ln(i)} \tag{21}$$

$$c_\sigma = \frac{\mu_{eff} + 2}{n + \mu_{eff} + 3},\ d_\sigma = 1 + 2.\max\left(0, \sqrt{\frac{\mu_{eff} - 1}{n+1}} - 1\right) + c_\sigma,\ c_c = \frac{4}{n+4} \tag{22}$$

$$\mu_{\text{cov}} = \mu_{eff},\ c_{\text{cov}} = \frac{1}{\mu_{\text{cov}}}.\frac{2}{\left(n + \sqrt{2}\right)^2} + \left(1 - \frac{1}{\mu_{\text{cov}}}\right)\min\left(1, \frac{2\mu_{eff} - 1}{(n+2)^2 + \mu_{eff}}\right) \tag{23}$$

In this evolutionary strategy, $1/c_\sigma$ and $1/c_c$ are considered as memory time constants and d_σ as damping parameter. Details about these parameters are discussed in [2].

CMA-based adaptive fuzzy controller design scheme

In this paper, rank-μ-update and weighted recombination version CMA are employed to design the adaptive fuzzy controller. In H$^\infty$ theory-based adaptive design methodology [4, 8], only the optimized values of singletons are obtained and the scaling gains are manually tuned by trivial method. The CMA-based design methodology not only tunes the singletons properly but also it can optimize the values of the scaling gains and can determine the optimum controller structure [5].

In CMA-based controller design, a candidate solution vector in solution space is a vector containing all required information to construct a fuzzy controller, e.g. (i) information about the positions of the MFs, (ii) values of scaling gains, (iii) positions of the output singletons, etc. The candidate solution vector (CSV) can be formed as [12]:

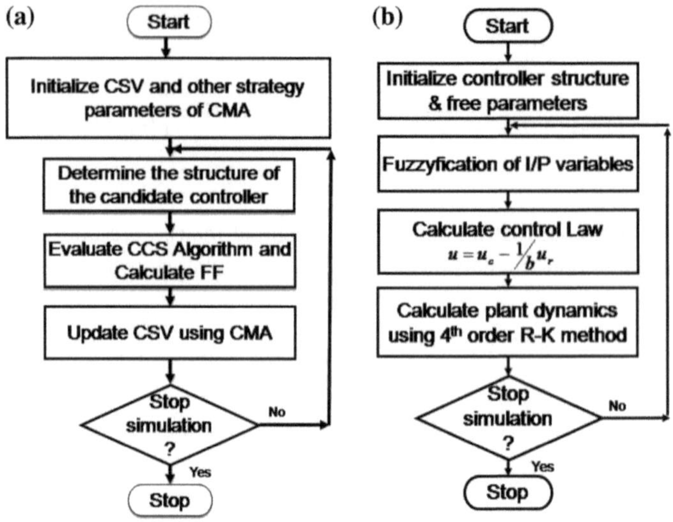

Fig. 2 Flowchart representation of **a** CMA-based AFLC design scheme and **b** CCS algorithm

$$z = [\text{center locations of the MFs} \mid \text{scaling gains} \mid\mid \text{learning rate} \mid \ldots \\ \ldots \mid \text{positions of the output singletons}] \tag{24}$$

Figure 2a shows the flowchart representation of CMA-based AFLC design algorithm. In this design process, a CSV is employed to simulate the system for each candidate controller by using the candidate controller simulation (CCS) algorithm [10], as shown in Fig. 2b, and the fitness function is evaluated for that CSV. The fitness function is chosen as the integral absolute error ($\text{IAE} = \sum_{k=0}^{PST} |e(k)|\Delta t_c$), where PST = plant simulation time, Δt_c = step size or sampling time.

4 Hybrid Stable Adaptive FLC Design

In hybrid design approach, H^{∞} theory-based local adaptation of output singleton of fuzzy controllers as presented in (9) and rank-μ-update and weighted recombination version CMA-based global optimization technique are combined to achieve a superior performance for the stable adaptive fuzzy controllers, and this design strategy is termed as hybrid adaptation strategy-based approach (HASBA) in this paper. In this hybrid design model, the H^{∞}-based local adaptation and CMA-based global optimization run concurrently to explore the solution space for achieving a robust stable tracking performance.

In this method, a candidate solution vector z is divided into two subgroups given as [5, 6, 12]:

$$\underline{Z} = [\underline{\psi} \,|\, \underline{\theta}] \tag{25}$$

where

$\underline{\psi}$ [centre locations of the fuzzy MFs | scaling gains | learning rate]
$\underline{\theta}$ [position of the fuzzy output singletons]

Here the decision variables comprising $\underline{\psi}$ have a non-linear influence, and the decision variables comprising $\underline{\theta}$ have a linear influence on the control signal from the AFLC [5, 7]. For each CSV chosen as a candidate controller, it is first subjected to the adaptation of the $\underline{\theta}$ portion according to (9) and the finally adapted values are then used to update the candidate controller. Then this updated CSV is subjected to an usual pass of the HASBA. In this process, $\underline{\theta}$ vector is subjected to both local and global search experiences in every update of CSV. The flowchart representation of this HASBA is shown in Fig. 3.

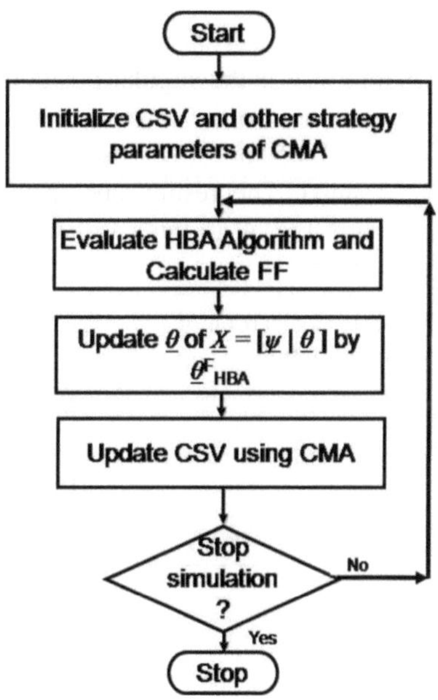

Fig. 3 Flowchart representation of HASBA algorithm

5 Simulation Case Study

The effectiveness of the proposed AFLC design methodologies is evaluated for a benchmark non-linear system like DC motor system with non-linear friction characteristics [7]. The non-linear plant model is simulated each utilizing fixed step fourth-order Runge–Kutta method with sampling time $\Delta t_c = 0.01$ s. A fixed structure (5 × 5 mf) zero-order T-S type fuzzy controller is used. The HBA algorithm is simulated for 210 s where 200 s is utilized for the adaptation purpose and then 10 s evaluation period with $\nu = 0$. In CMA and HASBA-based schemes, also the plant is simulated for 10 s duration in evaluation after a 200 number of CMA iterations. These CMA-based algorithms are run for 10 times each to calculate the stochastic variation of the results. An external disturbance d is applied to the plant in both case studies where d is a square wave of random amplitude within the range [−1, 1] and a period of 0.5 s. [5].

The controlled plant under consideration is a second-order DC motor containing non-linear friction characteristics described by the following model [5, 7]:

$$\left.\begin{array}{l} \dot{x}_1 = x_2 \\ \dot{x}_2 = -\frac{f(x_2)}{J} + \frac{C_T}{J} u + d \\ y = x_1 \end{array}\right\} \tag{26}$$

where $y = x_1$ is the angular position of the rotor (in rad), x_2 is the angular speed (in rad/sec) and u is the current fed to the motor (in A). $C_T = 10$ Nm/A, $J = 0.1$ kgm^2 and $f(x_2) = 5\tan^{-1}(5x_2)$ Nm. The control objective is to track a reference signal $y_m = \sin(t)$. The results are provided in Table 1, and sample performances are shown in Fig. 4. In this case study, HASBA design scheme is superior in terms of IAE value.

Table 1 Comparison of simulation results

Control strategy	IAE values		
	Best IAE	Avg. IAE	Std. Dev.
HBA	0.5526	–	–
CMABA	0.6024	0.6698	0.0548
HASBA	0.4902	0.5340	0.0451

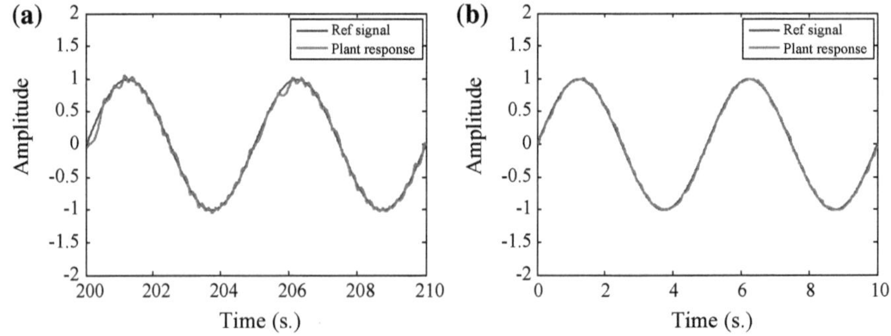

Fig. 4 Evaluation period response of **a** HBA control scheme and **b** HASBA control scheme

6 Conclusion

In this paper, rank-μ-update and weighted recombination version CMA algorithm-based optimization approach and H$^\infty$-based robust adaptive control technique have been concurrently hybridized for designing the AFLCs. The proposed hybrid design model has evolved as a superior technique for designing the stable AFLCs with external disturbances when compared with the design schemes discussed in this paper. The stability of closed-loop system and the convergence of the plant output to a desired reference signal in the presence of external disturbances are guaranteed precisely in the proposed design methodology. The main advantage of these proposed methods is that it requires no a priori knowledge about the controlled plant, and the inherent approximation error of the system is greatly reduced. The proposed design scheme can also be exploited to design the variable structure stable AFLCs [5, 6].

References

1. Hansen N., Ostermeier A., Gawelczyk A.: On the Adaptation of arbitrary normal mutation distributions in evolution strategies: the generating set adaptation. In: Proceedings of the 6th International Conference on Genetic Algorithms, San Francisco, CA, pp. 57–64 (1995)
2. Auger, A., Hansen, N.: A restart CMA evolution strategy with increasing population size. Congr. Evol. Comput. **2**, 1769–1776 (2005)
3. Suganthan, P.N., Hansen, N., Liang, J.J., Deb, K., Chen, Y.-P., Auger, A., Tiwari, S.: Problem definitions and evaluation criteria for the CEC 2005 special session on real-parameter optimization. Technical Report. Nanyang Tech. University. Singapore (2005)
4. Chen, B.S., Lee, C.H., Chang, Y.C.: H$^\infty$ tracking design of uncertain nonlinear SISO systems: adaptive fuzzy approach. IEEE Trans. Fuzz. Syst. **4**, 32–43 (1996)
5. Das, Sharma K., Chatterjee, A., Rakshit, A.: a hybrid approach for design of stable adaptive fuzzy controllers employing Lyapunov theory and particle swarm optimization. IEEE Trans. Fuzzy Syst. **17**, 329–342 (2009)
6. Das, Sharma K., Chatterjee, A., Rakshit, A.: Design of a hybrid stable adaptive fuzzy controller employing Lyapunov theory and harmony search algorithm. IEEE Trans. Control Syst. Technol. **18**, 1440–1447 (2010)
7. Fischle, K., Schroder, D.: An improved stable adaptive fuzzy control method. IEEE Trans. Fuzzy Syst. **7**, 27–40 (1999)
8. Tong, S., Li, H.X., Wang, W.: Observer-based adaptive fuzzy control for SISO nonlinear systems. Fuzz. Sets Syst. **148**, 355–376 (2004)
9. Hansen, N., Ostermeier, A.: Completely derandomized self-adaptation in evolution strategies. Evol. Comput. **9**, 159–195 (2001)
10. Hansen, N., Muller, S.D., Koumoutsakos, P.: Reducing the time complexity of the derandomized evolution strategy with covariance matrix adaptation (CMA-ES). Evol. Comput. **11**, 1–18 (2003)
11. Hansen, N., Kern, S.: Evaluating the CMA evolution strategy on multimodal test functions. In: Parallel Problem Solving from Nature—PPSN VIII. Lecture Notes in Computer Science, vol. 3242, pp. 282–291 (2004)
12. Das Sharma, K., Chatterjee, A., Matsuno, F.: A Lyapunov theory and stochastic optimization based stable adaptive fuzzy control methodology. In: Proceedings of the SICE Annual Conference 2008, Japan, pp. 1839–1844 (2008)

A Genetic Algorithm-Based Clustering Approach for Selecting Non-redundant MicroRNA Markers from Microarray Expression Data

Monalisa Mandal, Anirban Mukhopadhyay and Ujjwal Maulik

Abstract During the last few years, different studies have been done to reveal the involvement of microRNAs (miRNAs) in pathways of different types of cancers. It is evident from the research in this field that miRNA expression profiles help classify cancerous tissue from normal tissue or different subtypes of cancer. In this article, miRNA expression data of different cancer types are analyzed using a novel multi-objective genetic algorithm-based feature selection method for finding reduced non-redundant set of miRNA markers. Three objectives, viz. classification accuracy, a cluster validity index call Davies–Bouldin (DB) index, and the number of miRNAs encoded in a chromosome of genetic algorithm is optimized simultaneously. The classification accuracy is maximized to obtain the most relevant set of miRNAs. DB index is optimized for clustering the miRNAs and choosing representative miRNAs from each cluster in order to obtain a non-redundant set of miRNA markers. Finally, the number of miRNAs is minimized to yield a reduced set of selected miRNAs. The performance of the proposed genetic algorithm-based method is compared with that of the other existing feature selection techniques. It has been found that the performance of the proposed technique is better than that of the other methods with respect to most of the performance metrics. Lastly, the obtained miRNA markers with their associated disease and number of target mRNAs are reported.

Keywords MicroRNA · Multiobjective optimization · Genetic algorithm Clustering · Davies–Bouldin index

M. Mandal (✉)
Department of Computer and Information Science, University of Science and Technology (NTNU), Trondheim, Norway
e-mail: monalisa.mandal@ntnu.no

A. Mukhopadhyay
Department of Computer Science and Engineering, University of Kalyani, Kalyani 741235, West Bengal, India
e-mail: anirban@klyuniv.ac.in

U. Maulik
Department of Computer Science and Engineering, Jadavpur University, Kolkata 700032, West Bengal, India
e-mail: umaulik@cse.jdvu.ac.in

S. Kar et al. (eds.), *Operations Research and Optimization*, Springer Proceedings in Mathematics & Statistics 225, https://doi.org/10.1007/978-981-10-7814-9_12

1 Introduction

MicroRNAs (miRNAs) are small noncoding RNA molecules which do not participate in protein synthesis. They are mostly involved in discouraging the process of translation from mRNA to proteins. Literally, miRNA dampens the production of proteins. Within the last few years, noncoding miRNAs have become a focus of interest as possible biomarkers for different diseases. Specifically, the changes in miRNA expression levels have been found to be highly related to different types of cancer [21, 25]. Extensive studies have been done for analyzing miRNA expression data [11]. However, there are several problems in miRNA expression study such as (1) a particular miRNA targets a number of mRNAs, (2) there are some miRNAs which have identical sequences, thus difficulty lies in interpretation of those miRNAs, (3) and finally different miRNAs having different sequences may have same target mRNAs. Moreover, to establish an association between miRNA expression and cancer is really a more complicated issue for the researchers. Furthermore, for a complex disease like cancer various transcripts are regulated by individual miRNAs and thus, its function in oncogenesis is completely dependent on the biological circumstances. Despite all these complexities, researchers get an obvious profit in terms of comparatively less size than mRNAs while experimenting on cancer biomarkers discovery.

Previously, microarray gene expression analysis has been performed for cancer-related gene selection and gene clustering [12, 16, 18, 20]. Currently, miRNA microarray expression [17, 21, 22, 25] analysis has gained popularity in association with cancer marker identification. A miRNA microarray dataset is organized as a matrix, in which each row represents a miRNA and each column represents a sample or an experimental condition. Each sample is associated with a class label (such as normal or malignant). The miRNAs which have differential expression pattern in two different classes of expressions are called differentially expressed miRNAs and are treated as miRNA markers. Hence to distinguish different subtypes of miRNAs related to different diseases, researchers have employed microarray analysis of differential expression.

Many supervised and unsupervised classification approaches are available in the literature [7, 14, 15] which have been utilized for classification or clustering of disease samples, respectively. In most of the cases, the existing approaches yield some top-ranked miRNAs or features which are often found to be redundant. In this article, to select a small set of non-redundant miRNA markers, a multiobjective clustering-based feature selection strategy encoded in a genetic algorithm (GA) has been proposed. The multiobjective GA is guided by non-dominated sorting [5, 19] and crowding distance measure [5, 19]. Clustering of miRNAs (features) and picking up the center (prototype) feature from each cluster ensures non-redundancy. The intention here is to achieve a good fscore (representing classification accuracy) while optimizing a clustering validity index [4] (for better clustering of miRNAs) and the number of encoded miRNA markers. Davies–Bouldin (DB) cluster validity index [4] used for clustering purpose. The yielded miRNAs of this proposed study are considered as miRNA markers.

Genetic algorithms (GAs) are widely known class of popular evolutionary algorithms [8]. GAs simulate reproduction with solutions represented as chromosomes. Usually, a binary encoding is used for presenting a candidate solution having one or more objectives and selection, crossover and mutation are done while optimizing the objectives. As multiple objectives have been considered, the GA has been modeled as multiobjective optimization (MOO) [5, 19] problem where the objectives may estimate different aspects of typically conflicting solutions. The concept of Pareto optimality in MOO arises to solve the inconsistency among the objectives. The Pareto optimal set contains all those solutions so that it is not possible to improve any solution in this set with respect to an objective without simultaneously worsening it in terms of another objective. As there exist various "trade-off" solutions of the problem with respect to different objective functions, the Pareto optimal set usually consists of multiple non-dominated solutions. In this article, we have used the popular non-dominated sorting GA-II (NSGA-II) [5] as the basis of developing the proposed multiobjective GA-based miRNA marker selection technique.

The rest of the article is organized as follows. The next section describes the proposed multiobjective approach for miRNA marker selection in detail. Section 3 discusses various existing feature selection techniques used for comparison purpose. In Sect. 4, the datasets used for experiments are described along with preprocessing techniques applied on the datasets. Subsequently, Sect. 5 introduces the metrics used for performance evaluation of the algorithms. In Sect. 6, the results of experiments are reported and discussed. Finally, Sect. 7 concludes the article.

2 Proposed Multiobjective Feature Selection Technique

In this section, we describe in detail the proposed multiobjective feature selection technique for obtaining relevant, non-redundant, and reduced set of miRNA markers. NSGA-II has been utilized as the underlying multiobjective GA-based optimization tool. Given a population of solutions P, an individual $x \in P$ (chromosome) represents a binary encoding of a candidate solution. While optimizing the objectives, in each iteration, a new population of different individuals is created. The fitnesses or objectives are calculated, and fittest solutions are kept in an archive. The steps from encoding, initialization, objective calculation, reproduction, archive updation, and final solution selection are described in the following subsections.

2.1 Encoding Scheme and Initialization

A chromosome encodes all the miRNAs contained by the dataset. Thus, a chromosome has m cells if there are m miRNAs in the data matrix. The cells contain values 0 or 1. If a cell of a chromosome contains value 1, it means that the corresponding miRNA can be treated as representative center of a cluster of miRNAs. Hence, each chromosome encodes a set of cluster centers. Since total number of 1's in a chromosome is different from another, the number of cluster centers encoded in a chromosome is different. Each cell of a chromosome is initialized with randomly generated with value 0 or 1. After that their corresponding fitness values are calculated. For each dataset, the proposed technique has been executed for 100 iterations. The other input parameters, i.e., population size, crossover probability, and mutation probability are set to 20, 0.9, and 0.1, respectively.

2.2 Computing the Objectives

Each chromosome encodes some cluster centers in form of cell having value. These miRNAs are used for objective computation. Three objective functions are considered here to be optimized simultaneously.

The first objective function is fscore as defined in Eq. 1 for the selected miRNAs. This is calculated by cross-validation using support vector machine (SVM) classifier [24]. Basically, fscore (Eq. 1) measure combines precision (Eq. 2) and recall through the harmonic mean of precision and recall:

$$fscore = \frac{2 \times Precision \times Recall}{Precision + Recall}, \tag{1}$$

$$Precision = \frac{tp}{tp + fp}, \tag{2}$$

$$Recall = \frac{tp}{tp + fn}. \tag{3}$$

In information retrieval, positive predictive value, which is also called as precision, is defined as in Eq. 2. Here *tp*, *tn*, *fp*, and *fn* represent the number of true positives, true negatives, false positives, and false negatives, respectively. Sensitivity, which is also known as recall, is defined as per Eq. 3. The objective is to maximize fscore which indicates better relevance of the set of miRNA markers. Since the algorithm is modeled as minimizing one, 1-fscore is minimized.

To have compact clusters, a cluster validity index is optimized as the second objective. Here we have considered DB index for cluster validation. The Davies–Bouldin (*DB*) index [4] is defined a function of the ratio of the within-cluster scatter to between-cluster separation. The scatter within the *i*th cluster, S_i, is found as

$$S_i = \frac{1}{|C_i|} \sum_{x \in C_i} D^2(z_i, x). \tag{4}$$

Here $|C_i|$ represents the number of points in cluster C_i. The separation between two clusters C_i and C_j, represented as d_{ij}, is defined as the distance between the corresponding cluster centers.

$$d_{ij} = D^2(z_i, z_j). \tag{5}$$

The *DB* index is then defined as

$$DB = \frac{1}{K}\sum_{i=1}^{K} R_i, \tag{6}$$

where

$$R_i = \max_{j,j \neq i}\left\{\frac{S_i + S_j}{d_{ij}}\right\}. \tag{7}$$

The value of *DB* index should be minimized to achieve proper clustering. A smaller value of *DB* index indicates that the chosen miRNA markers are non-redundant. Therefore, the objective is to minimize the value of *DB* index for obtaining suitable clustering.

Finally, less number of tight clusters is another objective. Hence, the number of cluster centers (number of 1 bits in a chromosome) is minimized as well. Therefore, the last objective value for a chromosome *ch* is defined as

$$NCC = \sum_{i=1}^{m} ch(i), \tag{8}$$

where $ch(i) = \{0, 1\}$ and m is the length of the chromosome.

2.3 Reproduction Using Selection, Crossover and Mutation

For generating new offspring solution from parent population, first, selection of best chromosomes in terms of fitness value is considered. But to maintain a bit variety in the populations, less suitable chromosomes should also be included. Here, for selection mechanism we have employed crowded binary tournament selection [5]. Therefore, as output, a set of chromosomes is selected for crossover.

The process of selecting two parent chromosomes from the current population based on the adopted selection mechanism and producing new offspring from those parents is called crossover. A single-point crossover has been used in which a crossover point is chosen randomly from both parent chromosomes, and at this point, two parent chromosomes are interchanged to produce two new offspring chromosomes. The process of alternating some characteristics of each chromosome in random manner to generate better solutions is called mutation. To get some more diversity in the offspring chromosomes, bit-flip mutation has been employed. Here, mutation is done by just replacing the randomly selected cell value with alternating value, i.e., 0 turns to 1 and vice versa depending on mutation probability.

2.4 Maintaining an Archive of Non-dominated Solutions

The non-dominated population is stored in an archive. Initially, the archive stores the non-dominated solutions from the initial population. The next-generation population is added to it. But the archive size should be fixed; hence, again non-domination sorting and crowded distance measure [5] are applied on this combined population. If still the size of the archive is larger than user-defined size (taken as 15 here), then it is reduced to the given length by eliminating chromosomes from the end (based on crowding distance). Non-dominated sorting and crowded distance measure are applied to improve the adaptive fit of the candidate solutions to get better diversity of the Pareto optimal front, respectively.

2.5 Selecting the Final Solution

The proposed feature selection technique yields a set of non-dominated solutions in the final generation each of which encodes a possible set of miRNA markers. As the relevance (represented by fscore) and non-redundancy (represented by DB index) are the two most important aspects of the set of miRNA markers, the solution with the minimum product of fscore and DB index is chosen as the final solution.

3 Adopted Comparative Methods

Over the years, many feature selection techniques have been developed. Here some of them from the existing literature are considered for comparison purpose. These are minimum redundancy maximum relevance (mRMR) scheme, statistical significance tests [1] like t-test and Ranksum test, graph-based feature selection, cluster-based feature selection, and Fisher score.

The statistical tests like t-test [13], Ranksum test [23], and Fisher score are first executed on the preprocessed datasets. The p values of the features (miRNAs) are sorted in ascending order, and the required numbers of features are taken for validation. The number of resultant features of our proposed approach is the input of mRMR schemes, t-test, Ranksum test, and Fisher score. In mRMR feature selection method [6, 10], the relevance of a miRNA is computed by mutual information [2] between the miRNA and its associated class labels, whereas redundancy is obtained as the mutual information among the selected miRNAs. The basic idea of mRMR is to identify the miRNAs which are relevant and mutually maximally dissimilar to each other simultaneously. Let S be the set of output miRNAs. The average minimum redundancy is given as per Eq. 9:

$$\text{Minimum } W = \frac{1}{|S|^2} \sum_{i,j \in S} I(i,j), \tag{9}$$

where $I(i,j)$ represents the mutual information between the ith miRNA and the jth miRNA and $|S|$ is the number of miRNAs in S. The power of differentiability of a miRNA is given by the mutual information $I(h, gi)$ which is computed as per Eq. 10. Thus, the maximum relevance condition is to maximize the average relevance of all miRNAs in S.

$$\text{Maximum } V = \frac{1}{|S|} \sum_{i \in S} I(h, i). \tag{10}$$

Therefore, the redundancy and relevance of a miRNA are to be minimized and maximized, respectively. Since both the conditions are equally important, two simplest combined criteria can be $Max(V - W)$ and $Max(V/W)$. Here we describe the mRMR for discrete variable using mutual information quotient (*mRMRmiq*) only. The mRMR with MIQ scheme is formulated as shown in Eq. 11.

$$mRMR(miq) = \max_{i \in \Omega_S} \left\{ \frac{I(i,h)}{\frac{1}{|S|} \sum_{j \in S} I(i,j)} \right\}. \tag{11}$$

Here Ω is the set of all features and $\Omega_S = \Omega - S$.

The state-of-the-art methods like a graph-based feature selection method [26] and cluster-based feature selection method [3] are also used for comparative analysis. The graph-based method uses a dominant-set clustering technique to cluster the feature vectors and find the optimal feature subset from each dominant set employing the multidimensional interaction information (MII) criterion. The cluster-based feature selection technique [3] employs partitioning for clustering of similar features

4 Datasets and Preprocessing

For evaluation of the proposed method, a publicly available real-life miRNA expression dataset has been collected and preprocessed. This data can be obtained from http://www.broad.mit.edu/cancer/pub/miGCM. The actual dataset has expression values of 217 mammalian miRNAs in different cancer tissue types. In the preprocessing step, at first we have extracted four datasets by separating the samples correspond to uterus, colon, kidney, and prostate cancers, respectively. Now each dataset is formed with 217 miRNAs and the number of samples belonging to different cancer types as described in Table 1.

Subsequently, these datasets are normalized to set the mean and standard deviation of each miRNA vector to 0 and 1, respectively.

Table 1 Number of normal and tumor samples in different miRNA datasets

Cancer types	Number of normal samples	Number of tumor samples
Uterus	9	10
Colon	5	11
Kidney	3	5
Prostate	8	6

These two-class datasets (whose columns are miRNAs and rows are samples) are further processed by calculating signal to noise ratio (SNR) [9] for each miRNA (column). The mean and standard deviation (SD) of samples belonging to class 1 and class 2 are computed first. The $|SNR|$ of each miRNA is defined as in Eq. 12.

$$|SNR| = \left| \frac{mean(class1) - mean(class2)}{SD(class1) + SD(class2)} \right| \tag{12}$$

After sorting in the decreasing order of obtained $|SNR|$, top 100 miRNAs are considered for the final application.

5 Evaluation Metrics

Here, as evaluation criteria the sensitivity, specificity, accuracy, precision, fscore, and area under ROC curve (AUC) have been computed. Linear SVM-based classifier has been considered for fivefold cross-validation and as a result false positives (*fp*), true negatives (*tn*), false negatives (*fn*), and true positives (*tp*) are obtained. Then using these four terms, sensitivity (Eq. 13), specificity (Eq. 14), and accuracy (Eq. 15) are determined as per the following equations.

$$sensitivity = \frac{tp}{tp + fn}, \tag{13}$$

$$specificity = \frac{tn}{tn + fp}, \tag{14}$$

$$accuracy = \frac{tp + tn}{tp + tn + fp + fn}, \tag{15}$$

The precision and fscore have already been defined in Eqs. 2 and 1, respectively. A receiver operating characteristic (ROC) curve shows the plot of sensitivity against

(1-specificity) for a binary classifier system, while its discrimination threshold between two classes is varied. The area under the ROC curve (AUC) is an approach to convert the performance of a classifier with respect to ROC curve into a single value signifying the expected performance.

6 Results and Discussion

To compare the proposed scheme with other comparative methods, first the proposed scheme is executed, then, using the output miRNAs, fivefold cross-validation has been done by linear support vector machine [24]. Thus, sensitivity, specificity, accuracy, precision, fscore, and AUC have been calculated. The number of output miRNAs is the input for the comparative algorithms which need the number of features as input such as mRMR schemes, t-test, Ranksum test, and Fisher score. Also, the outcome of these comparative methods is tested by fivefold cross-validation.

From Table 2, it is evident that for uterus dataset; although sensitivity is less than mRMR schemes, for other comparative methods the proposed method scores better. Except Fisher score, the specificity, and AUC of the proposed method are better than other methods. In case of other performance metrics, the performance of the proposed method is best among all the methods. In case of colon dataset, except specificity and fscore, the score of the proposed scheme for all the other performance metrics is better or same with other comparative strategies. Even the differences with Fisher score and mRMR for specificity and fscore are too small. When the kidney dataset is considered, it is clear that the proposed method outperforms other techniques although in some cases the outcome of the proposed method is similar to other schemes. It can also be noticed from the prostate dataset, sensitivity, accuracy, fscore, and AUC for the proposed algorithm are 1.0000, 1.0000, 1.0000, and 0.9837, respectively, and these scores are the best among all the schemes. But for specificity and precision, it is poorer slightly. Overall, it can be concluded that the proposed schemes undoubtedly the best performer among all.

The expression values of miRNA markers are presented through heatmaps in Fig. 1. The blocks of the heatmap are described as the expression levels of a miRNA in terms of colors. The shades of red represent high-expression levels, the shades of green represent low-expression levels, and the colors toward black represent the middle-level expression values. It is evident from Fig. 1 that the miRNA markers for each tumor subtype have either high-expression values (up-regulated) or low-expression values (down-regulated) mostly across all the samples. The regulation information of the miRNA markers is given in Table 3. Also, the table describes the probe-ID, miRNA, number of target mRNAs. Here we have used mirbase database for finding the number of target mRNAs of each miRNA.

Table 2 Performance analysis for the four real-life miRNA datasets

Dataset	Algorithm	Sensitivity	Specificity	Accuracy	Precision	Fscore	AUC
Uterus dataset	Proposed	0.9000	0.8889	0.8947	0.9000	0.9000	0.9778
	mRMR(miq)	1.0000	0.7629	0.8781	0.8800	0.8703	0.9591
	t-test	0.8000	0.7778	0.7895	0.8000	0.8000	0.9444
	Ranksum test	0.9145	0.7981	0.8373	0.8800	0.8575	0.9474
	Cluster-based	0.8170	0.8865	0.8730	0.8170	0.8634	0.9701
	Graph-based	0.8241	0.8537	0.7834	0.8378	0.8106	0.9441
	Fisher Score	0.8234	0.9001	0.8947	0.8612	0.8231	0.9811
Colon dataset	Proposed	1.0000	0.8000	0.9333	0.9091	0.9524	0.9000
	mRMR(miq)	1.0000	0.7429	0.8954	0.9091	0.9723	0.9000
	t-test	0.9809	0.7791	0.7157	0.3080	0.8448	0.8102
	Ranksum test	0.9000	0.7981	0.8667	0.7428	0.9000	0.8600
	Cluster-based	0.8130	0.7865	0.8321	0.8001	0.8444	0.8539
	Graph-based	0.8410	0.7571	0.7831	0.8399	0.7881	0.8172
	Fisher Score	1.0000	0.8104	0.8921	0.8879	0.9343	0.8880
Kidney dataset	Proposed	1.0000	0.9887	1.0000	0.9890	1.0000	0.9837
	mRMR(miq)	0.9792	0.9887	0.7477	0.7633	0.9800	0.9565
	t-test	0.9709	0.8979	0.7157	0.7308	0.9829	0.9524
	Ranksum test	0.9667	0.8544	0.7273	0.7428	0.8943	0.9443
	Cluster-based	0.9312	0.9091	0.7023	0.7110	0.8102	0.8999
	Graph-based	0.8755	0.9355	0.8788	0.9232	0.9266	0.9482
	Fisher Score	0.9399	0.9292	0.9286	0.9583	0.9697	0.9468
Prostate dataset	Proposed	1.0000	0.9887	1.0000	0.9890	1.0000	0.9837
	mRMR(miq)	0.8333	1.0000	0.8713	0.9967	1.0000	0.9792
	t-test	0.8571	0.9779	0.9396	0.9331	0.8667	0.9691
	Ranksum test	0.8904	0.9792	0.9518	0.9649	0.9438	0.9444
	Cluster-based	0.7833	0.8215	0.6762	0.7915	0.8425	0.7898
	Graph-based	0.7917	0.8139	0.6974	0.7827	0.8333	0.8102
	Fisher Score	0.9006	0.9118	0.8924	0.9228	0.9172	0.9417

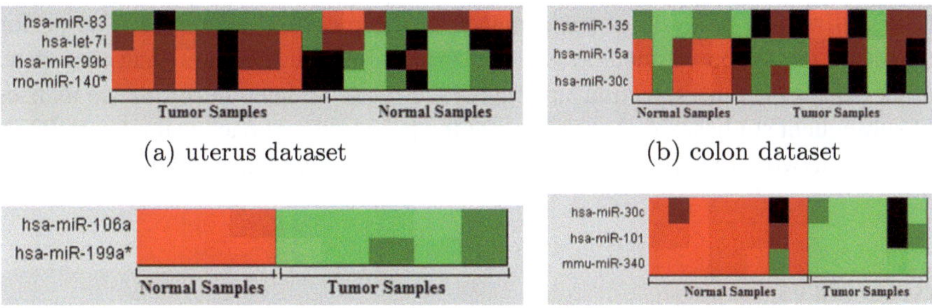

(a) uterus dataset (b) colon dataset

(c) kidney dataset (d) prostate dataset

Fig. 1 miRNA markers for different datasets identified by the proposed method

Table 3 Identified miRNA markers by the proposed method

Dataset	Probe-id	miRNA	Number of target mRNAs	Regulation
Uterus	EAM380	$rno-miR-140*$	405	Down
	EAM339	$hsa-miR-99b$	9	Down
	EAM183	$hsa-let-7i$	4340	Up
	EAM230	$hsa-miR-183$	3282	Down
Colon	EAM271	$hsa-miR-30c$	670	Down
	EAM222	$hsa-miR-15a$	5150	Down
	EAM203	$hsa-miR-135$	2654	Up
Kidney	EAM303	$hsa-miR-199a*$	1450	Down
	EAM186	$hsa-miR-106a$	5152	Down
Prostate	EAM369	$mmu-miR-340$	53	Down
	EAM311	$hsa-miR-101$	2576	Down
	EAM271	$hsa-miR-30c$	670	Down

7 Conclusion

The gene regulatory information is very vital and the biomolecule which regulates or dis-regulates those genes are also important in bioinformatics domain. This article gives an insight into the differential patterns contained by the miRNA which basically obstruct the process of translation. These miRNA may participate in cancer growth and progression by preventing important genes from being translated into proteins. Therefore, to recognize the marker miRNAs, a multiobjective GA-based feature selection technique has been proposed here. Three objective functions representing relevance, non-redundancy, and size of the set of miRNA markers are optimized simultaneously. Four real-life miRNA microarray expression datasets have been utilized for analysis of performance of the proposed technique. The performance of the proposed method is found to be significantly better than that of the other existing feature selection strategies in terms of adopted performance metrics. In subsequent studies, we plan to analyze the proposed approach using the miRNA-gene-TF network to explore the complete view of the regulatory information.

References

1. Bandyopadhyay, S., Mallik, S., Mukhopadhyay, A.: A survey and comparative study of statistical tests for identifying differential expression from microarray data. IEEE/ACM Trans. Comput. Biol. Bioinform. **11**(1), 95–115 (2014)
2. Cover, T., Thomas, J.: Entropy, Relative Entropy and Mutual Information. Elements of Information Theory, Wiley (2006)
3. Covoes, T.F., Hruschka, E.R., de Castro, L.N., Santos, A.M.: A cluster-based feature selection approach. In: International Conference on Hybrid Artificial Intelligence Systems, pp. 169–176 (2009)
4. Davies, D.L., Bouldin, D.W.: A cluster separation measure. IEEE Trans. Pattern Anal. Mach. Intel. **1**(2), 224–227 (1979)
5. Deb, K., Pratap, A., Agrawal, S., Meyarivan, T.: A fast and elitist multiobjective genetic algorithm: NSGA-II. In: IEEE Transactions on Evolutionary Computation, pp. 182–197 (2002)
6. Ding, C., Peng, H.: Minimum redundancy feature selection for microarray gene expression data. J. Bioinform. Comput. Biol. **3**(2), 185–205 (2005)
7. Gasch, A.P., Eisen, M.B.: Exploring the conditional coregulation of yeast gene expression through fuzzy k-means clustering. In: Genome Biol. **3**(11), 0059.1–0059.22 (2002)
8. Goldberg, D.E.: Genetic Algorithms in Search. Optimization and Machine Learning. Addison-Wesley, New York (1989)
9. Golub, T.R., Slonim, D.K., Tamayo, P., Huard, C., Gassenbeek, M., Mesirov, J.P., Coller, H., Loh, M.L., Downing, J.R., Caligiuri, M.A., Bloomeld, D.D., Lander, E.S.: Molecular classification of cancer: class discovery and class prediction by gene expression monitoring. Science **286**, 531–537 (1999)
10. Kamandar, M., Ghassemian, H.: Maximum relevance, minimum redundancy band selection for hyperspectral images. In: 19th Iranian Conference on Electrical Engineering (ICEE) (2011)
11. Lu, J., Getz, G., Miska, E.A., Alvarez-Saavedra, E., Lamb, J., Peck, D., Sweet-Cordero, A., Ebert, B.L., Mak, R.H., Ferrando, A.A., Downing, J.R., Jacks, T., Horvitz, H.R., Golub, T.R.: MicroRNA expression profiles classify human cancers. Nature **435**(7043), 834–838 (2005)
12. Mandal, M., Mukhopadhyay, A.: A graph-theoretic approach for identifying non-redundant and relevant gene markers from microarray data using multiobjective binary PSO. Plos One **9**(3), e90949 (2014)
13. Mankiewicz, R.: The Story of Mathematics. Princeton University Press (2000)
14. Maulik, U., Bandyopadhyay, S., Mukhopadhyay, A.: Multiobjective Genetic Algorithms for Clustering–Applications in Data Mining and Bioinformatics. Springer, ISBN 978-3-642-16615-0 (2011)
15. Mukhopadhyay, A., Bandyopadhyay, S., Maulik, U.: Multi-class clustering of cancer subtypes through SVM based ensemble of paretooptimal solutions for gene marker identification. PLoS One **5**(11), e13803 (2010)
16. A. Mukhopadhyay and M. Mandal. Identifying non-redundant gene markers from microarray data: a multiobjective variable length PSO-based approach. IEEE/ACM Trans. Comput. Biol. Bioinform. **pp**(99) (2014)
17. Mukhopadhyay, A., Maulik, U.: An SVM-wrapped multiobjective evolutionary feature selection approach for identifying cancer-microRNA markers. IEEE Trans. NanoBioSci. **12**(4), 275–281 (2013)

18. Mukhopadhyay, A., Maulik, U., Bandyopadhyay, S.: An interactive approach to multiobjective clustering of gene expression patterns. IEEE Trans. Biomed. Eng. **60**(1), 35–41 (2013)
19. Mukhopadhyay, A., Maulik, U., Bandyopadhyay, S.: A survey of multiobjective evolutionary clustering. ACM Comput. Surv. (CSUR) **47**(4), 61:1–61:46 (2015)
20. Ruiza, R., Riquelmea, J.C., Aguilar-Ruizb, J.S.: Incremental wrapper-based gene selection from microarray data for cancer classification. Pattern Recognit. **39**(12), 2383–2392 (2010)
21. Sun, J.-G., Liao, R.-X., Qiu, J., Jin, J.-Y., Wang, X.-X., Duan, Y.-Z., Chen, F.-L., Hao, P., Xie, Q.-C., Wang, Z.-X., Li, D.-Z., Chen, Z.-T., Zhang, S.-X.: Microarray-based analysis of microRNA expression in breast cancer stem cells. J. Exp. Clin. Cancer Res. **29**(174) (2010)
22. Thomson, J.M., Parker, J., Perou, C.M., Hammond, S.M.: A custom microarray platform for analysis of microRNA gene expression. Nat. Methods **1**(1), 47–53 (2004)
23. Troyanskaya, O., Garber, M., Brown, P., Botstein, D., Altman, R.: Nonparametric methods for identifying differentially expressed genes in microarray data. Bioinformatics **18**, 1454–1461 (2002)
24. Vapnik, V.: Statistical Learning Theory. Wiley, New York, USA (1998)
25. Wu, D., Hu, Y., Tong, S., Williams, B.R., Smyth, G.K., Gantier, M.: The use of mirna microarrays for the analysis of cancer samples with global mirna decrease. RNA **19**(7), 876–888 (2013)
26. Zhang, Z., Hancock, E.R.: A graph-based approach to feature selection. In: International Workshop on Graph-Based Representations, Pattern Recognition, pp. 205–214 (2011)

Ageing and Priority-Based Scheduling for Uplink in WiMAX Networks

Deepa Naik, Himansu Rathi, Asish Dhara and Tanmay De

Abstract WiMAX is an emerging technology for next-generation wireless networks which supports a large number of users in an economic way. To achieve quality of service (QoS) requirements, an efficient and reliable scheduling algorithm is needed. The existing approaches in the literature have been proven to provide the best performance in allocating bandwidth to subscriber stations to maximize the throughput and ensure the constraints of delay in each class of traffic. In these approaches, starvation of lower priority class was not considered, which is of great significance in reality. In this paper, we have considered the starvation of lower priority classes to provide QoS requirement to each class of traffic in an acceptable way, and a bandwidth scheduling algorithm is proposed. A comparative study between the proposed scheduling algorithm and the existing scheduling algorithm shows the better performance in terms of maximizing the network throughput in given network, while minimizing the starvation of lower priority classes.

Keywords WiMAX (world interoperability for microwave access network) QoS (quality of service) · UGS (unsolicited grant service) · RTPS (real-time polling service) · nRTPS (non-real-time polling service) · BE (best effort)

D. Naik (✉) · H. Rathi · A. Dhara · T. De
Department of Computer Science and Engineering,
National Institute of Technology, Durgapur, India
e-mail: naiksavantdeepa@gmail.com

H. Rathi
e-mail: himansurathi@gmail.com

A. Dhara
e-mail: ashishere15@gmail.com

T. De
e-mail: tanmayd12@gmail.com

S. Kar et al. (eds.), *Operations Research and Optimization*, Springer Proceedings in Mathematics & Statistics 225, https://doi.org/10.1007/978-981-10-7814-9_13

1 Introduction

Fixed WiMAX based on IEEE 802.16 standard is cost effective for fixed wireless networks. IEEE 802.16e is the new version to the fixed WiMAX mobility. WiMAX provides high data rate mobile wireless services for metropolitan areas. WiMAX coverage range is up to thirty-mile radius and data rates between 1.5 to 75 Mbps theoretically. The WiMAX defines the two modes of operation namely mesh and point-to-multipoint (PMP) mode. In mesh mode, each subscriber station (SS) can communicate to each other and to the base station (BS). In point-to-multipoint mode, subscriber stations can communicate only through base station. The base station is responsible for providing the QoS to each of its subscriber stations.

The scheduling algorithm must be simple to implement because real-time application supported by subscriber station needs the quick response from the centralized base station. So the time complexity of scheduling algorithm must be simple. Lots of research on scheduling algorithm has been investigated in [1, 2]. The issues in allocating resource are more challenging. In our proposed scheduling algorithm, base station uses the ageing and priority-based scheduling (APBS) to schedule the traffic of different classes. The downlink scheduler at the base station schedules the entire bandwidth among the subscriber stations depending on the grant per connection type.

The rest of the paper is organized as follows: Sect. 2 presents the previous works. In Sect. 3, problem is defined in formal notation. The proposed approach is presented in Sect. 4. The complexity analysis and results are described in Sect. 5 and Sect. 6, respectively. Finally, conclusion and future work are drawn in Sect. 7.

2 Previous Works

Our study focuses on centralized scheduling in point-to-multipoint (PMP) mode. In the literature, many scheduling algorithms have been proposed for downlink scheduling. There are different types of scheduling algorithm, namely traditional, dynamic scheduling. In [3] traditional scheduling algorithms uses the same Schelling technique as in computer operating system. It shares the equal network resources to all the subscribers without concern about the priority. However, this technique is not suitable for WiMAX scheduling, where the traffic demand of varying size arrives in random fashion. The authors Sagar et al. in [4] proposed the weight round robin (WRR) for scheduling the bandwidth to each queue. But it is not suitable for WiMAX scheduling because the main drawback of WRR is that when the traffic has a variable packet size, WRR provides an incorrect percentage of bandwidth allocation.

The dynamic schedulers are adaptive in bandwidth allocation. The authors Fathi et al. [5] proposed joint scheduling and call admission control (CAC) technique for scheduling packets. In [6] proposed the joint routing and scheduling algorithm for

WiMAX network with the provision for fairness in bandwidth allocation and increase in throughput.

Best of our knowledge, no work has been conducted to overcome the starvation of lower ordered traffic request, while maintaining the quality of service to each class of traffic.

3 Problem Formulation

The WiMAX network consists of base station (BS), subscriber station (ss) and users. The base station is connected to subscriber station in point-to-point mode. The user under the coverage of base station is directly connected through wireless links. The user not under the coverage area of base station is connected through subscriber stations. Given a network topology, which is a directed graph G(V,E), where V is a set of subscriber stations/user and E is the set of bidirectional wireless link between the base station and subscriber station/user at the particular instant of time. 'C' is the capacity of each link between the base station to subscriber station/users. The base station is responsible for allocating network resource to subscriber stations. The connection requests from all class of traffic are represented by notation(S, B.requested, $Class_{tupe}$, B, $stat_{timeofrequest}$, $request.duration$), where S, $B.requested$ and $star_{timeofrequest}$, $request.duration$ represent the source node, bandwidth demanded and arrival time of each request and duration of the request, respectively. The centralized uplink bandwidth scheduling with the prime objective is to maximize the network throughput and meanwhile eliminate the starvation due to the higher class traffic is considered.

Assumptions:
We use the following assumptions in this work.

- Traffic demands is static in nature. i.e. traffic demands are known in advance.
- The total number of traffic is $R_i = N$.
- w_{Mn} is 1 if the relay station M associated to base station n, $M \in ss$ and $j \in BS$ 0 otherwise.
- R_i, where M_i^{ss} is the number of subscriber station(M)/user including the traffic class i in the network.
- $B_{k,i}$ bandwidth request of kth subscriber/user with traffic class belong to i th class of traffic.
- β_n is 1 if the BS is installed at site n, $n \varepsilon BS$ 0 otherwise.
- γM is 1 if the ss is installed at site M, $M \varepsilon ss$ 0 otherwise.
- We have reserved 60% of bandwidth for UGS traffic in terms of bytes per seconds (Bps), and 20% is reserved for RTPS, and 10% is reserved for nRTPS and BE. However these are not rigid limits and are flexible if requests of another class are missing or are lesser.
- α value is assumed as 0.5.

Objective:

The objective here is to maximize the network throughput, while meeting the quality of service to each class of traffic requirement of WiMAX standard. The objective function to maximize the network throughput is defined as follows:

$$R_i = \sum_{k=1}^{M_i^{ss}} B_{k,i} \tag{1}$$

$$Maximize \quad \sum_{i}^{N} R_i * p \tag{2}$$

$$p = \begin{cases} 1, & \text{if } R_i \text{ is established.} \\ 0, & \text{otherwise.} \end{cases}$$

Subjected to:

Base station to relay station constraints

$$\sum_{\forall n \varepsilon BS} w_{Mn} = \gamma_M \quad where \quad \forall M \varepsilon ss \tag{3}$$

$$w_{Mn} \leq \beta_n \quad where \quad \forall M \varepsilon ss \quad n \varepsilon BS \tag{4}$$

Equation 3 ensures the constraint that each relay station is connected to only one base station. Equation 4 assures the base station to relay station connection. The number of base station is less than or equal to the relay station installed.

$$\sum_{i=1}^{n} T_{ss_i} \leq T \tag{5}$$

$$\sum_{i=1}^{n} T_{ss_i} \times \beta \leq T \tag{6}$$

$$\beta = \begin{cases} 1, & \text{if } ss_i \text{ used time slot } T_{ss_i} \text{ and associated with base station BS.} \\ 0, & \text{otherwise.} \end{cases}$$

ss_i refers to the number of subscriber station attached. *BS* refers to the number of base station. T_{ss_i} refers to the time slot allocated to each subscriber station. Equation 5 indicates the number of time slot allocated to each subscriber station should be less than or equal to the total number of time slot(T) available with base station *BS*.

4 Proposed Approach

In this section, the proposed algorithm (APBS) is presented with the objective to maximize the throughput of given network. Our algorithm is categorized in two stages. The first stage of our algorithm is to assign the base station to subscriber stations/nodes under the coverage area. Base station executes the uplink scheduler at every frame interval of time (t) and sends the grant requests to subscriber station in uplink map message. In the second stage, each subscriber station associated with four different class of traffic, namely UGS, RTPS, nRTPS and BE are checked for the expire time and stored in the list L_{req}. The base station maintains four types of priority queues. The request is stored in respective priority queues (UGS, RTPS, nRTPS, BE). The scheduler function is called to schedule the requests of each base stations. This type of scheduling is known as grant per connection (GPC). The algorithm, namely Ageing and Priority-Based Scheduling (ABPS), is depicted in Algorithm 1, to schedule the traffic of different classes. The downlink scheduler at the base station schedules the entire bandwidth among the subscriber stations. Scheduling technique considers the two factors, while serving the request. The scheduler has the age and deadline associated with each class of traffic. The scheduler calculates the weight associated with each request in each queue and stores it in the priority queue of respective class. The scheduler schedules each request before the deadline. The early deadline requests are served first. When alpha is 0, then the priority of all the requests is same and only the time in which they arrive is the only measure that we use to sort the requests; hence this turns out to be FCFS scheduling. And for any other value of alpha, we have taken both age and deadline of the requests into consideration, so it turns out to be APBS algorithm.

$$Age = current_{timeofsystem} - start_{timeoftherequest} \tag{7}$$

$$Deadline = End_{timeofrequest} - current_{timeofsystem} \tag{8}$$

$$Weight = \frac{\alpha \times (age + 1)}{(deadline + 1)} \tag{9}$$

α can vary between 0 and 1.

First, the UGS traffic is admitted in priority queue and calculate the weight associated with each request. The UGS connection is served depending on the bandwidth requirement. The request is served in queue depending on the deadline. The request with early deadline is served first. In this case, $\sum_{i=1}^{n} B_{UGS}$ are the total request with early deadline in UGS class, are served first. The remaining request of UGS class (in priority queue) are served in the next slot of frame allocation time. Thereafter, the RTPS connections are served by ordering them according to queue size information. In this case, $\sum_{i=1}^{n} B_{RTPS}$ are the total request with early deadline in RTPS class. The request which are waiting for a long time in a priority queue of RTPS are served according to their weight factor. The remaining request of RTPS class whose dead line is remaining are served in the next time slot of frame allocation. In our algorithm, we have fixed the percentage of bandwidth allocated to each class of traffic. If there is no higher priority traffic flows, the reserved bandwidth can be utilized by the lower class traffic including the intermediate queue.

4.1 Example

We have considered an WiMAX network connected to subscriber and mobile nodes shown in Fig. 2 to illustrate the working principle of APBS algorithm. There are just one base station and two subscriber stations that are connected to the base station. There are four nodes/end users, namely N0, N1, N2 and N3. The nodes N0 and N1 are connected directly to the base station. The N2 and N3 are connected via the subscriber stations. We have currently four requests for each class with bandwidth demand in kilo bytes. The traffic pattern are shown in Table 1 (and are graphically depicted in Fig. 1).

We use the following example to explain our algorithm. We have set the frame size as 100 KBps (kilo bytes per second) and frame duration as 0.5 s.

This frame allocation in Table 2 shows very simple distribution of the frame size. An important thing to watch here is that the UGS class is allocated just 60% of the total frame size despite being the highest priority class. Hence, other classes also get their requests served.

Fig. 1 Scheduler for uplink at base station

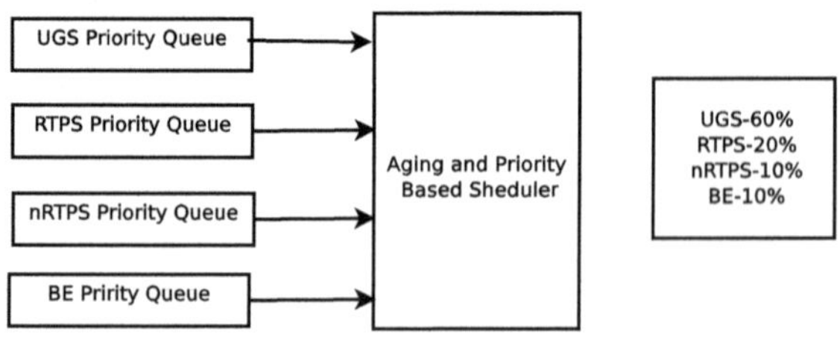

Fig. 2 The WiMAX network structure

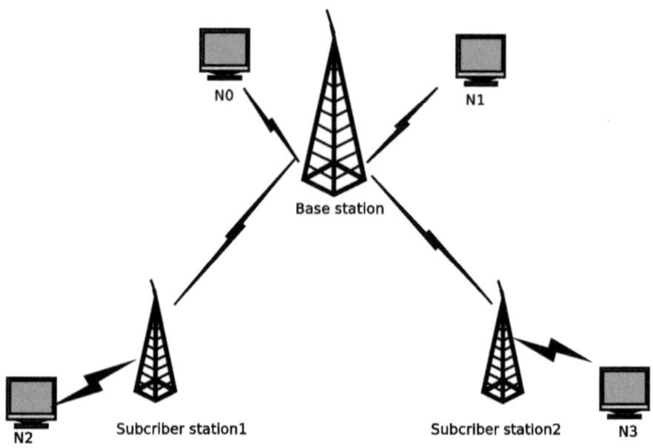

Table 1 Bandwidth demanded from different class of traffic

Request	Node	$Class_{type}$	$B.requested(KB)$	S.time(s)	Duration(s)
R_0	0	UGS	120	0	1
R_1	1	RTPS	50	0.5	2
R_2	2	RTPS	36	0	1
R_3	3	nRTPS	27	1	2
R_0	0	nRTPS	10	0.5	1
R_1	1	BE	3	0	1
R_2	2	BE	8	1	2
R_3	3	BE	5	0.5	1

Table 2 After running the scheduling algorithm, time t = 0s

Request	$Class_{type}$	$B.requested(KB)$	Age	Deadline	Weight	B.allocated
R_0	UGS	120	0	1	0.25	60
R_2	RTPS	36	0	1	0.25	20
R_5	BE	3	0	1	0.25	3

Table 3 Frame allocation for time t = 0.5s

Request	$Class_{type}$	$B.requested$(KB)	Age	Deadline	Weight	B.allocated
R_0	UGS	60	0.5	0.5	0.5	60
R_2	RTPS	16	0.5	0.5	0.5	16
R_1	RTPS	50	0	2	0.33	4
R_4	nRTPS	10	0	1	0.25	10
R_4	BE	5	0	1	0.2	5

Table 4 Frame allocation for time t = 1s

R	$Class_{type}$	$B.requested$(KB)	Age	deadline	Weight	B.allocated
R_1	RTPS	46	0.5	1.5	0.3	46
R_3	nRTPS	27	0	2	0.167	27
R_6	BE	8	0	2	0.1672	8

This frame shows the competitive frame allocation among requests of same class. Here, we can see that there are two requests from RTPS class. However, the one with greater weight value gets served first. The frame allocation at $t = 0.5$ as shown in Table 3. There is also competition for frame allocation between traffic requests from same class. Here, we can see that there are two requests from RTPS class. However, the one with greater weight value gets served first.

This frame allocation in Table 4 shows the reallocation property among the classes in terms of frame size for each class. We can observe that 46 KBps of RTPS request is satisfied despite the limit for RTPS request being 20 KBps initially.

It is because of the reallocation of frame size. Hence, RTPS is able serve up to 50 KBps (initially limited to 20 KBps) request in the above frame, and nRTPS is able to serve up to 27 KBps (initially limited to 10 KBps). All requests are served in an efficient manner.

$$RTPSF.size = \left[\frac{RTPS \quad F.size{+}RTPS \quad frame\%}{Remaining \quad classes \quad F.size\%} * F.size \quad leftout \right]$$

Example: RTPS frame size $= 20 + \left[\frac{20}{20+10+10} \times 60 \right] = 50$ KB.

Similarly for nRTPS $= 10 + \left[\frac{10}{20+10+10} \times 60 \right] = 25$ KB.

Also after RTPS leaves out 4 KB:

nRTPS new size $= 25 + \left[\frac{10}{10+10} \times 4 \right] = 27$ KB. RTPS size actually reflects the extra bandwidth available for RTPS class, when no higher class traffic request are made.

Algorithm 1: Ageing and Priority-Based Scheduling (APBS) in WiMAX network

---

Input: P is a instance of of traffic requests. The P consists of $[S, B.requested, Class_{type}, start_{timeofrequest}, request.duation]$ and $BS = [ss_1......ss_n]$ attached
Output: Magazining the network throughput of given network.

step 1: Allot all subscriber stations to Nearest base station.
step 2:for $i = 1$ *to stime* **do**
  step 2.1: for *framestart to frameend* **do**
    step 2.2: $current_{timeofsystem} = i + j$
    step 2.3: if $(request.startTime <= current_{timeofsystem})$ $and(start_{timeofrequest} + request.duration >= current_{timeofsystem})and$ $(B.requested! = B.maxAllocated)$ **then**
      step 2.4: Add the request to list L_{req}
      step 2.5: Check to which node the request belong, assign the node object to the request.
      step 2.5.1: Get the current co-ordinates of the nodes.
      step 2.5.2: Assign nodes to nearest station(whether BS/SS) on the distance and range.
    end
    step 3: for $k = 1$ *to* BS_k **do**
      step 3.1: for *each request of* L_{req} **do**
        step 3.2: Arrange them in queues in the base station.
          `/* where Priority Queue of all classes. */`
      end
      3.3: call scheduling(Priority queues, $current_{timeofsystem}$);
      step 3.4: Proceed with the request in the queue. Add the scheduled request to get final request of all base station
    end
  end
end

Algorithm 2: Scheduling algorithm in WiMAX network

Input: A Priority Queues of all classes $Class_{types}$
Output: Scheduling the bandwidth among the higher to lower priority classes to avoid starvation of lower priority classes.
step 1: for $i = 1$ *to* $|Class_{typen}|$ **do**
    `/* For each class in (UGS,RTPS,nRTPS,BE)*/`
    step 1.1: for req_j *to* $class_i$ **do**
        Calculate weight of request using the formula.
        step 1.2: $Weight_j = (\alpha \times (age + 1))/(deadline + 1)$.
        step 1.3: $age_j = current_{timeofsystem} - req_j.star_{timeofrequest}$
        step 1.4: $deadline_j = req_j.End_{timeofrequest} - req_j.currentstart_{timeoftherequest}$
        step 1.5: Schedule request according to weight in priority queue.
    end
end
step 2: Initialize all size of $class_i$.
step 2.1: while req_j *in priority queue* $class_i$ **do**
    step 2.2: Check allocable request frame size for the class.
    `/* For each class in (UGS,RTPS,nRTPS,BE)*/`
    step 2.3: if $(req.Allocation_i \leq size.class_i)$ **then**
        step 2.3.1: $size.class_i$-=$B.requested_i$
        step 2.3.2: $B.allocation_i$+=$B.requested_i$
        else
            step 2.4: $B.allocation_i$+=$size.class_i$.
            step 2.5: $size.class_i = 0$;
        end
    end
end
step 3: if $size.class_i > 0$ **then**
    step 3.1: Distribute the remaining bandwidth among all classes.
    $RTPSF.size = \left[\frac{RTPSF.size+RTPSframe\%}{RemainingclassesF.size\%} * F.size \quad leftout \right]$
end

5 Complexity Analysis

By algorithmic analysis, the total complexity of the algorithm is $O((no.BS) * ((no.R_N) * (ss_n) + (no.R_N) * log(no.R_N)))$. The algorithm is efficient and is feasible. We also notice that the complexity of the algorithm is eventually independent of the number of subscriber/relay stations. The ABPS algorithm based on multiple queues performs much better than the traditional first come first serve (FCFS) algorithm. We assume that number of requests is going to be much greater than number of base stations and subscriber stations. Hence, complexity reduces to $O(BS * (ss_n * R_N + R_N logR_N))$, where BS is the number of base stations, ss_n is the number of nodes/end users, and R_N is the number of requests generated.

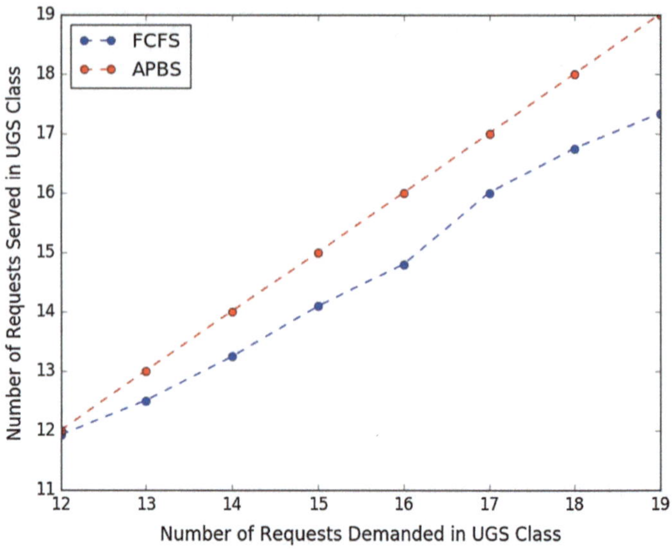

Fig. 3 Relationship of total number of requests demanded versus total number of requests served in UGS class

6 Results

In this section, the simulation results are observed under several scenarios by varying the traffic request of different classes. Total traffic request generated by subscriber stations under categories UGS:RTPS:nRTPS and BE are in ratio of 20:20:60, and requests are generated between the range of 64 KBps–128 KBps, 30–64 KBps and 1 KBps–30 KBps, receptively.

Figures 3, 4, 5 and 6 show the results of APBS with FCFS scheduling algorithm. They demonstrate the relationship between the number of the number of requests served in all four class with the number of requests demanded in all four class of traffic. The proposed algorithm outperforms the existing algorithm FCFS.

Figures 7, 8, 9 and 10 depict the results of APBS with FCFS scheduling algorithm. The proposed algorithm performed better in terms of serving the total bandwidth demanded in all classes compared to the existing algorithm FCFS.

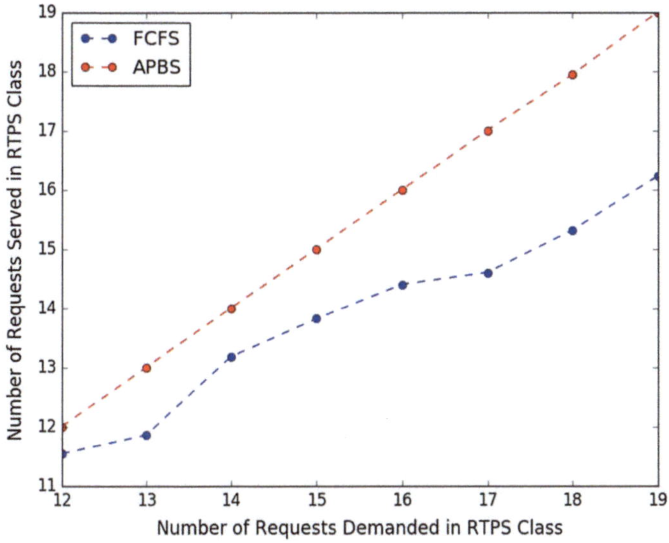

Fig. 4 Relationship of total number of requests demanded versus total number of requests served in RTPS

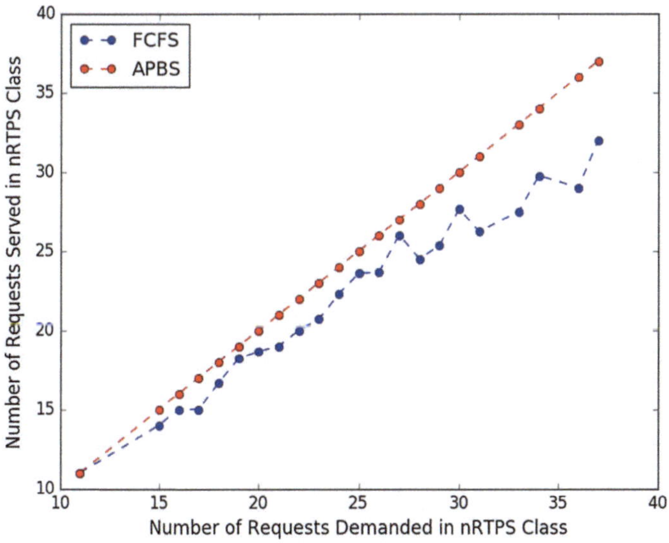

Fig. 5 Relationship of total number of requests demanded versus total number of requests served in nRTPS class

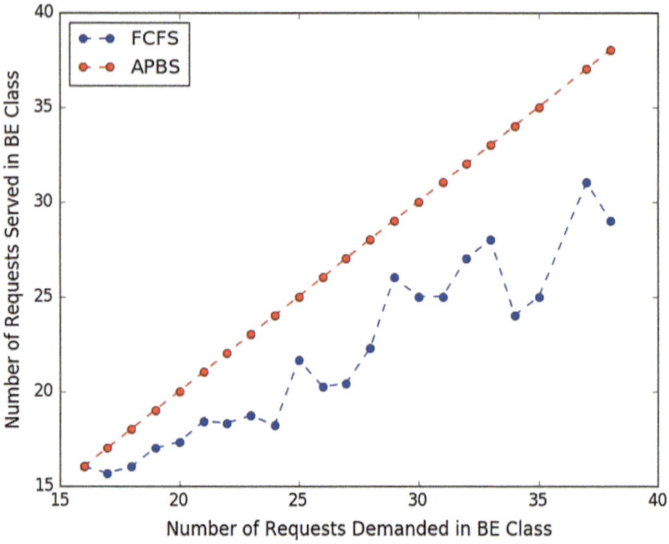

Fig. 6 Relationship of total number of requests demanded versus total number of requests served in BE class

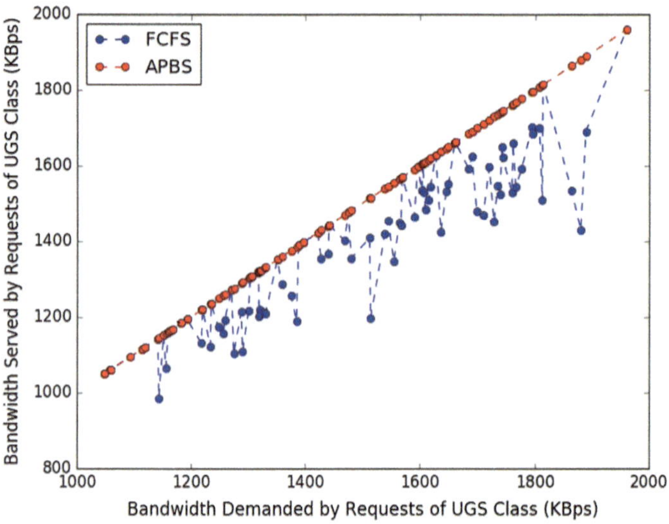

Fig. 7 Relationship of total bandwidth demanded versus total bandwidth served in UGS class

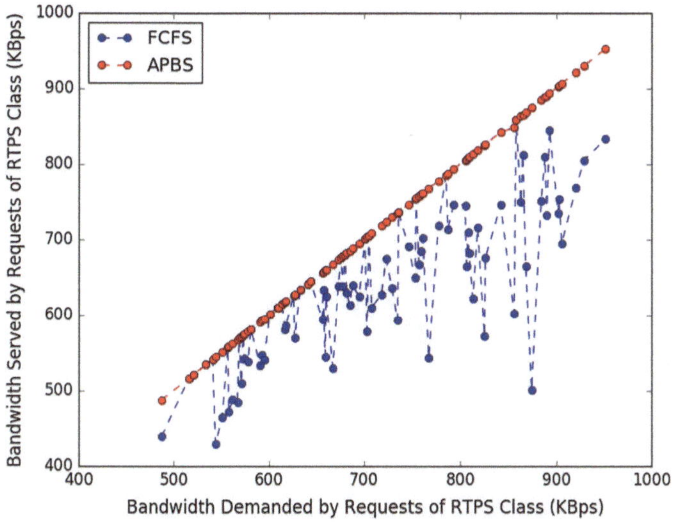

Fig. 8 Relationship of total bandwidth demanded versus total bandwidth served in RTPS class

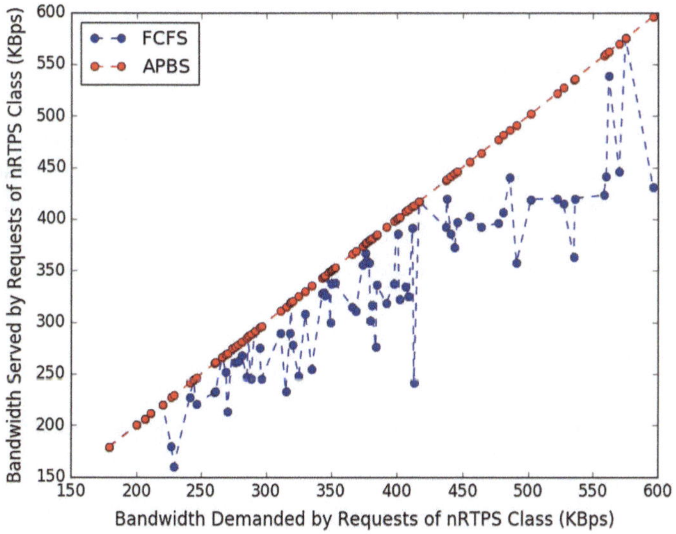

Fig. 9 Relationship of total bandwidth demanded versus total bandwidth served in nRTPS class

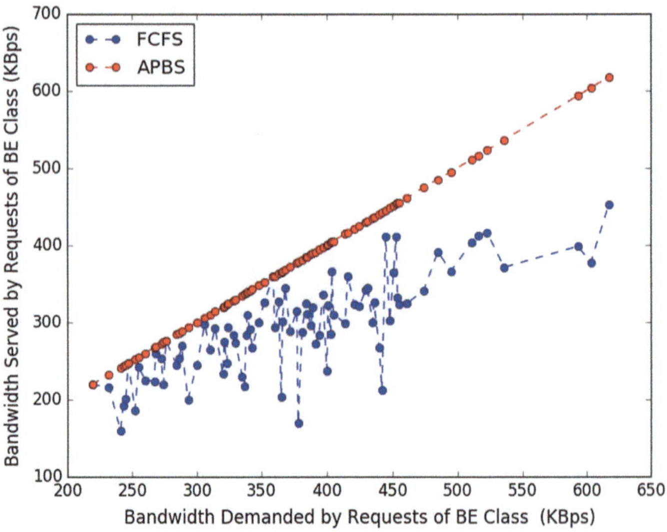

Fig. 10 Relationship of total bandwidth demanded versus total bandwidth served in BE class

7 Conclusion and Future Work

In this work, we have designed algorithm, Ageing and priority-Based Scheduling (APBS) to eliminated the bandwidth starvation due to the higher class monopolizing the lower class traffic. We have also significantly reduced starvation of requests, particularly of the less priority classes (like best effort traffic). The proposed algorithm can be extended to dynamic traffic demands in the hybrid networks in future.

References

1. Chen, J.C.: Efficient batch-arrival queueing model for QoS-aware packet scheduling in WiMAX networks. In: 2010 IET International Conference on Frontier Computing. Theory, Technologies and Applications, pp. 221–226 (2010)
2. Alsahag, A.M., Ali, B.M., Noordin, N.K., Mohamad, H.: Fair uplink bandwidth allocation and latency guarantee for mobile wimax using fuzzy adaptive deficit round robin. J. Netw. Comput. Appl. **39**, 17–25 (2014)
3. Cicconetti, C., Erta, A., Lenzini, L., Mingozzi, E.: Performance evaluation of the IEEE 802.16 MAC for QoS support. IEEE Trans. Mob. Comput. **6**(1), 26–38 (2007)
4. Yadav, A., Vyavahare, P., Bansod, P.: Review of WiMAX scheduling algorithms and their classification. J. Inst. Eng. (India): Ser. B **96**(2), 197–208 (2015)
5. Sagar, V., Das, D.: Modified EDF algorithm and WiMAX architecture to ensure end-to-end delay in multi-hop networks. In: TENCON IEEE Region 10 Conference, pp. 1–6 (2008)
6. Fathi, M., Rashidi-Bajgan, S., Khalilzadeh, A., Taheri, H.: A dynamic joint scheduling and call admission control scheme for IEEE 802.16 networks. Telecommun. Syst. **52**(1), 195–202 (2013)

Recognition System to Separate Text Graphics from Indian Newspaper

Shantanu Jana, Nibaran Das, Ram Sarkar and Mita Nasipuri

Abstract Identification of graphics from newspaper pages and then their separation from text is a challenging task. Very few works have been reported in this field. In general, newspapers are printed in low quality papers which have a tendency to change color with time. This color change generates noise that adds with time to the document. In this work we have chosen several features to distinguish graphics from text as well as tried to reduce the noise. At first minimum bounding box around each object has been identified by connected component analysis of binary image. Each object was cropped thereafter and passed through geometric feature extraction system. Then we have done two different frequency analysis of each object. Thus we have collected both spatial and frequency domain features from objects which are used for training and testing purpose using different classifiers. We have applied the techniques on Indian newspapers written in roman script and got satisfactory results over that.

Keywords Edge detection · Connected component analysis · Text graphics separation · FFT · 2D-DWT · Water marking · News paper segmentation

1 Introduction

Over a decade, a number of works have been done on text graphics separation from document images. But specifically on newspaper document few works were proposed. Those systems convert textual part of scanned newspaper documents in machine editable format for further processing. Such applications are available online but accuracy level is very poor. R. Garg proposed an Expectation Maximization based optimization framework which is an iterative method to find out maximum likelihood estimates of parameters in statistical model [1]. They first

S. Jana (✉) · N. Das · R. Sarkar · M. Nasipuri
CMATER Laboratory, Department of Computer Science and Engineering,
Jadavpur University, Kolkata 700032, India
e-mail: shantanujana.research@gmail.com

S. Kar et al. (eds.), *Operations Research and Optimization*, Springer Proceedings in Mathematics & Statistics 225, https://doi.org/10.1007/978-981-10-7814-9_14

eliminated background by binarizing the document image then applied the algorithm. But binarization of the document may destroy valuable grayscale information relevant to graphics. The most used documents in text graphics separation works are geographical maps. P. P. Roy described a technique to segment text and symbols from colormaps [2]. They used long line identification technique to remove symbols from map. This technique works fine for map documents but not fruitful to recognize graphics from news paper documents. Layout analysis, connected component labeling are noticeable features considered by Mollah et al. [3], Rege [4], Strouthopoulos et al. [5]. A clustering based technique described by Garg is very effective for these works [6]. Cao and Tan [7], Science [8], Vieux et al. [9] used cluster based techniques. Chinnasarn [10] used a modified kFill algorithm to remove noise from the document. This type of noise is basically salt and pepper noise where unwanted isolated black pixels are removed. In many cases of noise removal, mathematical morphology operations described by Haralick, S. E. Poland give best result [11, 12]. Mathematical morphology is an algebraic approach based on set theory. Two basic morphological operations are erosion and dilation. R. Verma has shown various noise removal techniques [13]. He focused on two types of noise, which are produced by photon nature of light and thermal energy of heat inside the image sensor. Noise reduction techniques described in this paper are effective for specific type of noise. Among those noises poisson noise is a kind of similar noise which is frequently identified in newspaper documents. Indian newspapers are printed in low graded paper materials which absorb water from moister and change its color gradually.

This type of noise is different from salt and pepper noise. The best procedure to extract such noise is to use a high pass filter. In our work we have used 2-d wavelet transform for this purpose. Newspaper graphics are combination of different grey shades where characters are all in single grey shade. Therefore a frequency transform can be used to detect this difference. Fast Fourier Transform (FFT) of such image objects has produced satisfactory result. Geometric features are not alone sufficient for this work because newspapers are printed in different fonts and styles.

The system we have developed in this work is highly efficient with an accuracy rate of the current system is 95.78%, achieved using Sequential Minimal Optimization (SMO) classifier on data set which are prepared at CMATER Laboratory. We have collected 27 article images from various Indian newspapers printed in roman script.

2 Connected Component Analysis

There are popular edge detection techniques compared by Kumar and Saxena [14]. Sobel, Canny, and other Gradient based classical edge operators, Laplasian based Marr Hildreth operators are efficient to detect edge in different perspective. In our work we needed an edge detection technique which rejects single discontinuity from connected component chain. We have calculated object's edge by 8- connected

US approves arms worth $1bn for Pak

Package Meant To Support Islamabad's Counterterrorism & Counter-Insurgency Ops

Chidanand.Rajghatta
@timesgroup.com

Washington: The Pentagon has notified US Congress about the state department's approval of a military arms package worth nearly $1 billion for Pakistan, ostensibly in support of its counterterrorism and counter-insurgency operations.

The list of arms and armaments notified for possible sale by the Defence Security Cooperation Agency in its communication to Congress on Monday includes 15 AH-1Z Viper attack helicopters, 1000 AGM-114 R Hellfire II missiles in containers, 32 T-700 GE 401C engines (30 installed and 2 spares) and 17 AN/AAQ-30 Target sight systems.

The package also includes 30 629F-23 ultra high frequency/very high frequency communication systems, 19 H-764 embedded global positioning system/inertial navigation systems, 32 helmet mounted display/optimized top owl, 17 … friend or foe, 17 AN/AAR-47 missile warning systems, 17 AN/ALE-47 countermeasure dispenser sets, 18 AN/APR-39C(V)2 radar warning receivers, 15 joint mission planning systems, and 17 M197 20mm gun systems.

The proposed sale "will contribute to the foreign policy and national security of the US by helping to improve … to US foreign policy and national security goals in South Asia," the DSCA said in its notification, adding that the helicopters and weapon systems will provide Pakistan with military capabilities in support of its counterterrorism and counter-insurgency operations in South Asia.

It also assured the Congress that the proposed sale of this equipment and support will not alter the basic military balance in the region where the US has also emerged as a large arms supplier to India.

Elaborating on the use of the arms package, the notification said the proposed sale will provide Pakistan with a precision strike, enhanced survivability aircraft that it can operate at high-altitudes. By acquiring this capability, it said, Pakistan will enhance its ability to conduct operations in North Waziristan Agency (NWA), the Federally Administered Tribal Areas (FATA), and other remote and mountainous areas in all-weather, day-and-night environments.

The sale, worth $952 million according to Pentagon estimates, is expected to go through without a hitch even though Pakistan lives a parlous, hand-to-mouth existence and US lawmakers and officials periodically vent their frustration about the country's two-timing, double-dealing ways in the war on terrorism and extremism.

Many of the military contractors involved are constituents of lawmakers who prize the creation of jobs and maintaining it in their districts above all else. The principal contractors of the Pakistan package will be Bell Helicopter, Textron in Fort Worth, General Electric in Lynn, The Boeing Company in Huntsville, and Lockheed Martin in Bethesda, Maryland.

ARMED AND DANGEROUS: The list of arms and armaments in the package includes 15 AH-1Z Viper attack helicopters (above), 1000 AGM-114 R Hellfire II missiles in containers, 32 T-700 GE 401C engines (30 installed and 2 spares) and 17 AN/AAQ-30 Target sight systems

'N-armed Pakistan a threat to S Asia'

A leading US daily on Tuesday described nuclear-armed Pakistan as "unquestionably the biggest concern" to stability in South Asia. The New York Times, in its editorial, said the world powers that worked on the Iran deal should now turn their attention to the "troubled region with growing nuclear risks of its own". It said the country has "wasted" billions on sophisticated weapons, that otherwise, should have been spent on its people PTI

Fig. 1 Sample newspaper image written in roman script for system testing

neighbourhood system. Let A(x) and B(x) are 8 connected neighbours, connected to the black pixel 'x'. Here 'A' denotes black pixels and 'B' denotes white pixels.

The edge region starts from one black pixel A(x) and grows looking connected A(x)s. During this process of growing, an A(x) is always connected with at least one B(x) shown in Fig. 3. Thus we have got the edge of each object. This algorithm starts from a black pixel and grows through the edge and stop somewhere when it finds a discontinuity. Thus it creates a black pixel chain detecting the edge or a portion of edge of objects, shown in Fig. 1. Depending on this edge we have drawn a minimum bounding box around each object, shown in Fig. 2. We have cropped each minimum bounding box from the image which denotes qualified objects for further processing.

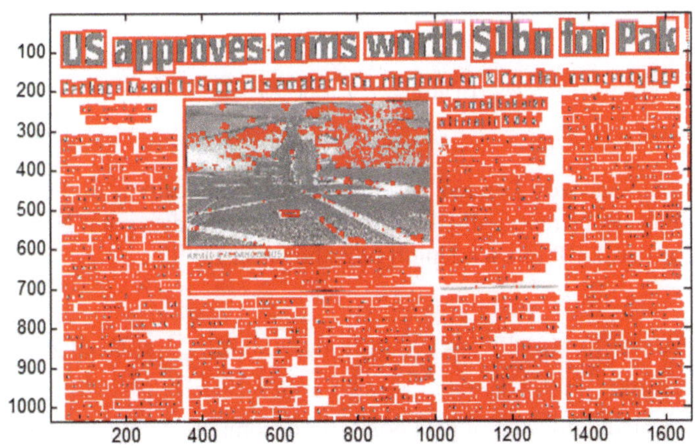

Fig. 2 Minimum bounding box has been drawn around each object

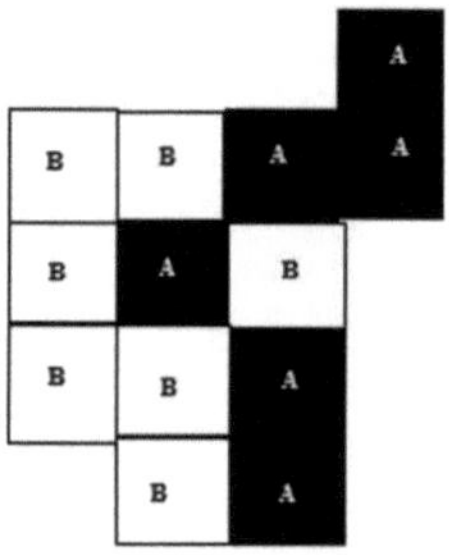

Fig. 3 8 connected neighborhood system for edge detection

3 Feature Extraction

3.1 Geometric Features

We have calculated height to width ratio and black pixel density of each object box. Height to width ratio is not an effective feature as the newspapers are printed in different fonts. To calculate black pixel density we assign a threshold ť to draw discernment boundary between white and black pixel (Fig. 3).

$\acute{r}_b = \frac{number\ of\ black\ pixel}{height \times width}$ *produce black pixel density and* $\acute{r}_h = \frac{height}{height + width}$ *produce height to width ratio.*

3.2 Fast Fourier Transform

FFT reduces number of computation of Discreet Fourier transform (DFT). A DFT of N data points of a continuous signal f(t) can be denoted by the Eq. (1).

$$F(j\omega) = \int_{0}^{(N-1)T} f(t)e^{-j\omega t} \quad (1)$$

Danielson Lancozos Lemma describes the decomposition of DFT into FFT. N Data points can be repeatedly divided by 2. A DFT of data point N can be written as a sum of two DFT of each length N/2 shown in Eq. (2).

$$F(n) = \sum_{K=0}^{N-1} x(k)e^{\frac{-j2\Pi kn}{N}} = \sum_{K=0}^{\frac{N}{2}-1} x(2k)e^{\frac{-j2\Pi kn}{\frac{N}{2}}} + W_N^n \sum_{K=0}^{\frac{N}{2}-1} x(2k+1)e^{\frac{-j2\Pi kn}{\frac{N}{2}}}. \quad (2)$$

Thus repeatedly division of data point can decompose an eight point DFT into 4 two point FFT shown in Figs. 4 and 6.

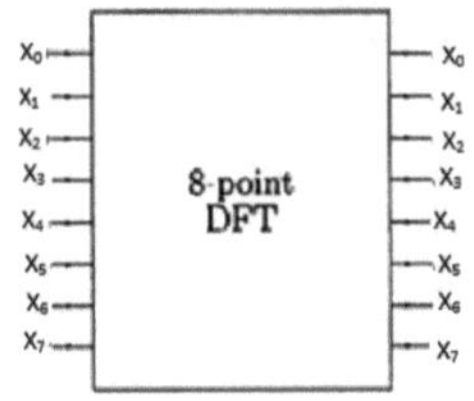

Fig. 4 An eight point DFT

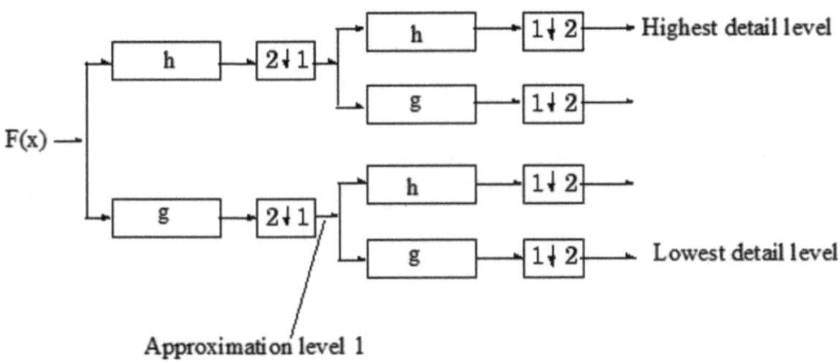

Fig. 5 2-D Wavelet transforms

Fig. 6 An eight-point DFT, divided into four two-point FFTs

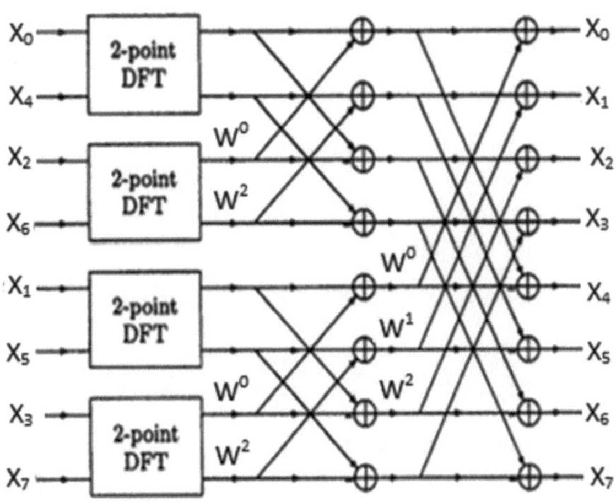

We have decomposed each object vector into 8-point, 30-point and 60-point Fast Fourier transform (FFT). In each case we took first and last three points of FFT. Among these cases we found that if we keep only all point DFT derived from whole object then it gives best result than FFT. We have extracted 12 features from DFT analysis (Fig. 5).

3.3 2-D Discrete Wavelet Transform

Wavelet means small wave. We can define a wavelet by Eq. (3).

$$CWT_x^{\varphi}(\tau, s) = \varphi_x^{\varphi}(\tau, s) = \frac{1}{\sqrt{|s|}} \int x(t)\varphi\left(\frac{t-\tau}{s}\right)dt \quad (3)$$

In Eq. (3)'s' is scaling factor. Changing this scaling factor, we can change the window size of frequency. Smaller window size creates smaller waves and gives higher time information. 'τ' is translation factor. 'τ' determines the window position in given time frame. Discrete wavelet transform (DWT) coefficients are sampled from Continuous wavelet transform (CWT) in discrete time function. DWT consists of scaling function and wavelet function which are associated with low pass and high pass filters respectively. Frequencies at different time scales are passed through the series of low pass and high pass filters shown in Fig. 5. 2-D wavelet transform decomposes an image into layers explained by Jiansheng to [15, 16]. Images feed to a high pass and low pass filters. Then each frequency is down sampled by factor of some positive integer. Down sampling simply reduces few frequency components from original frequency. Each level of decomposition creates an approximation level. In every approximation level, we get image details level. Low pass filter creates low details level and high pass filter creates high details level shown in Figs. 7, 8 and 9.

Fig. 7 Original graphics

Fig. 8 Lowest detailing of Fig. 7

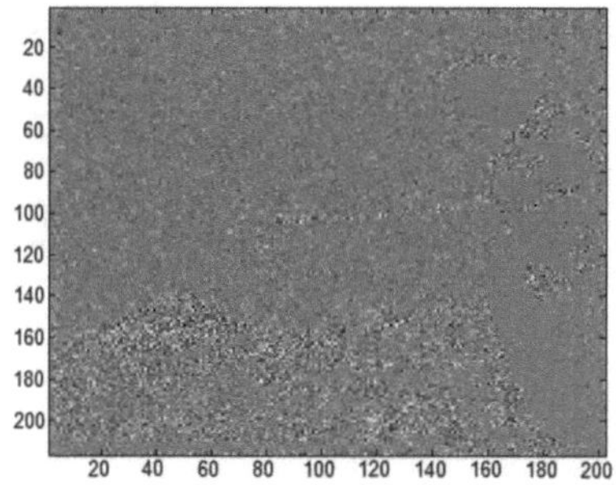

Fig. 9 Highest detailing of Fig. 7

We didn't win IPL twice playing only on turning tracks, says Gambhir

Archiman.Bhaduri
@timesgroup.com

Kolkata: Kolkata Knight Riders skipper Gautam Gambhir feels that it is the domestic players, rather than the foreigners, who make the difference in the Indian Premier League (IPL).

"It is always the domestic players who make the difference. Majority of the players in an IPL team are Indians and the team's fortune depends much on how they perform," he said on the eve of IPL-8 opener between KKR and Mumbai Indians here on Wednesday.

Talking about his own side, the Delhi opener said, "In this regard, we are very solid. We have the likes of Suryakumar Yadav, Manish Pandey and Robin Uthappa, who are seasoned campaigners. So, we have lot of experience."

Gambhir disagreed with the notion that the team management picked up a bunch of spinners in the squad fearing that Sunil Narine would not be cleared to play in IPL-8.

"They were never picked as back-up bowlers. We want different types of spinners and have based the entire team on that because, if you see, most of the wickets in India turn and start gripping especially in the later half of the tournament. That's the strategy," he explained.

Gambhir didn't agree that KKR is a side that depends solely on turning tracks.

"You don't win the IPL twice playing only on turning tracks. We have proved that we are a very consistent side, winning 14 games (including five in Champions League T20) on the trot. So people should stop staying that KKR win only on turning tracks," he said.

The skipper hoped rookie leg-spinner KC Cariappa will develop as a bowler playing alongside the likes of Narine and Brad Hogg.

"To be honest, we did not pick Cariappa hoping that he will be a match-winner for us. But, obviously he is different. We want to groom and develop young players. I haven't seen him bowl in a match so far, so we will see over the course of time how he shapes up," he stated.

"It would be nice to have spinners who can complement Sunil. It will always help the likes of Cariappa and Kuldeep Yadav to have Sunil and Brad Hogg around. It will help develop Indian cricket also.

"People and commentators talk a lot about the role of franchises. It's about walking the talk and we have done that. I think we have done that pretty well over the last few years and we'll continue to do that," he added.

KKR skipper Gautam Gambhir takes a break from training at the Eden Gardens on Tuesday

Fig. 10 Input image with connected characters

For our work, we have calculated up to approximation level 2 and down sampled the frequency by the factor of 2. We only have taken second highest details level to extract features. We have considered 10 features from second highest details. We have used 'Daubechies' wavelet and its filters.

4 Experimental Results and Discussion

For experimentation, we have selected 27 articles from Indian newspapers. All articles were scanned at 200 dpi resolution by HP Deskjet F380 scanner. We have collected 39 graphics and 181 text regions from 27 pages. From each image object we have extracted 24 features. We have trained the system with 125 data set and tested it with 95 data set. We have used four well known classifiers, which are Bayes Net, Support vector machine (SVM), Multilayer perceptron (MLP), Sequential minimal optimization (SMO). Recognition rates are shown in Tables 1 and 2. Figure 11 is output image of the system that we have built. System removes graphics from the newspaper page and stores it in different location. From Table 2

Table 1 Recognition rate achieved using cross validation on 220 data set

	Cross validation		
	3 fold	5 fold	7 fold
Bayes Net (%)	86.36	86.81	86.36
SVM (%)	82.27	82.27	82.27
MLP (%)	93.18	91.36	92.27
SMO (%)	94.54	93.18	95

US a p oves arms worth $1bn fo Pak

ackage Meant To Support Islama ad's Counterterrorism & Counter-Insurgency Ops

Chidanand.Rajghatta
@timesgroup.com

Washington: The Pentagon has notified US Congress about the state department's approval of a military arms package worth nearly $1 billion for Pakistan, ostensibly in support of its counterterrorism and counter-insurgency operations.

The list of arms and armaments notified for possible sale by the Defence Security Cooperation Agency in its communication to Congress on Monday includes 15 AH-1Z Viper attack helicopters, 1000 AGM-114 R Hellfire II missiles in containers, 32 T-700 GE 401C engines (30 installed and 2 spares) and 17 AN/AAQ-30 Target sight systems.

The package also includes 30 (29)F-23 ultra high frequency/very high frequency communication systems, 19 H-764 embedded global positioning system/inertial navigation systems, 32 helmet mounted display/optimized top owl, 17 friend or foe, 17 AN/AAR-47 missile warning systems, 17 AN/ALE-47 countermeasure dispenser sets, 18 AN/APR-39C(V)2 radar warning receivers, 15 joint mission planning systems, and 17 M197 20mm gun systems.

The proposed sale "will contribute to the foreign policy and national security of the US by helping to improve ... to US foreign policy and national security goals in South Asia," the DSCA said in its notification, adding that the helicopters and weapon systems will provide Pakistan with military capabilities in support of its counterterrorism and counter-insurgency operations in South Asia.

It also assured the Congress that the proposed sale of ... will not alter the basic military balance in the region where the US has also emerged as a large arms supplier to India.

Elaborating on the use of the arms package, the notification said the proposed sale will provide Pakistan with a precision strike, enhanced survivability aircraft that it can operate at high altitudes. By acquiring this capability, it said, Pakistan will enhance its ability to conduct operations in North Waziristan Agency (NWA), the Federally Administered Tribal Areas (FATA), and other remote and mountainous areas in all-weather, day-and-night environments.

The sale, worth $952 million according to Pentagon estimates, is expected to go through without a hitch even though Pakistan lives a perilous, hand-to-mouth existence and US lawmakers and officials periodically vent their frustration about the country's two-timing, double-dealing ways in the war on terrorism and extremism.

Many of the military contractors involved are constituents of lawmakers who prize the creation of jobs and maintaining it in their districts above all else. The principal contractors of the Pakistan package will be Bell Helicopter, Textron in Fort Worth, General Electric in Lynn, The Boeing Company in Huntsville, and Lockheed Martin in Bethesda, Maryland.

ARMED AND DANGEROUS: The list of arms and armaments in the package includes 15 AH-1Z Viper attack helicopters (above), 1000 AGM-114 R Hellfire II missiles in containers, 32 T-700 GE 401C engines (30 installed and 2 spares) and 17 AN/AAQ-30 Target sight systems

'N-armed Pakistan a threat to S Asia'

A leading US daily on Tuesday described nuclear-armed Pakistan as "unquestionably the biggest concern" to stability in South Asia. The New York Times, in its editorial, said the world powers that worked on the Iran deal should now turn their attention to the "troubled region with growing nuclear risks of its own". It said the country has "wasted" billions on sophisticated weapons, that otherwise, should have been spent on its people. PTI

Fig. 11 Output image after removing graphics from the page (Fig. 11 was produced using SVM classifier)

we find that SMO classifier gives best result. We have got this result tuning the classifier, choosing kernel type 'polykernal' and complexity parameter (c) = 2. We have used weka 3. 6 tool to train classifiers. To calculate black pixel density, we have chosen value of threshold t' is 100. We have chosen boxes below 13 × 13 size as characters because such small sizes of graphics are rare in newspaper documents. From Table 1 also it is found that SMO gives best result than other classifiers but we may conclude a decision form cross validation folds that, it is training data set that is creating such differences in results. Creating a training dataset with more numbers of images can give better result. It is observed from the output image shown in Fig. 11 that from same word *"approves"* one 'p' was recognized as character and another 'p' was recognized as graphics where both 'p' ware in same style and size. This type of error is reducing recognition rate of system. This error was occurred due to the addition of some gray shade with character 'p' at the time

Table 2 Recognition rate achieved using 125 training data set and 95 test data set

	Bayes Net	SVM	MLP	SMO
Recognition Rate (%)	90.52	85.26	92.63	95.78

Table 3 Recognition rate achieved for 125 training data set and 95 test data set using only FFT feature

	Bayes Net	SVM	MLP	SMO
Recognition Rate (%)	88.42	85.26	82.4	88.42

of scanning the document. That is the reason that FFT has failed to recognize the first ‘p’ from the word *“approves”* as character. Table 3 shows the recognition rate of system when we are using FFT as only feature.

In some cases in an image two or more characters ware found connected by printing mistake by press. In those cases two characters together are creating single minimum bounding box which are rejected by the height to width ratio feature and system consider those characters as graphics. These types of errors are reducing overall recognition rate of the system. In Fig. 10 the word “twice” has such connected characters. In the word ‘tw’ is connected and considered as graphics by the system.

5 Conclusion

This work aims to design a system to separate graphics from texts in the newspaper document on the basis of frequency transformation. Proper selection of FFT data points and wavelet level may increase recognition rate. Changing few implementation details, this work can be applied on logo, graphics recognition from visiting cards and camera captured nameplates. In those cases, analyzing shape of contour can be a vital feature. This is a robust system we have developed which is efficient for recognition on a defined dataset for a specific purpose. A larger good training dataset can destroy limitations.

Acknowledgements The authors are thankful to the Center for Microprocessor Application for Training Education and Research (CMATER) and Project on Storage Retrieval and Understanding of Video for Multimedia (SRUVM) of Computer Science and Engineering Department, Jadavpur University, for providing infrastructure facilities during progress of the work. The current work reported here, has been partially funded by University with Potential for Excellence (UPE), Phase-II, UGC, Government of India.

References

1. Garg, R., Bansal, A., Chaudhury, S., Roy, S.D.: Text graphic separation in Indian newspapers. In: Proceedings of 4th International Work Multiling. OCR-MOCR’13, August 24, p. 1 (2013)
2. Roy, P.P., Vazquez, E., Lladós, J., Baldrich, R., Umapada, P.: A System to Segment Text and Symbols from Color Maps. In: 7th International Workshop, GREC 2007, 5046, pp. 245–256 (2008). https://doi.org/10.1007/978-3-540-88188-9

3. Mollah, A.F., Basu, S., Nasipuri, M., Basu, D.K.: Text/Graphics Separation for Business Card Devices, pp. 263–270 (2009)
4. Rege, P.P., Chandrakar, C.A.: Text-Image Separation in Document Images Using Boundary/ Perimeter Detection (2011)
5. Strouthopoulos, C., Papamarkos, N., Atsalakis, A.E.: Text Extraction in Complex Color Documents, vol. 35, pp. 1743–1758 (2002)
6. Garg, R., Hassan, E., Chaudhury, S., Gopal, M.: A CRF Based Scheme for Overlapping Multi-Colored Text Graphics Separation," In: 2011 International Conference on Document Analysis and Recognition, no. c (2011)
7. Cao, R., Tan, C.L.: Separation of Overlapping Text from Graphics, pp. 44–48 (2001)
8. Science, C., Kent, L., Rd, R., Abe, N.: A Clustering-Based Approach to the Separation of Text Strings from Mixed Text Graphics Documents, pp. 706–710 (1996)
9. Vieux, R., Domenger, J., Talence, F.: Hierarchical Clustering Model for Pixel-Based Classification of Document Images, no. Icpr, pp. 290–293 (2012)
10. Chinnasarn, K.: Removing Salt-and-Pepper Noise in Text/Graphics Images, IEEE, pp. 459–462
11. Haralick, R.M., Sternberg, S.R., Zhuang, X.: Image Analysis Using Mathemetical Morphology, IEEE Trans. Pattern Anal. Mach. Intel. (4), pp. 532–550 (1987)
12. Kowalczyk, M., Koza, P., Kupidura, P., Marciniak, J.: Application of Mathematical Morphology Operations for Simplification and Improvement of Correlation of Images in Close-Range Photogrammetry, The International Archives of the Photogrammetry, Remote Sensing and Spatial Information Sciences, vol. XXXVII, part B5. Beijing (2008)
13. Verma, R., Ali, J.: A Comparative Study of Various Types of Image Noise and Efficient Noise Removal Techniques, Int. J. Adv. Res. Comput. Sci. Soft. Eng. **3**(10), 617–622 (2013)
14. Kumar, M., Saxena, R.: Algorithm and Technique on Various Edge Detection: A Survey, vol. 4, no. 3, pp. 65–75 (2013)
15. To, E.: The, A DWT, DCT and SVD Based Watermarking, vol. 4, no. 2, pp. 21–32 (2013)
16. Jiansheng, M., Sukang, L., Xiaomei, T.: A Digital Watermarking Algorithm Based on DCT and DWT, In: Proceedings of the 2009 International Symposium on Web Information Systems and Applications (WISA'09) Nanchang, P. R. China, May 22–24, vol. 8, no. 2, pp. 104–107 (2009)

Ant Lion Optimization: A Novel Algorithm Applied to Load Frequency Control Problem in Power System

Dipayan Guha, Provas Kumar Roy and Subrata Banerjee

Abstract In this article, an attempt has been made to find an effective solution of load frequency control problem in power system employing a powerful and stochastic optimization technique called "Ant Lion Optimization" (ALO). The proposed algorithm is inspired by the interaction strategy between ants and ant lions in nature. To appraise the effectiveness of ALO algorithm, a widely used two-area multi-unit multi-source power plant equipped with distinct PID-controller is investigated. The integral time absolute error-based objective function has been defined for fine tuning of PID-controller gains by ALO algorithm. To judge the acceptability of ALO algorithm, the simulation results are compared with some recently published algorithms. The simulation results presented in this paper confirm that the proposed ALO algorithm significantly enhanced the relative stability of the power system and can be applied to the real-time problem.

Keywords Load frequency control · Multi area power system
High voltage DC · Ant lion optimization · Transient analysis

D. Guha (✉) · S. Banerjee
Department of Electrical Engineering, National Institute of Technology, Durgapur, West Bengal, India
e-mail: guha.dipayan@yahoo.com

S. Banerjee
e-mail: bansub2004@yahoo.com

P. K. Roy
Department of Electrical Engineering, Kalyani Government Engineering College, Kalyani, West Bengal, India
e-mail: roy_provas@yahoo.com

S. Kar et al. (eds.), *Operations Research and Optimization*, Springer Proceedings in Mathematics & Statistics 225, https://doi.org/10.1007/978-981-10-7814-9_15

1 Introduction

The main interest of power system operation and control is to maintain a balance between total generation with load demand plus losses associated in the system. Any mismatch between the generation and demand causes serious deviation of system frequency from its tolerance value. A considerable change in the frequency may lead to asynchronization between the nearby control areas and produces high magnetizing current in electric motors, transformers, etc. Thus to have a reliable and stable operation of power system, the control of generation and, hence, frequency is the most important subject in the power system dynamics. In the past, fly ball mechanism of speed governor was employed to maintain the equilibrium condition in the power system. But recent research founds that the conventional method is not sufficient to hold the stability of power system because of rapid advancement of the power system, high power consumption, and excessive load demand. An intelligent controller in the name of "***load frequency controller (LFC)***" has been employed in a power system in coordination with speed governor to overcome the above-mentioned problem. In the power system, LFC is used as a regulator to monitor the net amount of change between generation and demand, and accordingly, regulate the valve position of the governor to control the steam flow through the turbine.

In the recent time, LFC has gained a huge attention and considerable research efforts in order to explore new control algorithms or to update the existing control theory. The literature review reveals that the work on LFC was proposed by Chon [1] in 1957; however, the use of optimal control theory in LFC area was coined by Elgerd and Fosha [2] in 1970. Tripathy et al. [3] have examined the effect of governor dead band (GDB) nonlinearity on the system dynamics of an interconnected power system. Yousef [4] has described an adaptive fuzzy logic control (FLC) strategy for LFC of a multi-area power system. Several classical controllers such as integral (I), proportional integral (PI), integral derivative (ID), proportional integral derivative (PID), integral double derivative (IDD) controllers have been proposed in [5] and their performances are compared for the LFC system. A neuro-fuzzy hybrid intelligent PI-controller has been applied to the LFC system by Prakash and Sinha [6]. In [7], the differential evolution (DE) algorithm was used to tune the PID-controller gains for multi-area power system network and show the superiority over the optimal controller. Guha et al. [8] in his most recent endeavor solved LFC problem using gray wolf optimization and shows the excellence of same over other existing control algorithms. The effective use of biogeography-based optimization (BBO) in LFC area is available in [9, 10]. In [11], the author has employed teaching-learning-based optimization (TLBO) to solve the LFC problem. A novel krill herd algorithm with optimal PID-controller is proposed for LFC in [12].

The aforementioned techniques effectively improved the system performances leaving behind some drawbacks which are further revised by the researchers. For example, in the FLC, the selection of rules in knowledge-based process, scaling factor, membership function is essential for effective implementation of same. In the neural network, large training data set is required for supervised learning. The neuro-fuzzy controller is insufficient to give the satisfactory performance when measurement noise and parametric variations are included in the analysis. The performance of DE is controlled by the mutation factor and crossover rate. Further, premature convergence degrades the search ability of DE. The control parameters of BBO are habitat modification probability, mutation probability, maximum emigration rate, maximum immigration rate, the step size of integration. If the parameters are not properly elected, then the algorithm may trap into suboptimal solution.

It is quite clear from the above discussion that the performance of the aforementioned techniques more or less is controlled by some input parameters. Further, in the line of "***no-free-lunch***" algorithm, no optimization algorithm is suitable for all optimization problems. Hence, it is very much essential to search a parameter-free algorithm that can solve a maximum number of optimization problems. Therefore, it may be justified to propose a new optimization scheme for the betterment of the existing results. Inspired from the above discussion, an attempt has been made in this paper to design and implement a population-based stochastic optimization technique called ant lion optimization (ALO) [13] for the effective solution to LFC problem. The main advantage of this technique is that its functionality is only controlled by the population size and maximum iteration count. To show the effectiveness, a multi-area multi-source power system with distinct PID-controller is investigated. The superiority of ALO algorithm has been established over DE, TLBO, and BBO algorithms for the same test system.

Rest of the paper is organized as follows: Sect. 2 describes the mathematical model of the test system with controller structure. Section 3 offers an overview of ALO algorithm. Section 4 highlights the algorithmic steps of ALO. Comparative transient responses and discussion are given in Sect. 5. Finally, Sect. 6 concludes the present study.

2 Problem Formulation

To demonstrate the ability of proposed ALO algorithm to cope with interconnected multi-source power system with distinct secondary controller, the simulation study is performed on a widely used interconnected two-area multi-unit multi-source power system having thermal, hydro, and gas power plants [7, 11, 14]. The transfer function model of concerned power system is shown in [7, 11, 14]. In Table 1, R_1, R_2, R_3 are the speed regulation parameters of thermal, hydro, and gas units in Hz/p.u. MW, respectively; u_{th}, u_{hy}, u_g are the controlled input to thermal, hydro, and gas power plants, respectively; T_{sg} is time constant of speed governor of thermal unit in sec; T_t is time constant of steam turbine of thermal unit in sec; K_r and T_r are the reheater gain and time constant of thermal unit, respectively; T_w is nominal starting time of water in penstock in sec; T_{RS} is hydro turbine speed governor reset time constant in sec; T_{RH} is hydro turbine speed governor transient droop time constant in sec; T_{GH} is hydro turbine speed governor main servo time constant in sec; X_C is lead time constant of gas turbine speed governor in sec; Y_C is lag time constant of gas turbine speed governor in sec; c_g is gas turbine valve positioner; b_g is gas turbine constant of positioner; T_F gas turbine fuel time constant in sec; T_{CR} gas turbine combustion reaction time delay in sec; T_{CD} gas turbine compressor discharge volume time constant in sec; B_1, B_2 are the bias factor; T_{12} is synchronizing time constant of tie-line in sec; T_{PS} is time constant of power system in sec; K_{PS} is the gain of power system in Hz/p.u. MW; $\Delta f_1, \Delta f_2$ are the incremental change in frequency of the respective area in Hz; ΔP_D is incremental load change in p.u. The nominal system parameters are collected from [11] and shown in Table 1. The step load increase of 1% in area-1 at $t = 0\text{s}$ is considered to assess the tuning efficacy of the proposed ALO algorithm.

The main concern in LFC study is to minimize the ACE as early as possible. In order to do that an optimized PID-controller has been designed employing ALO algorithm. In optimal control theory, different performance indices are defined for optimal tuning of controller parameters. The integral time absolute error (ITAE) integrates the absolute error multiplied by time over a specific time horizon, and therefore, the errors which exist after a long time will be much more heavily affected than those at the start of the response. This motivates authors to select ITAE-based ACE of the respective area as an objective function for fine tuning of LFC gains. The proposed objective function is defined in (1).

$$J = \int_{t=0}^{T_{Final}} |ACE_i| * t * dt \quad i = 1, 2, 3 \tag{1}$$

where T_{Final} is the final simulation time in sec. The defined LFC problem can be defined as a constraint optimization problem bounded by the control parameters. Thus, the said problem can be expressed as follows:

Minimize J

Subjected to: $K_{p,\min} \le K_p \le K_{p,\max}$; $K_{i,\min} \le K_i \le K_{i,\max}$; $K_{d,\min} \le K_d \le K_{d,\max}$

where $K_{PID,\min}$, $K_{PID,\max}$ are the minimum and maximum gains of PID-controller, respectively. In this case, the controller settings are selected between [0, 6] [11, 14].

3 Ant Lion Optimization Algorithm (ALO)

Ant lion (doodlebugs: the larvae of ant lion), shown in Fig. 1a, belongs to the "Myrmeleontidae" (scientific name) family and "net-winged insects." The complete lifespan of ant lion mainly consists of two phases: larvae and adult. In a lifespan of 3 years, the ant lion can take 3–5 weeks for adult phase and mostly undergo as a larvae. The ant lion go through the metamorphosis process in cocoon to become adults. They mostly hunt in larvae, and adult phase is only for reproduction. The name originates from their distinct hunting mechanism, as shown in Fig. 1b and their favorite prey (ant).

Table 1 Nominal values of system parameters [11]

Parameter	Values	Parameter	Values	Parameter	Values	Parameter	Values	Parameter	Values
f	60 Hz	T_{12}	0.0433	T_{rh}	28.75 s	$R_1 = R_2 = R_3$	2.4	T_{rs}	5 s
T_r	10 s	K_T	0.54348	T_{cr}	0.01 s	$K_{P1} = K_{P2}$	68.9566	$B_1 = B_2$	0.4312
K_r	0.5	K_G	0.13044	T_{fc}	0.23 s	$T_{P1} = T_{P2}$	11.49 s	T_t	0.3 s
T_{sg}	0.08 s	K_H	0.32608	T_{cd}	0.2 s	c_g	1	X_C	0.6 s
T_w	1 s	T_{gh}	0.2 s	b_g	0.5	Y_C	1 s		

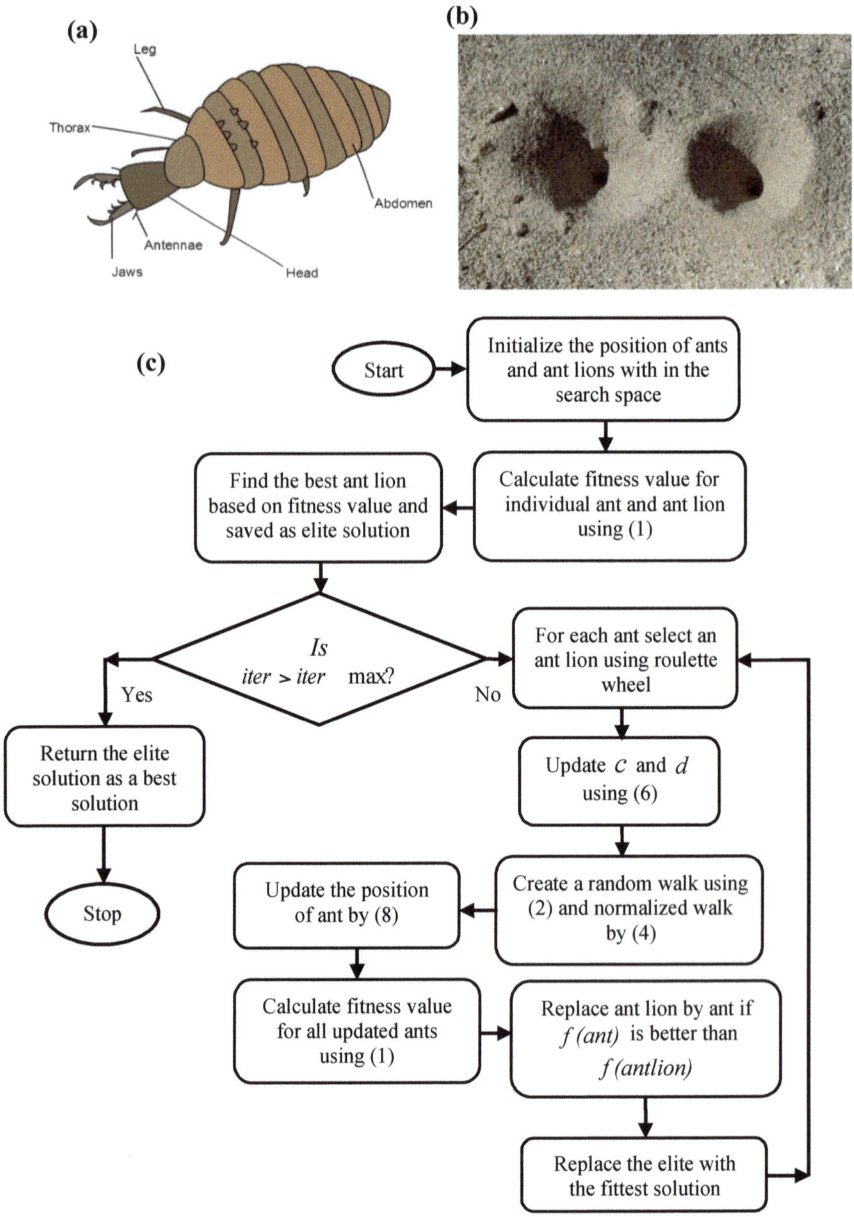

Fig. 1 **a** Ant lion, **b** hunting mechanism, **c** general flowchart of ALO algorithm

3.1 Mathematical Modeling of ALO Algorithm

The main inspiration of ALO algorithm is the interaction strategy between ant lion and ant in trap. As stated earlier that the hunting process of ant lion consists of five different steps namely random walk of ant, building traps, entrapment of ants in the traps, catching prey, and rebuilding of traps.

To model ALO algorithm, the ants are allowed to move randomly within the search space and ant lions are allowed to hunt them and become the fittest solution. The random walk of ants during the searching of foods can be modeled by (2).

$$x(t) = [0, cumsum(2r(t_1) - 1), cumsum(2r(t_2) - 1), - - -, cumsum(2r(t_n) - 1)] \tag{2}$$

where *cumsum* is the cumulative sum, n defines the total no. of iteration, t is the steps of random walk, and $r(t)$ is a stochastic function and defined in (3).

$$r(t) = \begin{cases} 1 & if, rand > 0.5 \\ 0 & if, rand \leq 0.5 \end{cases} \tag{3}$$

where *rand* is a random number uniformly generated between $[0, 1]$.

3.1.1 Random Walk of Ant

Random movement of ant during the searching of food in nature is monitored by (2). Since the ant walk is completely stochastic in nature, it is very much essential to check whether the ants are within the defined search area or not. To ensure the position of ants in the search area, normalized random walk as defined in (4) updates the current position of ant in each step of generation.

$$x_i^t = \frac{(x_i^t - a_i) * (d_i - c_i^t)}{(d_i^t - a_i)} + c_i \tag{4}$$

where a_i and b_i are the minimum and maximum of random walk of ith ant, respectively, c_i^t and d_i^t are the minimum and maximum of ith ant at tth iteration, in order.

3.1.2 Trapping Ants in Ant Lions Pits

As discussed earlier that the ant lions are gets hidden in pit to trap the ants and following mathematical relationship is defined to model such hunting mechanism in ALO algorithm.

$$\begin{cases} c_j^t = antlion_j^t + c^t \\ d_j^t = antlion_j^t + d^t \end{cases} \quad (5)$$

where $antlion_j^t$ indicates the position of selected jth ant lion at tth iteration, c^t and d^t are the minimum and maximum, in order, value of all variables at tth iteration, c_j^t indicates the minimum of all variables for ith ant, and d_j^t indicates the maximum of all variables for ith ant.

3.1.3 Building of Trap

In order to model the hunting capability of ant lion, a "roulette wheel" is employed in [13]. In roulette wheel selection algorithm, a set of weights or numbers represents the probability of selection of each individual in a group of choices. This mechanism ensures high possibility to the fitter ant lions for catching ants.

3.1.4 Sliding Ants into the Pit: Exploitation

Having knowledge of above discussion, it is identified that the ant lions are now at the position to build trap in proportional to their fitness value. But the ants are not easily caught into the pit, they try to escape themselves from the tarp. At this stage, ant lions cleverly throw sands toward the ant to trap into pits. To model such technique, the radius of ants during the random walk is decreased and defined by (6).

$$\begin{cases} c^t = \dfrac{c^t}{10^W\left(\frac{t}{T}\right)} \\ d^t = \dfrac{d^t}{10^W\left(\frac{t}{T}\right)} \end{cases} \quad (6)$$

where T is the maximum number of iteration and w is a constant and selected based on current iteration (t), i.e., $w = 2$ when $t > 0.1T$; $w = 3$ when $t > 0.5T$; $w = 4$ when $t > 0.75T$; $w = 5$ when $t > 0.9T$; $w = 6$ when $t > 0.95T$. Actually, w is defined to show the accuracy level of exploitation.

3.1.5 Catching Prey and Rebuilding Pit

This is the final stage of hunting process of ant lion in ALO algorithm where preys (ant) are caught in the ant lion's jaw. The ant lion pulls ant inside the sand and consumes its body. To mimic this strategy, it is anticipated that catching of prey is occured when ants become the fittest solution than its corresponding ant lions. Mathematically, it is defined in (7)

$$antlion_j^t updateant_j^t \quad if \quad f\left(ant_i^t\right) > f\left(antlion_i^t\right) \tag{7}$$

3.1.6 Elitism

Elitism is an important attribute of any optimization technique that permits the algorithm to holds best possible solution obtained at any stage of generation. In ALO algorithm, the best ant lion computed so far is saved as elite and defined by (8).

$$Ant_i^t = \left(R_{Antlion}^t + R_{elite}^t\right)/2 \tag{8}$$

where R_{elite}^t is the random walk of ant around elite and $R_{antlion}^t$ is the random walk of ant around ant lion at *t*th iteration. The general flowchart of proposed ALO algorithm is shown in Fig. 1c. For more details regarding the ALO algorithm, readers are referred to [13].

4 ALO Applied to LFC Problem

In this section, the algorithmic steps of ALO algorithm for solving the LFC problem in power system are discussed. The steps are enumerated as follows:

Step 1. Initialize the position of ant and ant lion within the search space with dimension of n_p* dim and calculate the fitness value, minimum ACE, for the individual solution using (1).

Step 2. Identify the best solution based on fitness value, as calculated in step 1, and marked the best solution as an elite solution.

Step 3. Create a random walk of ants toward the ant lion, employing roulette wheel mechanism, by performing the following pseudocodes:

Step 4. Update the position of ants, i.e., PID-controller gains, by (8).

Step 5. Calculate the fitness value, i.e., minimum ACE, for all the updated ants using (1).

Step 6. Replace the current position of ants by the updated solution if $f(ant) < f(antlion)$ and update the elite solution.

Step 7. Go to step 3 until the termination criterion is met and print the optimal solution.

$input: \dim, lb, ub, antlion, current\_iter$
$output: random\_walk(RW)$

Check if the bounds are scalar

$if \;\; size(lb,1) == 1 \,\&\&\, size(lb,2) == 2$
$lb = ones(1,\dim) * lb;$
$ub = ones(1,\dim) * ub;$
end

Check if bounds are horizontal or vertical

$if \;\; size(lb,1) > size(lb,2)$
$lb = lb';$
$ub = ub';$
end

Calculation of normalized random walk of ants:

Firstly, the value c^t and d^t are updated using (6) for different values of w as discussed in Section 3.2.4.

Next, calculate the normalized ant position by (2)-(4).

$if \;\; rand < 0.5$
$lb = lb + antlion;$
$else$
$lb = -lb + antlion;$
end
$if \;\; rand \ge 0.5$
$ub = ub + antlion;$
$else$
$ub = -ub + antlion;$
end
$for \;\; j = 1:\dim$
$x = [0 \;\; cumsum(2*(rand(\max\_iter,1) > 0.5) - 1)];$
$a = \min(x); b = \max(x); c = lb(j); d = ub(j);$
$$x_{norm} = \frac{(x-a)*(d-c)}{(b-a)} + c;$$
end

5 Results and Discussion

In this section, simulation results are presented and compared with the existing algorithms in order to evaluate the effectiveness of proposed ALO algorithm in LFC area. The study has been done in MATLAB/SIMULINK environment. The model of the test system, as shown in [7, 11, 14], is developed in SIMULINK domain, whereas the codes for ALO algorithm has been separately written in the *.m file*. The simulation was done on a personal computer having 2.4 GHz core i3 processor with 2 GB RAM in MATLAB R2009 environment. As it is already stated that the performance of ALO is independent on the prior initialization of control parameter, therefore only population size $n_p = 50$ and maximum iteration count $iter_{\max} = 100$ have been defined for the successful operation of ALO. The step load perturbation of 1% of nominal value is considered to assess the dynamic stability of the concerned test system.

The simulation is performed under two phases; initially, the test system is investigated with AC transmission line. The ALO algorithm has been employed to design PID-controller using the defined fitness function. At the end of optimization, optimal controller gains and minimum fitness value are presented in Table 2. In order to make a fair comparison, the test system is also studied with DE [7], TLBO [11], and BBO [14] algorithms, and the results are shown in Table 2. It is clearly noticed from Table 2 that the error criterions are further decreased with ALO algorithm that justifies the tuning superiority of ALO over DE, TLBO, and BBO. The signals of the closed-loop system are shown in Fig. 2. For comparison, the results with DE, TLBO, and BBO are also presented in Fig. 2. It is clearly viewed from Fig. 2 that the number of oscillations, overshoot, and settling time of frequency and tie-line power deviation is drastically reduced with ALO-tuned PID-controller. A comparative study on settling time is given in Table 3. It can be easily inferred from Fig. 2 and Table 3 that the proposed controller significantly improved the degree of relative stability.

The convergence behaviors of proposed ALO algorithm with PID-controller structure are shown in Fig. 3. It is viewed from Fig. 3 that proposed algorithm smoothly approaches to the global optimal solution. It is further noticed from Fig. 3 that ALO takes 70–80 iterations to reach the optimum point, which further justified the choice of maximum iteration of 100 for present study.

In the second phase of analysis, DC transmission line in parallel with AC line is included in the system model. The same procedure as described in Sect. 4 is followed to tune the PID-controllers. The optimal controller value with minimum ITAE value is deployed in Table 4. The comparative transient behavior of the test system is painted in Fig. 4. The setting time of Δf and ΔP_{tie} are noted down from Fig. 4 and shown in Table 5. Critical observation of Fig. 4 and Table 5 reveals that proposed ALO-based PID-controller outperforms over DE, TLBO, and BBO-tuned PID-controllers for the identical test system.

Table 2 Optimum values of controller gains and minimum fitness value without DC tie-line

EAs	Controller gains									Objective function			
	Thermal			Hydro			Gas			ISE ($\times$ 10^{-2})	ITSE ($\times$ 10^{-4})	ITAE ($\times$ 10^{-4})	IAE
	K_{p1}	K_{i1}	K_{d1}	K_{p2}	K_{i2}	K_{d2}	K_{p3}	K_{i3}	K_{d3}				
DE [7]	0.779	0.2762	0.6894	0.5805	0.2291	0.7079	0.5023	0.9529	0.6569	52.551	28.23	44.7442	0.1172
TLBO [11]	4.1468	4.0771	2.0157	1.0431	0.6030	2.2866	4.7678	3.7644	4.9498	11.121	15.43	30.1024	0.1134
BBO [14]	2.5991	5.7884	5.7402	0.3195	5.9200	5.6377	2.5158	4.1542	2.0376	1.2594×10^{-3}	0.2517	38.4	0.0264
ALO	5.9749	5.9940	4.6417	5.9823	0.0119	3.0820	5.8659	5.9240	0.7605	**6.2336×10^{-4}**	**4.6009×10^{-2}**	**17.4**	**0.0105**

Boldface shows the best value
EA Evolutionary algorithm

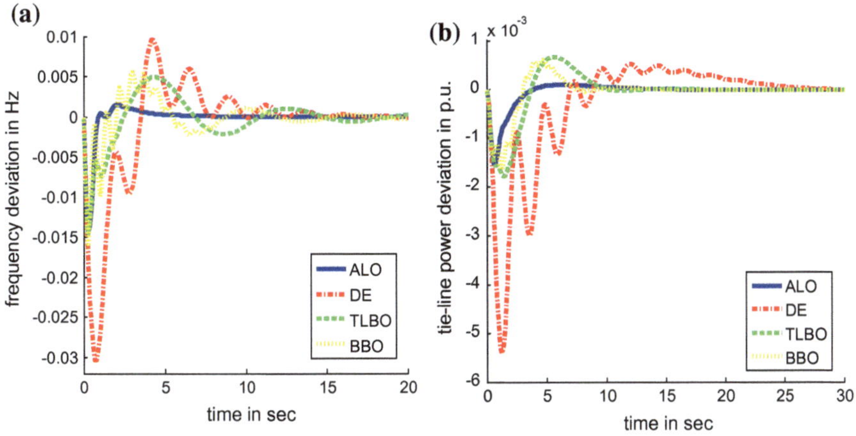

Fig. 2 Transient responses with AC tie-line only **a** change in frequency, **b** change in tie-line power

Table 3 Settling time of frequency and tie-line power excluding DC tie-line

Parameters	ALO	% of improvement by ALO	DE [7]	% of improvement by ALO	TLBO [11]	% of improvement by ALO	BBO [14]
Δf_1	**5.35**	61.98	14.06	69.93	17.79	62.67	14.33
Δf_2	**9.96**	34.43	15.19	60.06	24.94	50.55	20.14
ΔP_{tie}	**12.97**	45.42	23.85	–	9.86	0.154	12.99

Boldface shows the best value

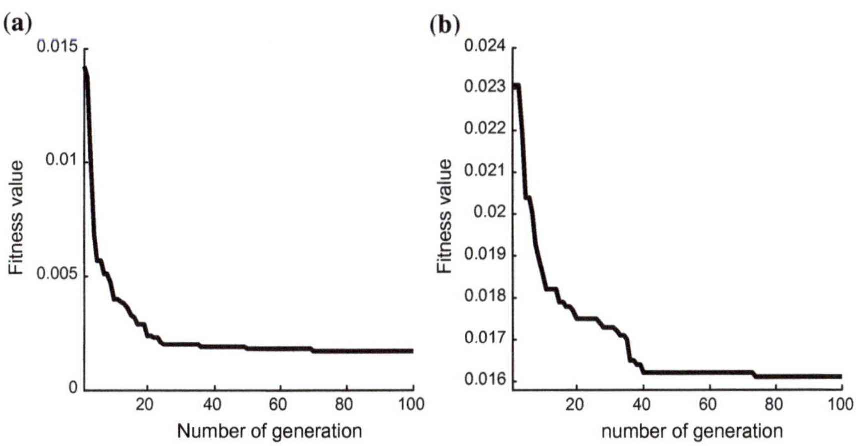

Fig. 3 Convergence characteristic of ALO, **a** with AC tie-line, **b** with AC–DC tie-line

Table 4 Optimum values of controller gains and fitness value considering AC–DC tie-line

EAs	Controller gains									Objective function			
	Thermal			Hydro			Gas			ISE ($\times 10^{-5}$)	ITSE ($\times 10^{-5}$)	ITAE	IAE
	K_{p1}	K_{i1}	K_{d1}	K_{p2}	K_{i2}	K_{d2}	K_{p3}	K_{i3}	K_{d3}				
DE [7]	1.6929	1.9923	0.8269	1.777	0.7091	0.4355	0.9094	1.9425	0.2513	19.22	34.79	0.1987	0.0384
TLBO [11]	5.0658	3.9658	2.4170	0.7032	0.0220	0.0264	8.7211	7.4729	2.4181	3.4388	2.1973	0.1029	0.0163
BBO [14]	5.5700	5.7741	2.1721	4.8648	0.1828	1.3833	5.7843	5.4653	1.2451	0.391	0.471	0.0191	0.0098
ALO	5.9210	5.9902	1.7548	5.9560	0.1300	0.0760	5.9654	5.9584	0.0390	**0.3872**	**0.3983**	**0.0160**	**0.0091**

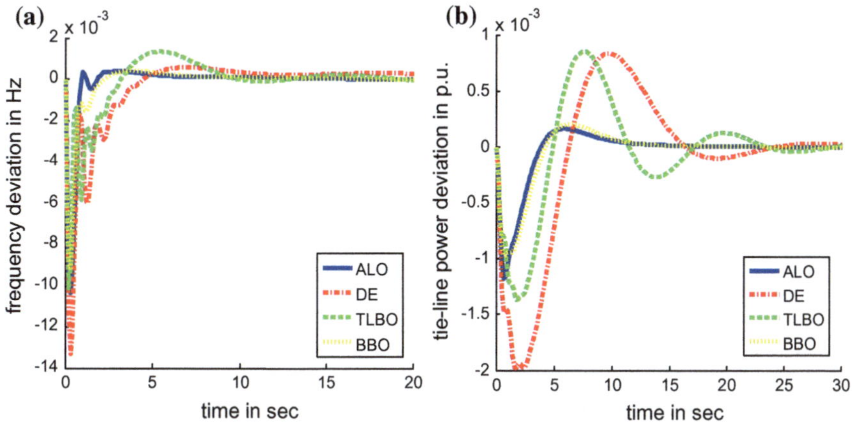

Fig. 4 Transient responses with AC–DC tie-line **a** change in frequency, **b** change in tie-line power

Table 5 Settling time of frequency and tie-line power after 1% SLP with AC–DC tie-line

Parameters	ALO	% of improvement	DE [7]	% of improvement	TLBO [11]	% of improvement	BBO [14]
Δf_1	**5.34**	37.6	8.56	11.74	6.05	75.4	21.7
Δf_2	**16.36**	32.2	24.13	12	18.59	46	30.3
ΔP_{tie}	**13.05**	11.35	14.72	10.3	14.55	35.87	20.35

Boldface shows the best results

6 Conclusion

ALO algorithm is suggested in this article to tune the parameters of PID-controllers for effective, simple, and robust solution of LFC problem. Two-area multi-unit multi-source power plants with and without AC–DC tie-line have been considered to appraise the acceptability of proposed algorithm in LFC domain. The ITAE-based ACE has been defined as a fitness function for fine tuning of PID-controller gains. The supremacy of ALO algorithm has been established by comparing the system performances with DE, TLBO, and BBO algorithm. Simulation results confirm that proposed ALO algorithm effectively increases the stability of the power system. The favorable impact of DC tie-line in parallel with AC tie-line on the system dynamics is also demonstrated in this article. However, the solution accuracy and enhancement of convergence speed of ALO algorithm can be further achieved by incorporating the quasi-oppositional-based learning into original ALO.

References

1. Cohn, N.: Some aspects of tie-line bias control on interconnected power systems. Am. Inst. Electr. Eng. Trans. **75**(3), 1415–1436 (1957)
2. Elgerd, O.I., Fosha, C.E.: Optimum megawatt-frequency control of multi-area electric energy systems. IEEE Trans. Power Appar. Syst. (PAS) **89**(4), 556–563 (1970)
3. Tripathy, S.C., Hope, G.S., Malik, O.P.: Optimisation of load-frequency control parameters for power systems with reheat steam turbines and governor dead band nonlinearity. IEE Proc. **129**(1), 10–16 (1982)
4. Yousef, H.: Adaptive fuzzy logic load frequency control of multi-area power system. Int. J. Electr. Power Energy Syst. **68**, 384–395 (2015)
5. Saikia, L.C., Nanda, J., Mishra, S.: Performance comparison of several classical controllers in AGC for multi-area interconnected thermal system. Int. J. Electr. Power Energy Syst. **33**, 394–401 (2011)
6. Prakash, S., Sinha, S.K.: Simulation based neuro-fuzzy hybrid intelligent PI-control approach in four-area load frequency control in interconnected power system. Appl. Soft Comput. **23**, 152–164 (2014)
7. Mohanty, B., Panda, S., Hota, P.K.: Controller parameters tuning of differential evolution algorithm and its application to load frequency control of multi-source power system. Int. J. Electr. Power Energy Syst. **54**, 77–85 (2014)
8. Guha, D., Roy, P.K., Banerjee, S.: Load frequency control of interconnected power system using grey wolf optimization. Swarm Evol. Comput. **27**, 97–115 (2016)
9. Guha, D., Roy, P.K., Banerjee, S.: Optimal design of superconducting magnetic energy storage based multi-area hydro-thermal system using biogeography based optimization. In: 2014 Fourth IEEE International Conference on Emerging Applications of Information Technology, ISI-Kolkata, pp. 52–57 (2014)
10. Guha, D., Roy, P.K., Banerjee, S.: Study of dynamic responses of an interconnected two-area all thermal power system with governor and boiler nonlinearities using BBO. In: 2015 Third International Conference on Computer, Communication, Control and Information Technology, pp. 1–6 (2015)
11. Barisal, A.K.: Comparative performance analysis of teaching learning based optimization for automatic load frequency control of multi-source power systems. Int. J. Electr. Power Energy Syst. **66**, 67–77 (2015)
12. Guha, D., Roy, P.K., Banerjee, S.: Application of Krill Herd algorithm for optimum design of load frequency controller for multi-area power system network with generation rate constraint. Adv. Intell. Syst. Comput. **404**, 245–257 (2015)
13. Mirjalili, S.: The ant lion optimizer. Adv. Eng. Softw. **83**, 80–98 (2015)
14. Guha, D., Roy, P.K., Banerjee, S.: Application of modified biogeography based optimization in AGC of an interconnected multi-unit multi-source AC-DC linked power system. Int. J. Energy Optim. Eng. **5**(3), 1–18 (2015)

ICMP-DDoS Attack Detection Using Clustering-Based Neural Network Techniques

Naorem Nalini Devi, Khundrakpam Johnson Singh and Tanmay De

Abstract DDoS comprises of one of the biggest problems in the network security. Monitoring the traffic is the fundamental technique used in order to discover the entity of probable irregularity in the traffic patterns. In this paper, we used SOM to divide the dataset into clusters, as analysis of clusters is easier than the whole dataset. We select the features such as mean inter-arrival time and mean probability of occurrence of the IP addresses that have the greater impact on the DDoS attack from the incoming packets. These features are given as input to the SOM to cluster the structure of similar member in a collection of unlabeled data. The comparison is made between pre-observed features from already trained datasets and features present in each cluster. MLP classifier is used to categorize the incoming clients as normal and attack. In this paper, we used CAIDA 2007 attack datasets and CAIDA 2013 anonymized trace datasets as pre-observed samples. The proposed method detects a DDoS attack with maximum efficiency of 97% and with a low false positive rate of 3.0%.

Keywords DDoS attack · SOM · ICMP · MLP · Clusters

1 Introduction

The explosive growth rate of interconnection among computer systems has created a lot of instability and network security problems. One of the severe threats faced by the network security is Distributed Denial of Service (DDoS) where legitimate

N. Nalini Devi (✉) · K. J. Singh
Department of CSE, NIT Manipur, Imphal, India
e-mail: nnchanchan@gmail.com

K. J. Singh
e-mail: johnkh34@gmail.com

T. De
Department of CSE, NIT Durgapur, Durgapur, India
e-mail: tanmayd12@gmail.com

S. Kar et al. (eds.), *Operations Research and Optimization*, Springer Proceedings in Mathematics & Statistics 225, https://doi.org/10.1007/978-981-10-7814-9_16

users are denied from serving the services, thereby exhausting the available resources of the servers. Intrusion detection system (IDS) [1, 2] is being designed and developed in order to protect computer networks against the repeated expansion of different types of threats along with the DDoS attack. In general, IDS can be of two approaches, anomaly detection and misuse (or signature) detection. Misuse detection [3] relies on prior knowledge of DDoS attack patterns. Although misuse detection system gives accurate results in detecting known attacks, its major drawback is that network attacks are covered by repeated expansion that required the latest knowledge base for all attacks. Anomaly detection has a greater advantage to detect unknown attacks and mainly focuses on comparing the normal behavior and abnormal behavior of the system. Processing and analyzing of large data samples to observe similarities for each and every data is not an easy task since it takes time and requires patience in grouping the data of similar properties which can also lead to an inaccurate grouping of data. Our paper proposes a new approach that designs a model to detect Internet Control Message Protocol (ICMP) DDoS attack in the network layer based on a class of neural networks known as self-organizing maps (SOMs), popularly known as Kohonen's self-organizing maps (KSOMs). Using of SOM can easily interpret and understand any given data samples and cluster the data more accurately based on their identical properties in few seconds. Using of SOM reduces time, and reduction of dimensionality and grid clustering makes it simpler to observe similarities in the data. Therefore, SOM gives a better technique than the existing techniques without SOM application, and it is experimentally proven that will be discussed in the further sections. The proposed method is used for determining the real-time incoming network traffic. For the estimation of our approach, we used the CAIDA 2007 attack dataset [4] and CAIDA 2013 anonymized trace dataset [4].

The paper is systematized as follows: Sect. 2 presents the related work that attempts to detect DDoS attack. Section 3 describes our proposed methodology, using of SOM algorithm, and also displays the classification of cluster datasets using MLP classifier. Section 4 presents the results and discussion of the proposed method. We conclude our paper in Sect. 5 and highlight some future work.

2 Related Work

With the rise in threat of DDoS attack on the Internet, Du et al. [5] proposed a credit-based accounting mechanism where sender needs to send their packets based on their credit points. When credit points exceed a pre-defined threshold, the client is well-behaving client. Otherwise, the client is an ill-behaving client, and it is blocked. Karanpreet et al. [6], Saurabh and Sairam [7], and Bhavani et al. [8] proposed IP traceback and post-attack analysis of traffic that helps in reducing the possibility of future attacks by disclosing the compromised systems and sometimes the actual attacker. Sang et al. [9] proposed an enhanced detection model using traffic matrix. Furthermore, researchers use genetic algorithm for optimization of parameters utilized in the traffic pattern and decide whether inbound traffic is an attack or normal by comparing the computed variance and threshold value set by the GA. Alan et al. [10] described the detection and mitigation of known and unknown DDoS attacks in real-time environment using ANN algorithm to recognize DDoS attack based on distinct patterns where it defends DDoS attacking traffic from reaching the target but allows legitimate traffic to pass through. Rashmi and Kailas [11] proposed the DDoS defense mechanisms and different countermeasures such as prevention technique and detection technique that are used for mitigating against the DDoS attack mostly dominated by an attacker that desires to build an authorized access to the victim. Hongbin et al. [12] proposed identifier/locator separation for defending a DDoS attack which proves to be more secure than the networks without identifier/locator separation. Sujatha et al. [13] differentiated flash crowd traffic from attack traffic by using AYAH Webpage that allows dynamic determination of whether a signature truly represents an attack or non-human users like robots or a legitimate human user. Liao et al. [14] proposed the detection of application-layer DDoS attack during a flash crowd event based on aggregate-traffic analysis and differentiation between user behaviors based on Web log analysis.

3 Proposed Methodology

Figure 1 represents the workflow of the proposed methodology for the classification of incoming network traffic. The workflow consists of the modules, viz. data clustering using SOM, training datasets, MLP classifier, and some unknown datasets. First, we collect the datasets, i.e., CAIDA attack 2007 and CAIDA 2013 anonymized trace datasets from the CAIDA data center. The received data are observed and analyzed. From the observed data, features such as mean inter-arrival time (MIAT) and mean probability of occurrence of IP (POIP) are chosen. Since the behavioral changes of these two features are relatively high for each IP comparing to the other features, i.e., time-to-live (TTL), sequence number, total length, etc. MIAT is calculated by choosing 20 instances from the arrival time and taking their mean. POIP is computed by fixing a time window of 20 ms each for 3 min and finding the average value of the results found in the 20-ms time windows. The selected parameters are presented as an input to the SOM clustering to separate the datasets into clusters. The cluster data are compared with the CAIDA datasets and subdivided the datasets into two training datasets, i.e., training dataset 1 and training dataset 2 based on their behaviors. The training datasets having identical properties with that of the CAIDA attack dataset are assigned as an attack and the other as normal. The two training datasets are given as input to the multilayer perceptron (MLP) for classification. The MLP classifier will classify the given input as an attack or normal. Lastly, an unknown dataset is tested with MLP to classify the input data as either attack or normal.

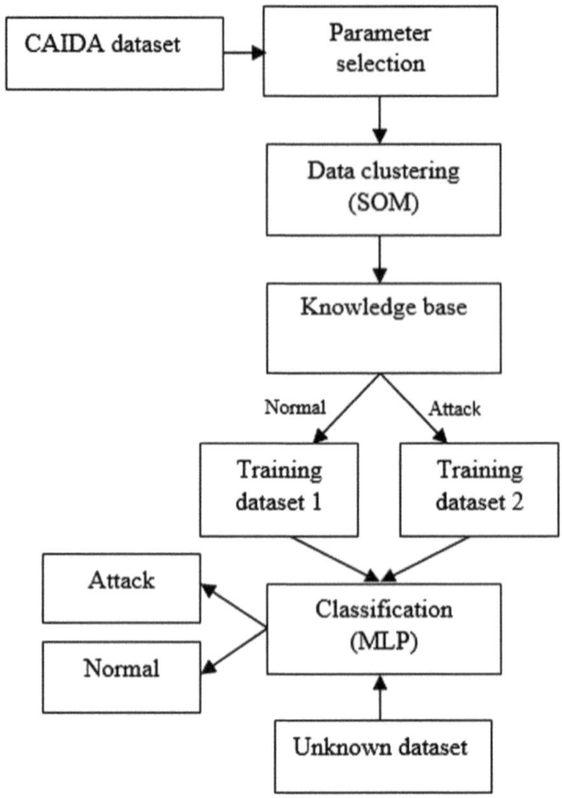

Fig. 1 Workflow of the proposed methodology for DDoS detection

3.1 Self-organizing Maps

SOM popularly known as KSOM [1, 15–18] is neurobiologically motivated. It is a type of artificial neural network that is trained using unsupervised learning. SOM uses a competitive learning mechanism. SOM organizes the neurons in the structure of the lattice. The lattice can be 1D, 2D, or even high-dimensional space. For practical applications, 1D and 2D are normally used. When the input patterns are fed to the output neurons, these input patterns will be acting as stimuli to those neurons which are output, and neurons in the output layer compete among themselves; then, one of the neurons will emerge as a winner whose weight vector lies closest to the input pattern. It is solely determined by calculating the Euclidean distance between the input vector and the weight vector. It is also referred to as the winner takes all mechanism since all the synaptic weights of the neighboring neurons are adjusted for the winning neuron. So if the same pattern is presented once more, then it has the greater chance of winning the competition. Neurons which are closed to the winning neuron tend to have an excitatory response, whereas inhibitory response is created for the neurons which are far apart. With SOM, dimensions will be reduced, and similarities will be displayed. In SOM, the topological relations are fixed from the beginning.

3.1.1 Learning Steps of SOM

Algorithm 1: Learning of SOM.

BEGIN

Step 1: Initialization of random weights with small random values.
Step 2: Using X as an input pattern.
Step 3: Finding of winning neuron (Best Matching Unit) can be determined using the Euclidean Dist. (1) [1] between the input data and weight of each neuron.

$$Dist. = \sqrt{\sum_{i=0}^{n} (X - W_{ij})} \tag{1}$$

Step 4: Winning neuron [1] updates all the weights of neighboring neurons using the Eq. (2).

$$\forall_j: W_{ij}(t) = W_{ij}(t-1) + \alpha(t)\eta(t').(X_i(t) - W_{ij}(t-1) \tag{2}$$

where,

α is the learning rate
η is the neighborhood function
t′ is the time that was spent in the current context

The neighborhood function t′ increases as η decreases.
Step 5: Repeat steps b, c and d until convergence.

END

3.1.2 Example of SOM Clustering

Figure 2 represents the SOM network architecture where input units are fully connected to the output units, and by randomly initializing the weight matrix, we determine the winning neuron using the Euclidean Dist. formula.

Weight matrix (0 < W<1) (using step 1)

$$\begin{cases} units1: \\ units2: \end{cases} \begin{bmatrix} 0.2 & 0.6 \\ 0.8 & 0.4 \end{bmatrix}$$ is the weight of the output unit.

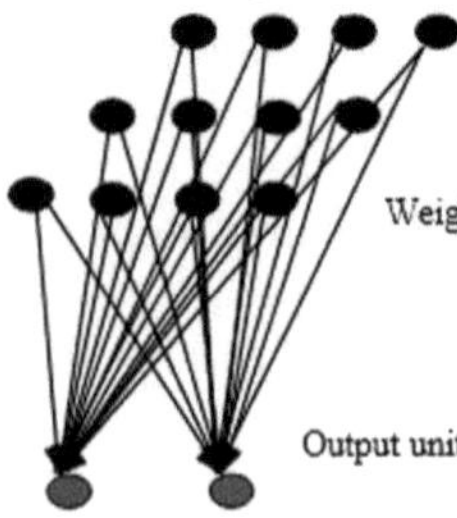

Fig. 2 SOM network architecture

Sample 1: (1.211, 0.052) (using step 2)

Dist. from unit 1 weight and unit 2 weight is calculated using Eq. (1).

Dist. from unit 1 weights = sqrt $((1.211-0.2)^2+(0.052-0.6)^2)=1.149$

Dist. from unit 2 weights = sqrt $((1.211-0.8)^2+(0.052-0.4)^2)=0.5385$ (winner)

Similarly, using the same procedure and weight matrix, we find the winning neuron for the remaining training samples as given in Table 1.

SOM clustering results as shown in Fig. 3 are computed using Neural Network Clustering Tool (nctool), a pre-defined function of MATLAB. In Fig. 3a, lines between the clusters give their relations. In Fig. 3b, dark colors between clusters indicate that the two clusters are closely related, whereas light colors are not closely related. Figure 3c gives the SOM input planes, and Fig. 3d presents SOM hits.

Table 1 Clustering of training samples

Training samples	Dist. from unit 1 weight	Dist. from unit 2 weight	Winning neuron	Class
(1.211,0.052)	1.149	0.5385	Unit 2	1
(1.516,0.009)	0.5702	0.8669	Unit 2	1
(0.016,0.03)	1.4426	0.8158	Unit 1	0
(0.013,0.04)	0.5903	0.8654	Unit 1	0
(1.187,0.008)	1.1504	0.5508	Unit 2	1
(0.016,0.03)	0.5989	0.8669	Unit 1	0
(2.379,0.04)	2.2498	1.6195	Unit 2	1
(1.743,0.022)	1.6477	1.0159	Unit 2	1
(3.503,0.008)	3.3556	2.7312	Unit 2	1
(0.01,0.06)	0.5724	0.8600	Unit 1	0
(0.017,0.09)	0.5418	0.8421	Unit 1	0
(0.015,0.03)	0.5992	0.8678	Unit 1	0
(3.423,0.012)	3.2761	2.6515	Unit 2	1
(3.171,0.006)	2.9970	2.4035	Unit 2	1

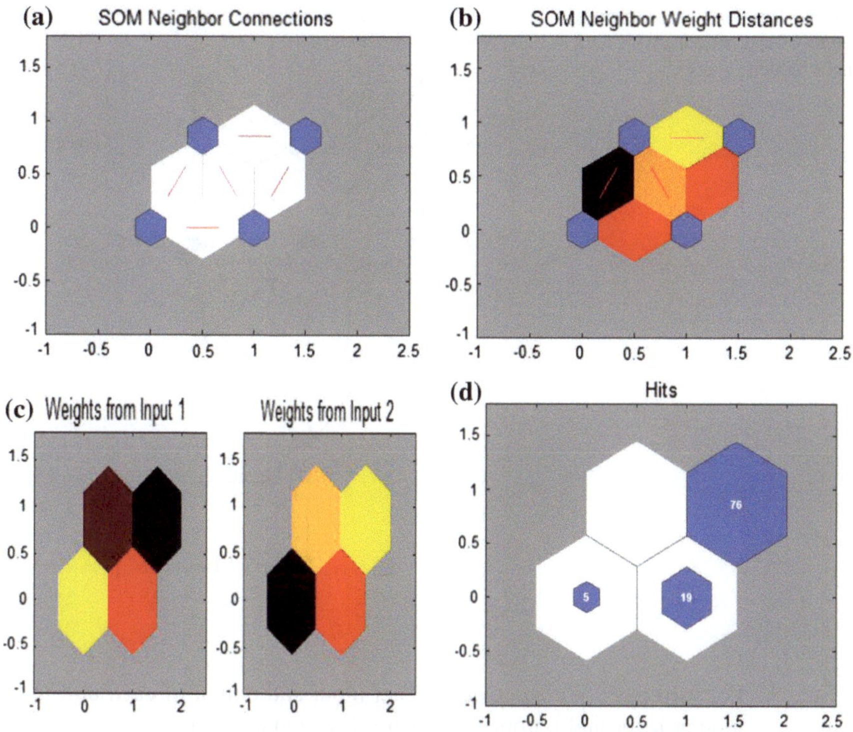

Fig. 3 SOM clustering results, **a** SOM neighbor connections, **b** SOM neighbor weight distances, **c** SOM input planes, and **d** SOM hits

3.2 Multilayer Perceptron

MLP [19, 20] is a feed-forward neural network model, consisting of multiple layers of nodes (neurons) in a directed graph where each layer is completely connected to the next layer. Every node is a neuron with a nonlinear activation function excluding the input nodes. The nodes are arranged in multiple layers, i.e., an input layer, hidden layer, and an output layer as shown in Fig. 4. MLP exercises supervised learning method known as back-propagation for training the network.

3.2.1 Training Steps of Back-Propagation Algorithm

Algorithm 2: Training of MLP.

BEGIN

Step 1: Initialization of weights and biases.
Step 2: Provide training samples.

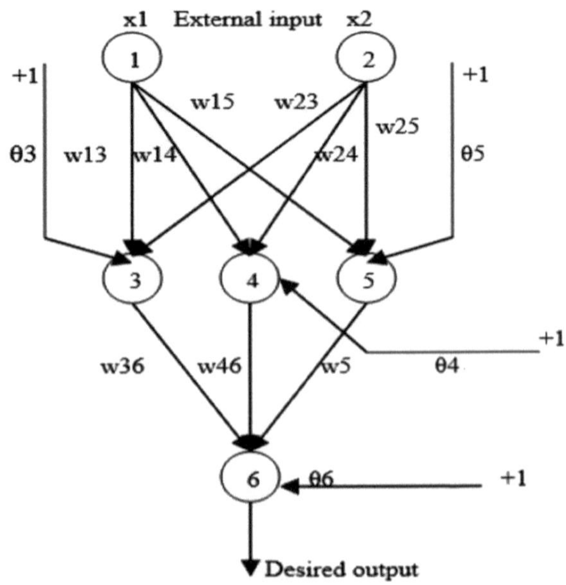

Fig. 4 Topology of a typical feed-forward network with three hidden layers

Step 3: Forward pass: We calculate the net input and output of each unit in the hidden layer and output layer using Eqs. (3) and (4) [21].

$$I_j = \sum_j \left(W_{ij}O_i + \theta_j\right) \tag{3}$$

where, I_j is the inputs present to the neuron. W_{ij} is the weight of the connection from unit i in the previous layer to unit j. O_i is the output of unit 1 from the previous layer. θ_j is the bias of the unit. Assuming that the activation function [21] in all three neurons are sigmoid functions given by

$$O_j = \frac{1}{1 + e^{(-I_j)}} \tag{4}$$

where O_j is the actual output of unit j (computed by the activation function).

Step 4: Backward pass: For a unit K in the output layer, the error is calculated by using the Eq. (5) [21].

$$Err_k = O_k(1 - O_k)(T_K - O_k) \tag{5}$$

where, O_K is the actual output of unit k. T_k is the true output based on known class label. $O_K(1 - O_k)$ is a derivative (rate of change) of the activation function. For a unit j in the hidden layer, the error is computed using Eq. (6) [21].

$$Err_j = O_j\left(1 - O_j\right) \sum_k Err_k W_{jk} \tag{6}$$

where, W_{jk} is the weight of the connection from unit j to unit k in the next higher layer. Err_k is the error of unit K.

Step 5: Update weights and biases to match the propagated errors. Weights are updated using the Eqs. (7) and (8) [21]. Where, α is the constant between 0.0 and 1.0 matching the learning rate, this learning rate is fixed for implementation.

$$\Delta W_{ij} = (\alpha) Err_j O_i \tag{7}$$

$$W_{ij} = W_{ij} + \Delta W_{ij} \tag{8}$$

Biases are updated using the Eqs. (9) and (10) [21].

$$\Delta\theta_j = (\alpha) Err_j \tag{9}$$

$$\theta_j = \theta_j + \Delta\theta_j \tag{10}$$

Step 6: At the end of the epoch → check if the stopping criteria are satisfied. The stopping criteria are given by ((Maximum number of iterations > Iteration threshold) || (Error Function < Error threshold))

END

Algorithm 3: Assigning threshold value.
In the paper, we set the threshold to 0.5.

```
BEGIN
If (Stopping criteria are satisfied)
  Stop training
Else
Continue training
    Epoch++
Go to Step 1
END
```

Training samples are presented to the input layers which are then propagated forward through the hidden layers and yield an output activation vector in the output layer. Training samples are classified using MLP classifier as shown in Table 5. Some unknown dataset is provided which is to be tested and classified based on their identity properties. The connection weights in the hidden layer are W_{ij}, and W_{jk} is the connection weights in the output layer. Temp is the result obtained from W_{jk} (connection weight in the output layer) * Y_j (output obtained from the hidden layer).

4 Results and Discussions

To test the efficiency of the classifier, we have used Neural Network Pattern Recognition Tool (nprtool), a pre-defined function of MATLAB. Using this pre-defined function, we find the accuracy and the corresponding ROC curve of the classifier. Unknown test samples are classified as an attack and normal using the Neural Network/Data Manager (nntool), which is also a pre-defined function of MATLAB. A confusion matrix [22] includes information on actual and predicted classification which is used to describe the efficiency of a classification model on a test dataset for which the values are known. The efficiency of the confusion matrix is computed using Eq. (11).

$$Accuracy = \frac{TP + TN}{TP + FN + FP + TN} \tag{11}$$

where

TP is the true prediction that an instance is (+)ve.
FN is the true prediction that an instance is (−)ve.
FP is the false prediction that an instance is (+)ve.
TN is the false prediction that an instance is (−)ve.

To provide a comprehensive evaluation, we compared the data of CAIDA dataset before SOM application and after SOM application as displayed in Fig. 5. In this comparison, data after SOM application have the higher proficiency on classifying the incoming traffics as normal and attack with an efficiency of 97% as shown in Table 2 than that of the data before SOM application. The corresponding receiver operating characteristic (ROC) curve is computed for each confusion matrix. The more the curve tends toward the left-hand corner and the top outline of the ROC space, more is the accuracy. The area below the curve is a measure of test efficiency. An unknown test sample which is given in Table 3 is classified using MLP classifier before SOM application, whereas test samples in Table 4 are classified using the same classifier after SOM application according to their behavioral changes. Based on the class that is classified, as shown in Tables 3 and 4, some test samples are misclassified in Table 3. However, in Table 4 test samples are more accurately classified. Hence, Table 4 gives a better performance in classifying unknown datasets, as either attack or normal.

4.1 Time Complexity

In the paper, we use both SOM and MLP algorithms for classification. The time complexity for SOM algorithm is given by T1, and the time complexity for MLP algorithm is given by T2.

T1 = O(S2), where S is the number of samples used in SOM.

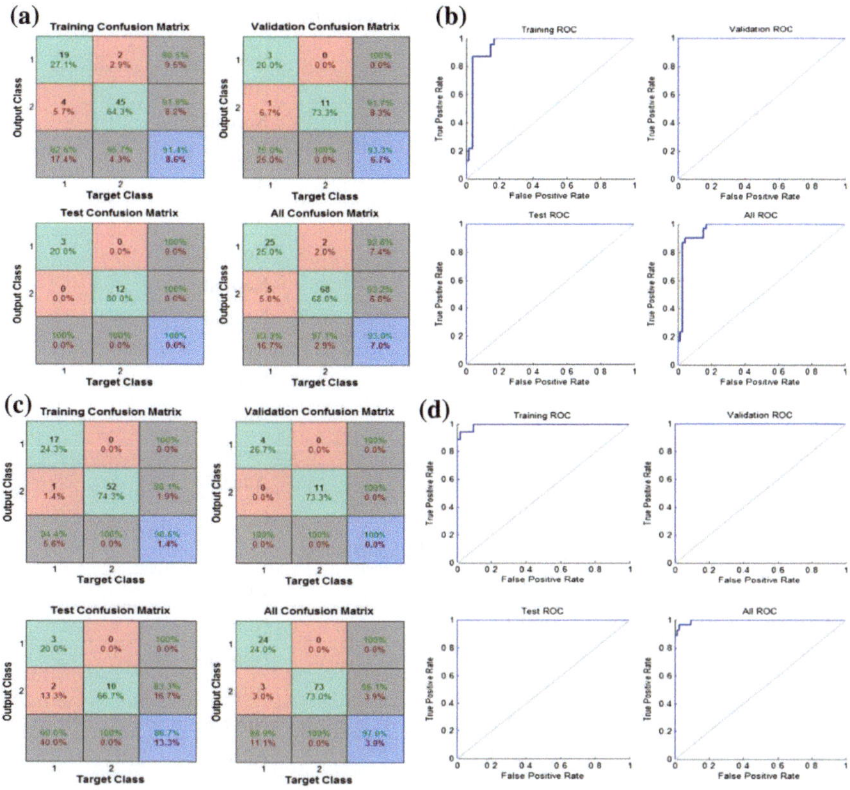

Fig. 5 MLP results, **a** and **c** confusion matrices before and after SOM application, **b** and **d** corresponding ROC curves of confusion matrices before and after application of SOM

Table 2 Accuracy comparison table of MLP before and after SOM application

Conditions	Accuracy (%)	TP	TN	FP	FN
Before SOM application	93	83.3	97.1	6.8	7.4
After SOM application	97	88.9	100	3.9	0

T2 = O(2^n), where n is the number of neurons used in the MLP structure. Therefore, the total time complexity for the proposed method is T1 + T2.

In Table 5, W_{ij} represents the weights between the input layer and hidden layer. Since the number of input attributes is two and the number of hidden neurons is three, we have only six weights from input layer to hidden layer. Similarly, W_{jk} represents the weights from hidden layer to output layer of MLP. Since we have three hidden neurons and one output neuron, the number of weights is three.

Table 3 A sample MLP classification before SOM application

MIAT	POIP	Class
1.502	0.012	Attack
1.187	0.008	Normal
0.016	0.03	Attack
1.743	0.022	Attack
0.01	0.06	Attack
3.423	0.012	Normal
2.379	0.04	Normal
3.503	0.008	Normal
0.013	0.04	Attack
3.171	0.006	Normal
0.017	0.09	Attack
0.015	0.03	Attack
0.021	0.18	Attack

Table 4 A sample MLP classification after SOM application

MIAT	POIP	Class
1.502	0.012	Normal
1.187	0.008	Normal
0.016	0.03	Attack
1.743	0.022	Normal
0.01	0.06	Attack
3.423	0.012	Normal
2.379	0.04	Normal
3.503	0.008	Normal
0.013	0.04	Attack
3.171	0.006	Normal
0.017	0.09	Attack
0.015	0.03	Attack
0.021	0.18	Attack

Table 5 A sample classification using MLP

Testing samples	W_{ij}	W_{jk}	$Temp = W_{jk} * Y_j$	Output	Class
(0.012,0.301)			−12.234049	0.000005	0
(1.019,0.001)			6.001845	0.997532	1
(0.01,0.255)	−7.484483		−12.106938	0.000006	0
(0.022,0.299)	7.923808		−13.026509	0.000002	0
(0.009,0.405)	0.181681	−8.306628	−11.653962	0.000009	0
(2.475,0.005)	−9.161539	−11.283950	6.006293	0.997543	1
(1.811,0.008)	10.763251	−6.006293	6.006283	0.997543	1
(0.017,0.117)	−0.040093		−7.948537	0.000353	0
(0.02,0.199)			−10.3813309	0.000031	0
(2.512,0.010)			6.006293	0.997543	1

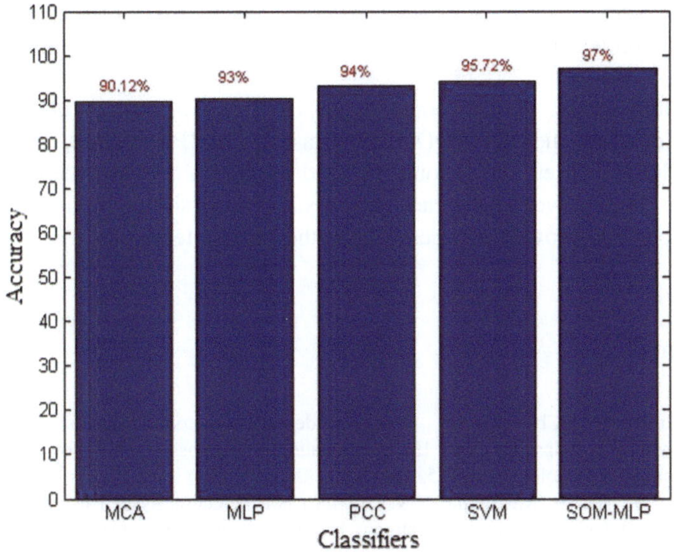

Fig. 6 Accuracy comparison graph of various classifiers

4.2 *Comparison with the Existing Methods*

To determine the superiority of the proposed method with the other existing methods, the efficiency of the proposed method before SOM and after SOM application is compared with other methods. Multivariate correlation analysis (MCA) [23] is proposed to accurately figure out the network traffic using ISCX 2012 IDS evaluation dataset which results in an accuracy of 90.12%. Pearson's correlation coefficient (PCC) [24] approach is used to discover known patterns and tested the method with generated dataset from a Web site of the World Cup 98 with an accuracy of 94%. KDD Cup 1999 dataset is trained using SVM [25] classifier and results in an accuracy of 95.72%. The comparison results as shown in Fig. 6 illustrate that our proposed detection system based on MLP classifier after SOM application achieves 97% on CAIDA dataset, which significantly exceeds the other methods based on accuracy.

5 Conclusions and Future Work

In this paper, CAIDA dataset is observed, analyzed, and clustered using SOM. Our approach is an attempt to detect the malicious incoming traffic and classify them into an attack and normal based on MLP classifier to prevent the illegitimate clients and permit only the legitimate clients to access the service. From the simulation results, it is found that the MLP classifier algorithm before SOM application results in lower detection accuracy of 93%, whereas results after SOM application give high detection accuracy of 97% in classifying the trained datasets into normal and attack.

As future works, we can compare the efficiency of various classifiers using different datasets such as KDD Cup datasets, DARPA dataset, ISCX dataset, CONFICKER dataset, etc. It could be used to determine which classifier using a particular dataset gives highest accuracy results in classifying the incoming packet as malicious and provide services only to the legitimate clients.

References

1. Aikaterini, M., Christos, D.: Detecting denial of service attacks using emergent self-organizing maps. In: 2005 IEEE International Symposium on Signal Processing and Information Technology, pp. 375–380. IEEE (2005)
2. Raman, S., Harish, K., Singla, R.K.: An intrusion detection system using network traffic profiling and online sequential extreme learning machine. (Elsevier) **42**(22), 8609–8624 (2015)
3. Monowar, H.B., Bhattacharyya, D.K., Kalita, J.K.: A multi-step outlier anomaly detection approach to network-wide traffic. (Elsevier) **348**, 243–271 (2016)
4. The CAIDA UCSD “DDoS Attack 2007” Dataset. http://www.caida.org/data/passive/ddos20070804dataset.xml. Accessed 23 Sept 2015
5. Ping, D., Akihiro, N.: Overcourt: DDoS mitigation through credit-based traffic segregation and path migration. (Elsevier) **33,** 2164–2175 (2010)
6. Karanpreet, S., Paramvir, S., Krishan, K.: A systematic review of IP traceback schemes for denial of service attacks. Comput. Secur. (2015). http://dx.doi.org/doi:10.1016/j.cose.2015.06.007
7. Saurabh, S., Sairam, A.S.: ICMP based IP traceback with negligible overhead for highly distributed reflector attack using bloom filters. (Elsevier) **42**, 60–69 (2014)
8. Bhavani, Y., Janaki, V., Sridevi, R.: IP traceback through modified probabilistic packet marking algorithm using Chinese remainder theorem. (Elsevier) **6**(2), 715–722 (2015)
9. Sang, M.L., Dong, S.K., Je, H.L., Jong, S.P.: Detection of DDoS attacks using optimized traffic matrix. (Elsevier) **63**, 501–510 (2012)
10. Alan, S., Richard, E.O., Tomasz, R.: Detection of known and unknown DDoS attacks using artificial neural networks. (Elsevier) 1–9 (2015)
11. Rashmi, V.D., Kailas, K.D.: Understanding DDoS attack & its effect in the cloud environment. (Elsevier) **49**, 202–210 (2015)
12. Hongbin, L., Yin, L., Hongke, Z.: Preventing DDoS attacks by identifier/locator separation. (IEEE) 60–65 (2013)
13. Sujatha, S., Radcliffe, P.J.: A novel framework to detect and block DDoS attack at the application layer. In: IEEE TENCON Spring Conference, pp. 578–582. IEEE (2013)
14. Liao, Q., Li, H., Kang S., Liu, Ch.: Feature extraction and construction of application layer DDoS attack based on user behavior. In: Proceedings of the 33rd Chinese Control Conference 28–30 July 2014, Nanjing, China (2014)
15. Luiz, F.C., Sylvio, B., Leonardo, D.S.J.M., Mario, L.P.: Unsupervised learning clustering and self-organized agents applied to help network management. **54**, 29–47 (2016)
16. Emiro de la, H., Eduardo de la, H., Andres, O., Julio, O., Antonio, M.A.: Feature selection by multi-objective optimization: application to network anomaly detection by hierarchical self-organizing maps. (Elsevier) **71**, 322–338 (2014)

17. Dennis, I., Xiaobo, Z.: A-GHSOM: an adaptive growing hierarchical self-organizing map for network anomaly detection. **72**(12), 1576–1590 (2012)
18. Dusan, S., Natalija, V., Aijun, A.: Unsupervised clustering of web sessions to detect malicious and non-malicious website users. (Elsevier) **5**, 123–131 (2011)
19. Gunasekhar, T., Thirupathi, R.K., Saikiran, P., Lakshmi, P.V.S.: A survey on DDoS attacks. Int. J. Comput. Sci. Inf. Technol. **5**, 2373–2376 (2014)
20. Arun Raj Kumar, P., Sevalkumar, S.: Distributed denial of service attack detection using an ensemble of neural classifier. (Elsevier) **34**(11), 1328–1341 (2011)
21. http://slideplayer.com/slide/3278185. Accessed 20 April 2016
22. Xinyang, D., Qi, L., Yong, D., Sankaran, M.: An improved method to construct basic probability assignment based on the confusion matrix for classification problem. (Elsevier) **340–341**, 250–261 (2016)
23. Zhiyuan, T., Aruna, J., Xiangjian H., Priyadarsi, N., Ren, P.L., Jiankun, H.: Detection of denial-of-service attacks based on computer vision techniques. IEEE Trans. Comput. http://eprints.eemcs.utwente.nl/25297/01/TC-2014-04
24. Theerasak, T., Shui, Y., Wanlei, Z., Beliakov, G.: Discriminating DDoS attack traffic from flash crowd through packet arrival patterns. In: The First International Workshop on Security in Computers, Networking and Communications, pp. 969–974
25. Sin, J.H., Min, Y.S., Yuan, H.C., Tzong, W.K., Rong, J.C., Jui, L.L., Citra, D.P.: A novel intrusion detection system based on hierarchical clustering and support vector machines. **38**(1), 306–313 (2011)

Bio-economic Prey–Predator Fishery Model with Intratrophic Predation, Time Delay in Reserved and Unreserved Area

D. Sadhukhan, B. Mondal and M. Maiti

Abstract In this paper, we have studied the dynamics of a fishery system by dividing the marine aquatic environment in two zones, one is free fishing zone where harvesting and predation are allowed and other is reserve zone which is used only for growing the small fishes to make the marine ecosystem stable. Here harvesting and predation are not allowed. In harvesting zone, there are predators which follow intratrophic predation. We also incorporate time delay in this intratrophic interaction. At the first part of the problem, we have studied the local stability and bionomic equilibrium of the system without time delay, and in the second part of the study, the stability and bifurcation of the model have been discussed taking delay parameter into account. Optimal harvesting policy with Pontrygian's maximal principle has also been discussed for the model. Finally, some numerical results and simulation are given to illustrate the model.

Keywords Prey–predator · Stability · Reserved area · Intratrophic predation Optimal harvesting

1 Introduction

Biological resources are renewable resources. Since from the time of Lotka [1] and Volterra [2], there are numerous studies considering the prey–predator interaction for fisheries and other renewable resources such as Nicholson et al. [3], Gurtin and Maccamy [4], De Angelis [5], Dekker [6], Landhal and Hansen [7], Kapur [8] and

D. Sadhukhan
Haldia Government College, Haldia, Purba Midnapore, WB, India
e-mail: dipankar.sadhukhan2@gmail.com

B. Mondal (✉)
Raja N.L.K. Womens College, Midnapore, WB, India
e-mail: abm_mondal@rediffmail.com

M. Maiti
Department of Applied Mathematics, Vidyasagar University, Midnapore, WB, India

S. Kar et al. (eds.), *Operations Research and Optimization*, Springer Proceedings in Mathematics & Statistics 225, https://doi.org/10.1007/978-981-10-7814-9_17

Smith Maynard [9]. Economic and Biological aspect of renewable resources management has been considered by Clark [10]. In recent years, the optimal management of renewable resources, which has a direct relationship to sustainable development, has been studied extensively by many authors. Again as the harvesting is a very important phenomenon for fishery in modern day life so there is need for some limit to use the natural resources for a long period of time. For the harvesting of natural resources, Clark [11] has studied few papers on optimal harvesting policy, Mesterton-Gibbons [12, 13], which presented a dynamic model for fishery resources and showed that the use of diesel-powered trawling may lead to the extinction of predator as well as prey species if the trawling efficiency in the catch of prey species is improved. Chaudhuri [14] studied the problem of combined harvesting of two competing fish species. Pitchford and Brindley [15] investigated a prey–predator model under intratrophic predations. Kar and Misra [16] have developed a prey–predator fishery model with influence of prey reserve and also noted that the fish population maintains at an equilibrium level in absence or presence of predator provided the population in the unreserved area lies in a certain interval. Dubay et al. [17] have proposed and analysed a fishery resources system with reserved area. He has also investigated optimal harvesting policy of this system. Dubay [18] has developed a model to study the role of a reserved zone on the dynamics of prey–predator system and also established that the reserved zone has a stabilizing effect on predator-prey interactions. Louartassi et al. [19] have studied the stability and static output feedback design for a model the fishery resource with reserve area. Chakraborty and Kar [20] have studied a bio-economic model of a prey–predator fishery with protected area and also discussed the system numerically and observed that marine protected area can be used as an effective management tool to improve resource rent under a number of circumstances.

In our work, we have considered prey–predator model with ratio dependant response function and also intratrophic predation under time delay. We have also assumed that no migration from unreserved zone to reserved zone and a combined harvesting in unreserved zone. We have discussed the local stability, global stability, bionomic equilibrium for the system without time delay and after that we have investigated the local stability with time delay and checked the bifurcation condition of the model taking time delay as bifurcation parameter. Optimal harvesting policy with Pontrygian's maximal principle has also been discussed for the model. Finally, some numerical results and simulation are given to illustrate the model.

2 Model Formulation

Consider a fishery habitat, in an aquatic ecosystem, with reserved and unreserved areas. In reserved area, it is considered that no harvesting and predation will take place while the unreserved area is the harvesting and predation zone. Let $x(t)$ and $y(t)$ be the respective population size of the prey in unreserved and reserved zone and let $z(t)$ be the biomass densities of the predator at time t. Let r, s are intrinsic growth rate of prey in unreserved and reserved area; K, L are the carrying capacity of prey in unreserved and reserved area and b, d are the birth and death rate of predator. Let the prey subpopulation of unreserved area migrate into reserved area at a rate σ_1 and prey subpopulation of reserved area migrate into unreserved area at a rate σ_2.

Also let 'E' be the combined harvesting effort for the fish population in unreserved areas and q_1, q_2 are catchability coefficient of prey and predator in unreserved area. Again, we assume that in each area prey population follows logistic growth. Therefore, with this condition in view the dynamics of the prey–predator system may be written in the form of a system of differential equation as:

$$\left.\begin{array}{c}\frac{dx}{dt} = rx\left(1 - \frac{x}{K}\right) - \sigma_1 x + \sigma_2 y - \frac{mxz}{x+az} - Eq_1 x\\ \frac{dy}{dt} = sy\left(1 - \frac{y}{L}\right) + \sigma_1 x - \sigma_2 y\\ \frac{dz}{dt} = bz + \frac{mxz}{x+az} + \frac{amzz(t-\tau)}{x+az} - \frac{mzz(t-\tau)}{x+az} - dz - Eq_2 z\end{array}\right\} \quad (1)$$

In this model, τ represents the time delay, which is to incorporate the interaction in between mature and juvenile predators in intratrophic predation. The model differs from the standard predator-prey model as here the food available to the predator depends on the linear combination of both the prey and predator densities. Explicitly as in [15] food available to $z = x + az$. Where a is the measure of intensity of intratrophic predation. If $a = 0$, there is no intratrophic predation and consequently the prey is the only food resource for the predator. If $a = 1$, the predator regards prey and predator alike and thus depends on both populations.

In our model, we consider that no migration will take place unreserved zone to reserved zone, so $\sigma_1 = 0$. Now let us take $\sigma_2 = \sigma$. Therefore, considering $\beta = m - am$ the model (1) reduces to

$$\begin{aligned}\frac{dx}{dt} &= rx\left(1 - \frac{x}{k}\right) + \sigma y - \frac{mxz}{x+az} - Eq_1 x\\ \frac{dy}{dt} &= sy\left(1 - \frac{y}{l}\right) - \sigma y\\ \frac{dx}{dt} &= bz + \frac{mxz}{x+az} - \frac{\beta zz(t-\tau)}{x+az} - dz - Eq_2 z\end{aligned} \quad (2)$$

In this model, the response function blows up at the point where $x = 0$ and $z = 0$ since there is an apparent division by zero. However, since both $\frac{xz}{x+az}(\geq 0)$ and $\frac{z^2}{x+az}(\geq 0) \to 0 + \text{as}\, x, z \to 0 + \cdots$ [15].

It is clear that $\frac{dx}{dt} = 0$ at $(0, 0, 0)$. Similar argument is applicable for $\frac{dz}{dt}$ at$(0, 0, 0)$. Now the problem is that the system (2) is to be analysed along with the initial conditions $x(0) > 0, y(0) \geq 0,$ and $z(0) > 0$.

3 Existence of Steady States

The equilibrium point of the system (2) is obtained by solving, $\frac{dx}{dt} = \frac{dy}{dt} = \frac{dz}{dt} = 0$.

Suppose $P(x^*, y^*, z^*)$ be the positive solution of the algebraic equations:

$$rx\left(1 - \frac{x}{k}\right) + \sigma y - \frac{mxz}{x + az} - Eq_1 x = 0 \tag{3}$$

$$sy\left(1 - \frac{y}{l}\right) - \sigma y = 0 \tag{4}$$

$$bz + \frac{mxz}{x + az} - \frac{\beta z z(t - \tau)}{x + az} - dz - Eq_2 z = 0 \tag{5}$$

Now from Eq. (4) we get

$$y = \frac{l(s - \sigma)}{s} \tag{6}$$

Again from (5)

$$z = \frac{b - d - Eq_2 + m}{\beta - a(b - d - Eq_2)} x \tag{7}$$

Therefore, with the help of (3), (6) and (7), the equation for x is

$$\frac{r}{k} x^2 + \left[\frac{m(b - d - Eq_2 + m)}{\beta + am} - r + Eq_1\right] x - \frac{\sigma l(s - \sigma)}{s} = 0 \tag{8}$$

As $s > \sigma$,, Eq. (8) has a unique positive solution x^*, if the following condition holds

$$\frac{m(b - d - Eq_2 + m)}{\beta + am} > r - Eq_1 \tag{9}$$

Therefore with the condition (9), z of (7) will be positive if

$$b - \frac{\beta}{a} < d + Eq_2 < b + m \tag{10}$$

4 Dynamical Behaviour of Steady States for $\tau = 0$

The variational matrix corresponding to the steady state $P(x^*, y^*, z^*)$ is

$$V(x^*, y^*, z^*) = \begin{bmatrix} \left(-\frac{r}{k}x^* - \frac{\sigma y^*}{x^*} + \frac{mx^* z^*}{(x^* + az^*)^2} \right) & \sigma & -\frac{mx^{*2}}{(x^* + az^*)^2} \\ 0 & -\frac{s}{l}y^* & 0 \\ \frac{(am - \beta)z^{*2}}{(x^* + az^*)^2} & 0 & \left(-\frac{amx^* z^*}{(x^* + az^*)^2} + \frac{a\beta z^*}{(x^* + az^*)^2} - \frac{\beta z^*}{(x^* + az^*)} \right) \end{bmatrix} \quad (11)$$

Therefore, the characteristic equation corresponding to the variational matrix (11) is

$$\mu^3 + a_1\mu^2 + a_2\mu + a_3 = 0 \quad (12)$$

where

$$a_1 = \left[\frac{r}{k}x^* + \frac{\sigma y^*}{x^*} - \frac{mx^* z^*}{(x^* + az^*)^2} + \frac{s}{l}y^* + \frac{amx^* z^*}{(x^* + az^*)^2} - \frac{a\beta z^*}{(x^* + az^*)^2} + \frac{\beta z^*}{(x^* + az^*)} \right],$$

$$a_2 = \frac{s}{l}y^* \left[\left\{ \frac{r}{k}x^* + \frac{\sigma y^*}{x^*} - \frac{amx^* z^*}{(x^* + az^*)^2} \right\} - \left\{ -\frac{amx^* z^*}{(x^* + az^*)^2} + \frac{a\beta z^{*2}}{(x^* + az^*)^2} - \frac{\beta z^*}{(x^* + az^*)} \right\} \right] + \frac{(am - \beta)x^{*2}z^{*2}}{(x^* + az^*)^2}$$
$$+ \left[\left\{ \frac{r}{k}x^* + \frac{\sigma y^*}{x^*} - \frac{amx^* z^*}{(x^* + az^*)^2} \right\} \left\{ \frac{amx^* z^*}{(x^* + az^*)^2} - \frac{a\beta z^{*2}}{(x^* + az^*)^2} + \frac{\beta z^*}{(x^* + az^*)} \right\} \right] \text{and}$$

$$a_3 = -\frac{s}{l}y^* \left[\frac{(am - \beta)x^{*2}z^{*2}}{(x^* + az^*)^2} + \left\{ \frac{r}{k}x^* + \frac{\sigma y^*}{x^*} - \frac{mx^* z^*}{(x^* + az^*)^2} \right\} \times \left\{ \frac{amx^* z^*}{(x^* + az^*)^2} - \frac{a\beta z^{*2}}{(x^* + az^*)^2} + \frac{\beta z^*}{(x^* + az^*)} \right\} \right]$$

$$(13)$$

Therefore, the condition for the system to be asymptotically stable in the first octant is that the eigenvalues corresponding to the variational matrix $V(x^*, y^*, z^*)$ have negative real parts, i.e. $Re(\mu) < 0$. So by Routh–Hurwitz criteria, the equivalent conditions are

$$a_1 > 0, a_1a_2 - a_3 > 0 \, and \, a_3 > 0 \quad (14)$$

where $a_1, a_2 and\, a_3$ are given by (13).

With the help of above conditions, we can find a small sphere with centre at $P(x^*, y^*, z^*)$ such that any solution $P(x, y, z)$ of the system (2), which is inside the sphere at time $t = t'$ will remain inside the sphere for all $t \geq t'$ and will tend to $P(x^*, y^*, z^*)$ as $t \to \infty$.

5 Bionomic Equilibrium

The biological equilibrium of the system (2) is given by the solution of $\frac{dx}{dt}=\frac{dy}{dt}=\frac{dz}{dt}=0$. Now for bionomic equilibrium (in which the net revenue obtained by selling the harvested species equals to the total cost of harvesting), we have to solve $\frac{dx}{dt}=\frac{dy}{dt}=\frac{dz}{dt}=0$ together with the equation in which economic rent is zero. So if c be the constant fishing cost per unit effort with p_1 and p_2, which are the constant prices per unit biomass of the landed prey and predator, respectively, from the unreserved area, then the economic rent, i.e. the revenue at any time t is given by

$$\pi(x,y,z,E)=(p_1q_1x+p_2q_2z-c)E \tag{15}$$

Therefore, the bionomic equilibrium is $B(x_b,y_b,z_b,E_b)$ where x_b,y_b,z_b and E_b are the positive solutions of $\frac{dx}{dt}=\frac{dy}{dt}=\frac{dz}{dt}=\pi=0$.

Thus,

$$\left.\begin{array}{c} x_b=\dfrac{c\{\beta-a(b-d-E_bq_2)\}}{p_1q_1\{\beta-a(b-d-E_bq_2)\}+p_2q_2(b-d-E_bq_2+m)} \\ y_b=\dfrac{l(s-\sigma)}{s} \\ z_b=\dfrac{(b-d-E_bq_2+m)}{\{\beta-a(b-d-E_bq_2)\}}x_b \end{array}\right\} \tag{16}$$

where E_b is the positive solution of the equation

$$\begin{aligned}&\{acq_1q_2(ap_1q_1q_2-p_2q_2^2)\}E^3+\left[acq_2(ap_1q_1q_2-p_2q_2^2)\left\{\frac{m(b+d-m)}{am+\beta}-r\right\}-(ap_1q_1q_2\right.\\&\quad-p_2q_2^2)\frac{l\sigma(s-\sigma)}{s}+cq_1(\beta-ab+ad)(ap_1q_1q_2-p_2q_2^2)+acq_1q_2\{p_1q_1(\beta-ab+ad)\\&\quad+p_2q_2(b+d-m)\}]E^2+\left[\frac{2rac^2q_2}{k}(\beta-ab+ad)+\{c(\beta-ab+ad)(ap_1q_1q_2-p_2q_2^2)\right.\\&\quad+acq_2\{p_1q_1(\beta-ab+ad)+p_2q_2(b+d-m)\}\}\left\{\frac{m(b+d-m)}{am+\beta}-r\right\}+cq_1(\beta-ab+ad)\\&\quad\times\{p_1q_1(\beta-ab+ad)+p_2q_2(b+d-m)\}-2(ap_1q_1q_2-p_2q_2^2)\{p_1q_1(\beta-ab+ad)\\&\quad\left.+p_2q_2(b+d-m)\}\frac{l\sigma(s-\sigma)}{s}\right]E+\frac{rc^2}{k}(\beta-ab+ad)^2+c(\beta-ab+ad)\{p_1q_1(\beta-ab+ad)\\&\quad+p_2q_2(b+d-m)\}\left\{\frac{m(b+d-m)}{am+\beta}-r\right\}+\{p_1q_1(\beta-ab+ad)+p_2q_2(b+d-m)\}^2\frac{l\sigma(s-\sigma)}{s}=0\end{aligned} \tag{17}$$

6 Local Stability Analysis for Delay Model

We now investigate the dynamics of the given system (2) for $\tau>0$. Let $P(x^*,y^*,z^*)$ be the only interior equilibrium of the system (2). Now taking $X=$

$x - x^*, Y = y - y^* and\, Z = z - z^*$ as the perturbed variables and removing the nonlinear term using equilibria conditions, we obtain the linear variational system as

$$\frac{dX}{dt} = \left[-\frac{r}{k}x^* - \frac{\sigma y^*}{x^*} + \frac{mx^*z^*}{(x^*+az^*)} \right] X + \sigma Y + \left[\frac{amx^*z^*}{(x^*+az^*)^2} - \frac{mx^*}{(x^*+az^*)} \right] Z,$$

$$\frac{dY}{dt} = \left[-\frac{sy^*}{l} \right] Y$$

and

$$\frac{dZ}{dt} = \left[\frac{mz^*}{(x^*+az^*)} - \frac{mx^*z^*}{(x^*+az^*)} + \frac{\beta z^{*2}}{(x^*+az^*)^2} \right] X + \left[-\frac{amx^*z^*}{(x^*+az^*)^2} + \frac{a\beta z^{*2}}{(x^*+az^*)^2} \right] Y - \left[\frac{\beta z^*}{(x^*+az^*)} \right] Z(t-\tau) \tag{18}$$

From the above-linearized system, we get the characteristic equation as

$$\Delta(\lambda, \tau) = \lambda^3 - A\lambda^2 + B\lambda + C + \left(D\lambda^2 - I\lambda + F\right)e^{-\lambda\tau} = 0 \tag{19}$$

where

$$A = -\frac{r}{k}x^* - \frac{\sigma y^*}{x^*} - \frac{sy^*}{l} + \frac{(a-1)mx^*z^*}{(x^*+az^*)} + \frac{a\beta z^{*2}}{(x^*+az^*)^2},$$

$$\begin{aligned} B = &\left[-\frac{r}{k}x^* - \frac{\sigma y^*}{x^*} - \frac{sy^*}{l} + \frac{mx^*z^*}{(x^*+az^*)^2} \right] \times \left[-\frac{amx^*z^*}{(x^*+az^*)^2} + \frac{a\beta z^{*2}}{(x^*+az^*)^2} \right] \\ &- \left[\frac{amx^*z^*}{(x^*+az^*)^2} - \frac{mx^*}{(x^*+az^*)} \right] \times \left[\frac{mz^*}{(x^*+az^*)} - \frac{mx^*z^*}{(x^*+az^*)^2} + \frac{\beta z^{*2}}{(x^*+az^*)^2} \right] \\ &+ \left[-\frac{r}{k}x^* - \frac{\sigma y^*}{x^*} + \frac{mx^*z^*}{(x^*+az^*)^2} \right] \times \left[-\frac{sy^*}{l} \right], \end{aligned}$$

$$\begin{aligned} C = &\left[\left\{ \frac{amx^*z^*}{(x^*+az^*)^2} - \frac{mx^*}{(x^*+az^*)} \right\} \times \left\{ \frac{mz^*}{(x^*+az^*)} - \frac{mx^*z^*}{(x^*+az^*)^2} + \frac{\beta z^{*2}}{(x^*+az^*)^2} \right\} \right. \\ &\left. - \left\{ -\frac{r}{k}x^* - \frac{\sigma y^*}{x^*} + \frac{mx^*z^*}{(x^*+az^*)^2} \right\} \times \left\{ -\frac{amx^*z^*}{(x^*+az^*)^2} + \frac{a\beta z^{*2}}{(x^*+az^*)^2} \right\} \right] \times \left[-\frac{sy^*}{l} \right], \end{aligned}$$

$$D = \frac{\beta z^*}{(x^*+az^*)},$$

$$I = \left[-\frac{r}{k}x^* - \frac{\sigma y^*}{x^*} - \frac{sy^*}{l} + \frac{mx^*z^*}{(x^*+az^*)^2} \right] \times \left[\frac{\beta z^*}{(x^*+az^*)} \right]$$

and

$$F=\left[-\frac{r}{k}x^{*}-\frac{\sigma y^{*}}{x^{*}}+\frac{mx^{*}z^{*}}{(x^{*}+az^{*})^{2}}\right]\times\left[-\frac{sy^{*}}{l}\right]\times\left[\frac{\beta z^{*}}{(x^{*}+az^{*})}\right]$$

For $\tau=0$, the characteristic Eq. (19) becomes

$$\lambda^{3}+(D-A)\lambda^{2}+(B-I)\lambda+(F+C)=0 \quad (20)$$

As Eq. (20) is same as Eq. (12), Eq. (20) has roots with negative real parts by the condition (14). Now for nonzero τ, if $\lambda=i\omega$ is a root of the characteristic Eq. (19), then we have

$$\Delta(i\omega,\tau)=-i\omega^{3}-D\omega^{2}(\cos\omega\tau-i\sin\omega\tau)+A\omega^{2}+iB\omega-i\omega I(\cos\omega\tau-i\sin\omega\tau)+F(\cos\omega\tau-i\sin\omega\tau)+C=0 \quad (21)$$

Separating real and imaginary parts, we get

$$\left.\begin{array}{l}-\omega^{2}D\cos\omega\tau+A\omega^{2}-\omega I\sin\omega\tau+F\cos\omega\tau+C=0\\-\omega^{3}+\omega^{2}D\sin\omega\tau+B\omega-I\omega\cos\omega\tau-F\sin\omega\tau=0\end{array}\right\} \quad (22)$$

From the above system (21), we obtain the sixth-order equation for ω as

$$\omega^{6}+\left(A-2B-D^{2}\right)\omega^{4}+\left(B^{2}+2AC-I^{2}+2DF\right)\omega^{2}+\left(C^{2}-F^{2}\right)=0 \quad (23)$$

The above Eq. (23) does not have any real roots if

$$\left(A-2B-D^{2}\right)>0,\left(B^{2}+2AC-I^{2}+2DF\right)>0\,and\left(C^{2}-F^{2}\right)>0 \quad (24)$$

then Eq. (22) does not have any real roots. So, there will be no purely imaginary roots for the characteristic Eq. (19). Again as the condition (14) ensures that all the roots of Eq. (12) have negative real parts, so by Rouche's theorem, it is clear that Eq. (23) also have the roots with negative real parts.

Theorem *A necessary and sufficient conditions for $P(x^{*},y^{*},z^{*})$ is to be locally asymptotically stable in presence of delay $\tau\geq 0$ if the following conditions are satisfied:*

(i) *the real parts of all the roots of $\Delta(\lambda,0)=0$ are negative;*
(ii) *for all real ω and any $\tau>0$, the following holds:*

$$\Delta(i\omega,\tau)\neq 0,\quad where\, i=\sqrt{-1} \tag{25}$$

Proof In Eq. (19) considering $\tau=0$, we get Eq. (20), which is same as (12). So first condition of the theorem is satisfied if (14) holds along with the conditions (9) and (10). Again from (24) the second condition of the theorem is obvious.

7 Bifurcation Analysis

In this section, we find out the condition for Hopf bifurcation considering the discrete time delay τ. For this, we start from the characteristic equation

$$\Delta(\lambda,\tau)=\lambda^3-A\lambda^2+B\lambda+C+\left(D\lambda^2-I\lambda+F\right)e^{-\lambda\tau}=0 \tag{26}$$

Since we are interested the bifurcation analysis around the interior equilibrium point $P(x^*,y^*,z^*)$ for the variation of delay parameter. The stability of the interior equilibrium $P(x^*,y^*,z^*)$ is determined by the sign of the real part of the characteristic root of (26).

Here λ is a function of time delay τ, so if we write $\lambda=\mu+i\nu$, then μ and ν are also functions of τ, that is, $\mu=\mu(\tau),\nu=\nu(\tau)$.

Now substituting,

$$\lambda(\tau)=\mu(\tau)+i\nu(\tau) \tag{27}$$

in the characteristic equation and separating real and imaginary parts we get

$$\begin{aligned}&\mu^3-3\mu\nu^2+\left(\mu^2-\nu^2\right)\left(De^{-\mu\tau}\cos\nu\tau-A\right)+2\mu\nu De^{-\mu\tau}\sin\nu\tau+\left(B-Ie^{-\mu\tau}\cos\nu\tau\right)\mu\\&\quad-\nu Ie^{-\mu\tau}\sin\nu\tau+Fe^{-\mu\tau}\cos\nu\tau+C=0\end{aligned} \tag{28}$$

and

$$\begin{aligned}&3\mu^2\nu-\nu^3+2\mu\nu\left(De^{-\mu\tau}\cos\nu\tau-A\right)-\left(\mu^2-\nu^2\right)2\mu\nu De^{-\mu\tau}\sin\nu\tau+\left(B-Ie^{-\mu\tau}\cos\nu\tau\right)\nu\\&\quad+\mu Ie^{-\mu\tau}\sin\nu\tau-Fe^{-\mu\tau}\sin\nu\tau=0\end{aligned} \tag{29}$$

A necessary condition for the change of stability near interior equilibrium point $P(x^*,y^*,z^*)$ is that the characteristic Eq. (26) has purely imaginary roots. As λ,μ and ν are functions of τ, the change of stability occurs at such values of τ such that $\mu(\tau)=0$ and $\nu(\tau)\neq 0$. Let $\hat{\tau}$ be the critical values of τ, such that $\mu(\hat{\tau})=0$ and $\nu(\hat{\tau})\neq 0$ then (28) and (29) become

$$\left.\begin{array}{l} -\hat{\nu}^2(De^{-\mu\tau}\cos\hat{\nu}\tau - A) - \hat{\nu}I\sin\hat{\nu}\tau + F\cos\hat{\nu}\tau + C = 0 \\ -\hat{\nu}^3 + \hat{\nu}^2 D\sin\hat{\nu}\tau + (B - I\cos\hat{\nu}\tau)\hat{\nu} - F\sin\hat{\nu}\tau = 0 \end{array}\right\} \quad (30)$$

Now we study the change of stability behaviour near the interior equilibrium point $P(x^*, y^*, z^*)$ when the value of the parameter τ passes through their critical value $\hat{\tau}$.

For this, we eliminate $\hat{\tau}$ from the above set of Eq. (30) and arranging we get

$$\hat{\nu}^6 + \left(A - 2B - D^2\right)\hat{\nu}^4 + \left(B^2 + 2AC - I^2 + 2DF\right)\hat{\nu}^2 + \left(C^2 - F^2\right) = 0 \quad (31)$$

Without any detailed analysis of the above Eq. (31), we assume that there exists at least one positive real root denoted by $\hat{\tau}$. At this point, it is quite clear that the non-existence of such a real positive root will terminate our further analysis. So, on the assumption that $\hat{\tau}$ is a positive real root of Eq. (31), we can find the critical values of the delay parameter $\hat{\tau}$ for $\nu = \hat{\nu}$ as

$$\tau = \frac{1}{\hat{\nu}} arc\tan\left[\frac{\hat{\nu}^5 D + (AI - BD - F)\hat{\nu}^3 + (BF + IC)\hat{\nu}}{(AD - I)\hat{\nu}^4 + IB\hat{\nu}^3 + (DC - AF)\hat{\nu}^2}\right] + \frac{n\pi}{\hat{\nu}} \quad (32)$$

where $n = 0, 1, 2, \ldots$.

One possible value of $\hat{\tau}$ denoted by $\hat{\tau}^0$ can be obtained from (32) for $n=0$ as follows:

$$\hat{\tau}^0 = \frac{1}{\hat{\nu}} arc\tan\left[\frac{\hat{\nu}^5 D + (AI - BD - F)\hat{\nu}^3 + (BF + IC)\hat{\nu}}{(AD - I)\hat{\nu}^4 + IB\hat{\nu}^3 + (DC - AF)\hat{\nu}^2}\right] \quad (33)$$

The expression (31) and (33) give the critical values ν and τ for which the characteristic root of (19) will have a pair of purely imaginary roots.

To verify the transversality condition of Hopf bifurcation, we examine the value of $\frac{d\mu}{d\tau}$ evaluated at $\tau = \hat{\tau}$ with the condition that $\mu(\hat{\tau}) = 0$ and $\nu(\hat{\tau}) \equiv \hat{\nu} \neq 0$. If $\frac{d\mu}{d\tau}$ is anon-vanishing quantity, stabilization cannot take place at the critical parametric value $\hat{\tau}$. We differentiate Eq. (30) with respect to τ and putting $\tau = \hat{\tau}$ and using $\mu(\hat{\tau}) = 0$ and $\nu = \hat{\nu}$ we get

$$\left.\begin{array}{l} \mathrm{A}'\left[\frac{d\mu}{d\tau}\right]_{\tau=\hat{\tau}} + \mathrm{B}'\left[\frac{d\nu}{d\tau}\right]_{\tau=\hat{\tau}} = \mathrm{C}' \\ -\mathrm{B}'\left[\frac{d\mu}{d\tau}\right]_{\tau=\hat{\tau}} + \mathrm{A}'\left[\frac{d\nu}{d\tau}\right]_{\tau=\hat{\tau}} = \mathrm{D}' \end{array}\right\} \quad (34)$$

where

$$\left.\begin{aligned} A' &= -3\hat{\upsilon}^2 + \hat{\upsilon}^2 D\hat{\tau}\cos\hat{\upsilon}\hat{\tau} + 2D\hat{\upsilon}\sin\hat{\upsilon}\hat{\tau} + B - F\cos\hat{\upsilon}\hat{\tau} + I\hat{\upsilon}\hat{\tau}\sin\hat{\upsilon}\hat{\tau} + F\hat{\tau}\cos\hat{\upsilon}\hat{\tau} \\ B' &= \hat{\upsilon}^2 D\hat{\tau}\sin\hat{\upsilon}\hat{\tau} - 2D\hat{\upsilon}\cos\hat{\upsilon}\hat{\tau} + 2A\hat{\upsilon} - I\sin\hat{\upsilon}\hat{\tau} - I\hat{\upsilon}\hat{\tau}\cos\hat{\upsilon}\hat{\tau} - F\hat{\tau}\sin\hat{\upsilon}\hat{\tau} \\ C' &= \left(F\hat{\upsilon} - D\hat{\upsilon}^3\right)\sin\hat{\upsilon}\hat{\tau} \\ D' &= \left(F\hat{\upsilon} - D\hat{\upsilon}^3\right)\cos\hat{\upsilon}\hat{\tau} - I\hat{\upsilon}^2\sin\hat{\upsilon}\hat{\tau} \end{aligned}\right\} \tag{35}$$

Solving (34) for $\left[\frac{d\mu}{d\tau}\right]_{\tau=\hat{\tau}}$ we obtain

$$\left[\frac{d\mu}{d\tau}\right]_{\tau=\hat{\tau}} = \frac{A'C' - B'D'}{A'^2 + B'^2} \tag{36}$$

Clearly, the sign of $\left[\frac{d\mu}{d\tau}\right]_{\tau=\hat{\tau}}$ is same as that of the sign of $A'C' - B'D'$ Now it is clear that $A'C' - B'D' \neq 0$.

Therefore, for the appropriate values of the parameters and with the help of Eq. (36) and $A'C' - B'D' \neq 0$, we see that $\left[\frac{d\mu}{d\tau}\right]_{\tau=\hat{\tau}} \neq 0$ and consequently the transversality condition of Hopf bifurcation is satisfied for $\tau = \hat{\tau}^0$ which is given in (33). Thus, the model system exhibits Hopf bifurcation of small amplitude periodic solutions as τ passes through its critical value $\tau = \hat{\tau}^0$.

8 Optimal Harvesting Policy

In this section, the present value J of continuous time-stream of revenues is given by

$$J - \int_0^\infty e^{-\delta t}\pi(x, y, z, E, t)dt \tag{37}$$

where $\pi(x, y, z, E, t) = (p_1q_1x + p_2q_2z - c)E, \delta$ denotes the annual discount rate. Now we maximize J subject to the system of Eq. (2) using Pontryagin's Maximal Principle [21]. The control variable $E(t)$ is subjected to the constraints, $0 \leq E(t) \leq E_{max}$, so that $V_t = [0, E_{max}]$ is the control set and E_{max} is a feasible upper limit for the harvesting effort.

The Hamiltonian for the problem is given by

$$\begin{aligned} H = e^{-\delta t}(p_1q_1x + p_2q_2z - c)E + \lambda_1\left\{rx\left(1 - \frac{x}{k}\right) + \sigma y - \frac{mxz}{x + az} - Eq_1x\right\} \\ + \lambda_2\left\{sy\left(1 - \frac{y}{l}\right) - \sigma y\right\} + \lambda_3\left\{bz + \frac{mxz}{x + az} - \frac{\beta z^2}{x + az} - dz - Eq_2z\right\} \end{aligned} \tag{38}$$

Here, Hamiltonian H depends linearly on E with coefficient $\sigma = e^{-\delta t}(p_1 q_1 x + p_2 q_2 z - c) + \lambda_1 q_1 x + \lambda_3 q_2 z$. Consequently, its maximum value is reached for extremes of E, i.e. the harvest rate must be either $0\,\text{or}\,E_{max}$. This observation leads to the rule that one must harvest as such as possible when the switching function $\sigma > 0$ and will not harvest at all when $\sigma < 0$. Furthermore, when $\sigma = 0$ the harvest rate is undetermined. In this case, three solutions for E are possible, namely $0, E_{max}$ or E^* which is the singular control, that maintains the condition $\sigma = 0$. Therefore, the optimal control path will be either "bang-bang" control or singular. Our job is to reach optimal solution optimally from the initial state $(x(0), y(0), z(0))$. This can be achieved by applying a "bang-bang" control (Pontryagin et al. [21]) to the system as presented below.

Define

$$\tilde{E}(t) = \begin{cases} E_{max} & for\,\sigma > 0 \\ 0 & for\,\sigma < 0 \end{cases}$$

Moreover, let T be the time at which the path $(x(t), y(t), z(t))$, which generated via the "bang-bang" control $E(t) = \tilde{E}(t)$, reaches the steady state $(x_\delta, y_\delta, z_\delta)$. Then, the optimal control policy is

$$E(t) = \begin{cases} \tilde{E}(t) for\, 0 \le t \le T \\ E^* \quad for \quad t > T \end{cases}$$

and the optimal path is given by the trajectory generated by the above optimal control. In view of the stability property of the interior equilibrium of the system (2), we can also reach the singular optimal solution through a suboptimal by choosing the control policy $E(t)$ to be equal to E^* for all t. The advantage of choosing the optimal path is that it leads to the optimal singular solution more rapidly than does the suboptimal path.

Now the adjoint equations are

$$\frac{d\lambda_1}{dt} = -\frac{\partial H}{\partial x}, \frac{d\lambda_2}{dt} = -\frac{\partial H}{\partial y} and \frac{d\lambda_3}{dt} = -\frac{\partial H}{\partial z} \tag{39}$$

Therefore using Eqs. (38), (39) and (2), we have

$$\begin{aligned} \frac{d\lambda_1}{dt} &= a_1\lambda_1 + b_1\lambda_2 + c_1\lambda_3 - p_1 q_1 E e^{-\delta t}; \\ \frac{d\lambda_2}{dt} &= a_2\lambda_1 + b_2\lambda_2 + c_2\lambda_3 \quad and \\ \frac{d\lambda_1}{dt} &= a_3\lambda_1 + b_3\lambda_2 + c_3\lambda_3 - p_2 q_2 E e^{-\delta t} \end{aligned} \tag{40}$$

where

$$a_1 = -r + \frac{2rx}{k} + \frac{amz^2}{x+az} + Eq_1,$$
$$b_1 = 0, c_1 = \frac{amz^2}{(x+az)^2};$$
$$a_2 = -\sigma, b_2 = \sigma - s + \frac{2sy}{l},$$
$$c_2 = 0\, and\, a_3 = \frac{mz^2}{(x+az)^2},$$
$$b_3 = 0, c_3 = \frac{\beta(2xz - mx^2 + amz^2)}{(x+az)^2} + (d + b - Eq_2)$$

Now we use the steady-state solution from the system (40), as we are concerned with optimal equilibrium and we consider $x, y\, \text{and}\, z$ as constants in the subsequent steps.

The solutions of the above system of linear differential equations are given by

$$\lambda_1 = A_1 e^{m_1 t} + A_2 e^{m_2 t} + A_3 e^{m_3 t} + \frac{M_1}{N} e^{-\delta t} \tag{41}$$

where $m_1, m_2\, \text{and}\, m_3$ are the roots of the cubic equation

$$\beta_0 m^3 + \beta_1 m^2 + \beta_2 m + \beta_3 = 0 \tag{42}$$

With, $\beta_0 = 1, \beta_1 = -(a_1 + b_2 + c_3),\ \beta_2 = \begin{vmatrix} a_1 & 0 \\ a_2 & b_2 \end{vmatrix} + \begin{vmatrix} b_1 & 0 \\ 0 & c_3 \end{vmatrix} + \begin{vmatrix} a_1 & c_1 \\ a_3 & c_3 \end{vmatrix}$ and

$$\beta_3 = -\begin{vmatrix} a_1 & 0 & c_1 \\ a_2 & b_2 & 0 \\ a_3 & 0 & c_3 \end{vmatrix}.$$

Here λ_i is bounded iff $m_i < 0\, for\, i = 1, 2, 3\, or\, Ai = 0\, for\, i = 1, 2, 3$.

The Hurwitz matrix is

$$\begin{pmatrix} \beta_1 & 1 & 0 \\ \beta_3 & \beta_2 & \beta_1 \\ 0 & 0 & \beta_3 \end{pmatrix} with\, \Delta_1 = \beta_1, \Delta_2 = \beta_1\beta_2 - \beta_3, \Delta_3 = \beta_3(\beta_1\beta_2 - \beta_3) \tag{43}$$

Therefore, the roots of the cubic equation are real negative or complex conjugate having negative real parts iff. Δ_1, Δ_2 and Δ_3 are all greater than zero.

But $\Delta_1 < 0$, so it is difficult to check whether $m_i < 0$, therefore we assume $A_i = 0$.

Then

$$\lambda_1(t) = \frac{M_1}{N} e^{-\delta t} \tag{44}$$

By similar process, we get

$$\lambda_2(t) = \frac{M_2}{N} e^{-\delta t} \tag{45}$$

and

$$\lambda_3(t) = \frac{M_3}{N} e^{-\delta t} \tag{46}$$

where

$$M_1 = E[p_1 q_1 (c_3 + \delta) - p_2 q_2 c_1] \tag{47}$$

$$M_2 = -E\delta[p_1 q_1 (a_2 c_3 + a_2 \delta - b_2 a_3) + p_2 q_2 (b_2 a_1 + b_2 \delta - a_2 c_3)] \tag{48}$$

$$M_3 = E[p_2 q_2 (a_1 + \delta) - p_1 q_1 a_3] \tag{49}$$

and

$$N = \delta^2 + (a_1 + c_3)\delta + (a_1 c_3 - a_3 c_1) \tag{50}$$

We find that the shadow prices $\lambda_i(t)e^{\delta t}, \mathrm{i} = 1, 2, 3$ of the three species remain bounded as $t \to \infty$ and hence satisfy the transversality condition at ∞. The Hamiltonian must be maximized for $E \in [0, E_{max}]$. Assuming that the control constraints $0 \le E \le E_{max}$ are not binding (i.e. the optimal equilibrium does not occur either at $(E = 0\, or, E = E_{max})$, so we consider the singular control.

Therefore,

$$\frac{\partial H}{\partial E} = e^{-\delta t}(p_1 q_1 x + p_2 q_2 z - c) - \lambda_1 q_1 x - \lambda_3 q_2 z = 0 \tag{51}$$

or,

$$e^{-\delta t} \frac{\partial \pi}{\partial E} = \lambda_1 q_1 x + \lambda_3 q_2 z \tag{52}$$

As we know from (15) that,

$$\frac{\partial \pi}{\partial E} = (p_1 q_1 x + p_2 q_2 z - c) \tag{53}$$

This Eq. (53) indicates that the total user cost of harvest per unit effort must be equal to the discounted value of the future profit at the steady-state effort level (*cf.* Clark [11]).

Now from (52) and (53), we get

$$e^{-\delta t}(p_1q_1x + p_2q_2z - c) = \lambda_1q_1x + \lambda_3q_2z \tag{54}$$

Substituting the values of λ_1 and λ_2, we get

$$\left(p_1 - \frac{M_1}{N}\right)q_1x + \left(p_2 - \frac{M_3}{N}\right)q_2z = c \tag{55}$$

The above Eq. (51) together with Eq. (2) gives the optimal equilibrium population densities as $x = x_\infty, y = y_\infty \text{and}\, z = z_\infty$. Now when $\delta \to \infty$, the above equation leads to the result

$$p_1q_1x_\infty + p_2q_2z_\infty = c \tag{56}$$

which gives that $\pi(x_\infty, y_\infty, z_\infty, E) = 0$.

Using (55), we get

$$\pi = (p_1q_1x + p_2q_2z - c)E = \frac{(M_1q_1x + M_3q_2z)E}{N} \tag{57}$$

As the each $M1$ and $M2$ is of $o(\delta)$ and N is of $\left(\delta^2\right)$, therefore π is of $o\left(\delta^{-1}\right)$. Thus, π is a decreasing function of $\delta(\geq 0)$. We therefore conclude that $\delta = 0$ leads to maximization of π.

9 Numerical Experiments

Let us choose the parametric values as $r = 10, s = 5, k = 200, l = 200, \sigma = 0.9, m = 0.6, a = 0.1, b = 10,$
$\beta = 0.54, d = 0.001, q_1 = 0.5, q_2 = 0.3, p_1 = 2, p_2 = 10, c = 50, E = 10 \text{ and } \tau = 0.$ With these parametric values having suitable units, we get non-trivial steady states (30.17, 164, 463.3) asymptotically stable, which shows pictorially in Fig. 1, and corresponding Bionomic equilibrium point is (22.64, 164, 9.12) with the value of $E = 30.31$.

We find that the value of the optimal harvesting effort E corresponding to the optimal equilibrium (389.312,164, 73.645) is 5.28 units Also for the optimal effort E_{min} = 0, E_{max} = 20.01 and E* = 5.28, time taken to reach the optimal equilibrium point through optimal path is 1.66 unit and along suboptimal path, the time is 2.08 unit (shown in Fig. 4).

Now we check the behaviour of the system for $(E = 10 \text{ and } \tau = 0.2)$ and for $(E = 10, \tau = 0.21)$, respectively, which are depicted following in the two figures.

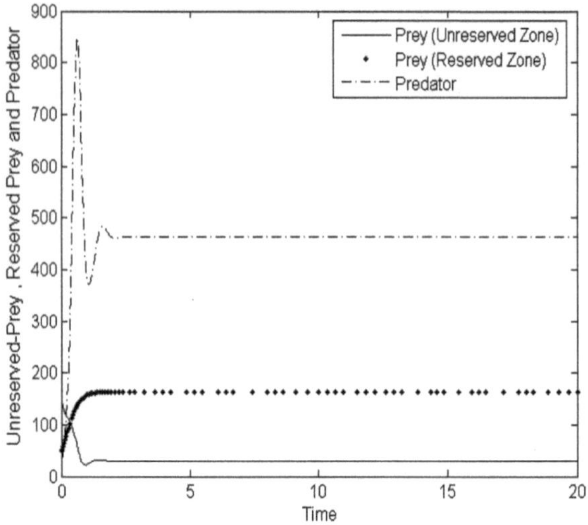

Fig. 1 Variation of the populations against time, beginning with $x = 130, y = 50$ and $z = 50$.

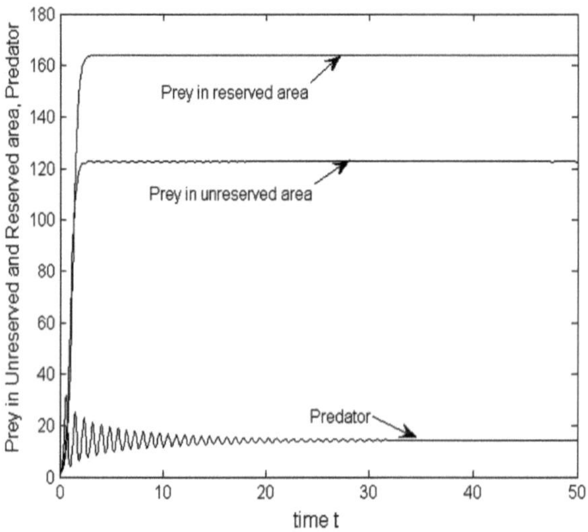

Fig. 2 Represent the stable behaviours of the populations for $\tau = 0.2$ and $\tau = 0.21$ with $E = 10$

10 Conclusion and Further Extension

In this work, a mathematical model has been formulated for an aquatic ecosystem consisting of two aquatic zones, one is predation and harvesting prohibited and other is free for harvesting and pre-predator interaction. We assumed that species

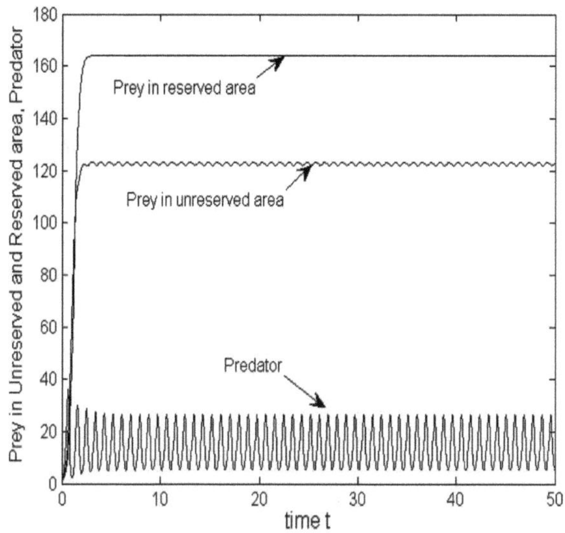

Fig. 3 Represent the unstable behaviours of the populations for $\tau = 0.2$ and $\tau = 0.21$ with $E = 10$

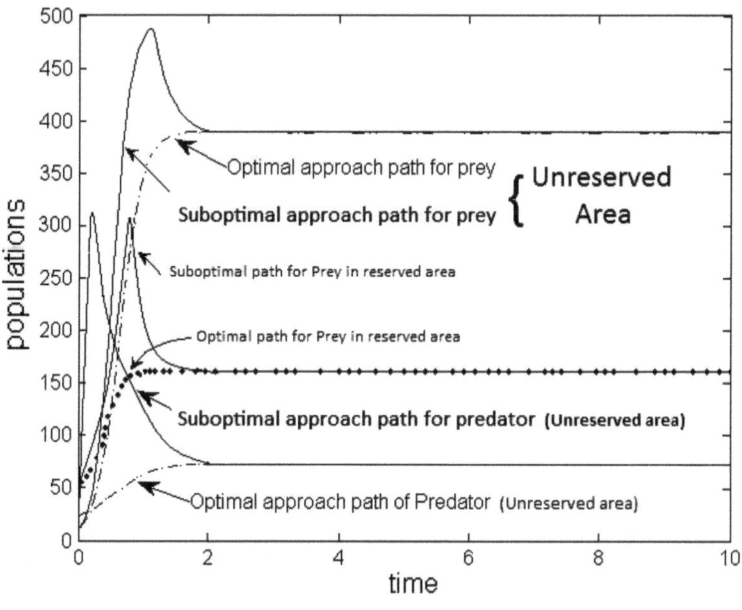

Fig. 4 Optimal and suboptimal approach paths with $E_{min} = 0$, $E_{max} = 20.01$, x(0) = 10, y (0) = 50 and z(0) = 20

follow the logistic growth law inside and outside of reserved zone and only migration takes place from reserved zone to unreserved zone but not vice versa.

Also, we incorporate intratrophic predation on predator populations using time delay.

Using Routh–Huwritz criteria, we derived the condition for local stability of the system and plotted the stability diagram in Fig. 1 for τ. Also, we have obtained the conditions for local stability with $\tau \neq 0$ along with the bifurcation for the system taking τ as bifurcation parameter. Also, we pictorially checked that as τ passes through its critical value $\tau = 20$, the stable behaviour of the system becomes unstable, which has been depicted in Figs. 2 (stable), 3 (unstable) and Fig. 4 represents the optimal approach path for the system under consideration.

Similar type models can be developed with finite time horizon and will be useful for the inland fisheries.

References

1. Lotka, A.J.: Elements of Physical Biology. Williams and Wilkins, Baltimore (1925)
2. Volterra, V.: Variazioni e fluttuazioni del numero di individui in species animali conviventi. Mem. Accd. Lincei. 31–113 (1926)
3. Nicholson, A.J., Bailey, V.A., Skellam, J.G., Brain, M.V., Proctor, J.R.: The simultaneous growth of interacting system. Acta Biotheor. **13**, 131–144 (1960)
4. Gurtin, M.E., Maccamy, R.C.: Non-linear age-dependent population dynamics. Arch. Rat. Mech. Anal. **54**, 281–300 (1974)
5. De Angelis, D.L.: Global asymptotic stability criteria for models of density-dependent population growth. J. Theor. Biol. **50**, 35–43 (1975)
6. Dekker, H.: A simple mathematical model of rodent population cycles. J. Math. Biol. **2**, 57–67 (1975)
7. Landhal, H.D., Hansen, B.D.: A three stage population model with cannibalism. Bull. Math. Biol. **37**, 11–17 (1975)
8. Kapur, J.N.: A stability analysis of continuous and discrete population models. Indian J. Pure Appl. Math. **9**, 702–708 (1976)
9. Smith Maynard, J.: Models in Ecology. Cambridge University Press, Cambridge (1971)
10. Clark, C.W.: Mathematical Bioeconomics: The Optimal Management of Renewable Resourses. Willey, New York (1976)
11. Clark, C.W.: Profit maximization and the extinction of annual species. Political Econ. **18**, 950–961 (1973)
12. Mesterton-Gibbons, M.: On the optimal policy for combined harvesting of independent species. Nat. Res. Model **2**, 109–134 (1987)
13. Mesterton-Gibbons, M.: On the optimal policy for combined harvesting of predator-prey. Nat. Res. Model **3**, 63–90 (1988)
14. Chaudhuri, K.S.: A bioeconomic model of harvesting a multispecies fishery. Ecol. Model **32**, 267–279 (1986)
15. Pitchford, J., Brindley, J.: Intratrophic predation in simple predator-prey models. Bull. Math. Biol. **60**, 937–953 (1998)
16. Kar, T.K., Misra, S.: Influence of prey reserve in a prey-predator fishery. Nonlinear Anal. **65**, 1725–1735 (2006)
17. Dubay, B., Chandra, P., Sinha, P.: A model for fishery resource with reserve area. Nonlinear Anal.: Real World Appl. **4**, 625–637 (2003)

18. Dubay, B.: A prey-predator model with reserved area. Nonlinear Anal.: Model. Control **12**(4), 479–494 (2007)
19. Louartassi, Y., Elalami, N., Mazoudi, E.: Stability analysis and static output feedback design for a model the fishery resource with reserve area. Appl. Math. Sci. **6**(66), 3289–3299 (2012)
20. Chakraborty, K., Kar, T.K.: Economic perspective of marine reserve in fisheries: a bionomic model. Math. Biosci. (2012)
21. Pontryagin, L.S., Boltyanskii, V.S., Gamkrelizre, R.V., Mishchenko, E.F.: The Mathematical Theory of Optimal Process. Pergamon Press, London (1962)

Part III
Management Applications

Applying OR Theory and Techniques to Social Systems Analysis

Tatsuo Oyama

Abstract The paper describes applications of Operations Research (OR) theory and techniques used to solve various types of social problems occurring in our social system. Social systems analysis has for quite some time been the main analytical and scientific approach used to investigate systems and to solve various problems related to modern social systems, including industry, business, the military, public administration, politics, and society in general. We will present here three major roles that operations research (OR) and social systems analysis (SSA) technique have played both practically and theoretically in the solution of social systems problems since it was developed almost 60 years ago. Firstly, we explain briefly OR, SSA, and public policy (PP) regarding what they are, how OR can be contributing to SSA and PP, and how traditional academic disciplines are related each other with the SSA. Secondly, we introduce several examples of the quantitative data analysis, which we have investigated in our school (National Graduate Institute for Policy Studies) to solve various types of social problems including population, traffic and accident, higher education policy, energy policy, and agriculture policy data analyses. Thirdly, we give mathematical modeling analysis with its application to the optimal location model analysis for integrating promotion branch offices in the local government. Fourthly, as an important role of OR as a theory building analysis technique, we explain two problems of apportionment problem and shortest path counting problem. Finally, in the summary section future perspectives of OR are given.

Keywords Operations research · Social systems analysis · Public policy Quantitative data analysis · Mathematical modeling analysis · Apportionment problem · Shortest path counting problem

---

T. Oyama (✉)
National Graduate Institute for Policy Studies, 7-22-1 Roppongi, Minato-ku, Tokyo 106-8677, Japan
e-mail: oyamat@grips.ac.jp

S. Kar et al. (eds.), *Operations Research and Optimization*, Springer Proceedings in Mathematics & Statistics 225, https://doi.org/10.1007/978-981-10-7814-9_18

1 OR, Social Systems Analysis and Public Policy

Operations Research (OR) is a scientific approach to solve various types of problems and make decisions to deal with these problems appropriately. Historically, OR appeared for the first time in the military sector during the World War II period in the 1950s, then it was applied to the private sector of business and industry, and subsequently began to be applied furthermore to the public sector. OR has been defined differently between the USA and the UK as follows. In the USA, the Operations Research Society of America defined OR in 1990 as follows.

(i) Scientific approach for decision making.
(ii) Making an optimal decision for designing and operating the system consuming a limited amount of resources under certain conditions.

In the UK, the Operational Research Society defined OR in 1962 as follows: "Operational research: applying mathematical models to investigate complex problems in industries, business, government, defense, and so on with respect to workers, machines, materials, budgets, and so on."

The characteristics of OR can be given as follows.

(i) Applying an interdisciplinary and scientific approach, and modeling the system through simplification and abstraction.
(ii) Providing the information to the top of the organization in order to make reasonable and desirable decisions regarding operation, planning, and management.
(iii) Trying to solve problems through "soft techniques" for management rather than through "hard techniques" for manufacturing.
(iv) Aiming for the long-term optimal operation of the whole system by applying mathematical programming modeling techniques with formulation, optimization, and analysis.

Social Systems Analysis tries to apply theories and methods for solving various types of societal problems occurring in our society and needing to be solved urgently. Generally, societal problems are complex, various, large scale, and diversified, which means that one unique discipline should not be sufficient to solve these problems, and thus an interdisciplinary approach is required. In the area of social systems analysis, we can say that there is no theory to define societal problems and to solve them, and also there is no principle to validate and justify these problems. Generally, in the academic field of social sciences, there is no repetition at all as societal problems can never occur under the same conditions and situations. A mathematical modeling approach can be applicable to solve these societal problems as OR theory and techniques can provide certain, e.g., optimization and/or simulation models. However, we know that these solutions obtained from applying these modeling approaches cannot be exact solutions for the original societal problems.

The basis of social systems analysis would be expressed by the census as we can say that almost all decisions made in both local and central governments would be based upon the census from the past in any country. The census was started in the USA and the UK in 1790 and 1801, respectively, while Japan started it very late in 1920. The word "statistics" originates from the German "Statistik." In the 1830s, so-called political arithmetic was initiated as "quantitative population data analysis" at London Statistics Institute. Since then, many distinguished scholars such as Jon Graunt, William Petty, Ernst Engel, and Adolphe Quetelet became interested in social phenomena such as crime, disease, politics, economics, industry, making great contributions in social science areas. Experimental observations, objective facts, and data itself are very important and can contribute to developing policy-related research, making decisions for the public sector by applying scientific methods to public policy-related problems. We find that the social sciences have been ramified and professionalized since then, and thus demographical statistics, sociology, economics, political science, public administration, and so on, including natural science and engineering and their application, have been developed. Saint Shimon, who is called the "Father of Technocracy," says that, "Science can solve not only technical problems but also social problems," and thus "technical experts" are needed to operate society.

In the twentieth century, big capitalistic businesses appeared in the USA and then they started to dominate the world economically and politically. "The American dream represented and expressed by liberty, democracy, wealth" has been realized internationally. In 1929, American President Franklin Roosevelt advocated the "New Deal Policy" consisting of NIRA, AAA, and TVA. In the 1950s, a new academic discipline called "The Policy Science" appeared in the Symposium Report at Stanford University in 1951, which is called the "Birth of Policy Sciences." In the 1960s, US President Lyndon B. Johnson proposed the "Great Society" Plan consisting of interdisciplinary researchers such as economists, OR researchers, mathematicians, computer engineers. People called this the "End-of-Ideology era." Thus, "social engineers" and "technocrats" have played major roles in all fields such as industry, university, government, and military, where all these people jointly participated to develop technology and for economic cooperation. We may say that in the "intellectual development system," the intellectually most advanced group, e.g., universities and research institutes, tries to develop science and technology and attain economic prosperity through obtaining governmental financial support and aggregating various types of research issues and research organizations.

OR may be said to be the theory and technique for conducting systems analysis or more specifically social systems analysis. Therefore, OR can be used to solve various types of societal problems appearing in the social system and making decisions in the area of public policy. Figure 1 shows the relationship among various academic disciplines focusing on social systems analysis.

We believe that OR could play three major analytical roles, through which we expect we can solve societal problems occurring in the public sector even though these solutions might not be exact, complete, and final decisive solutions. Three

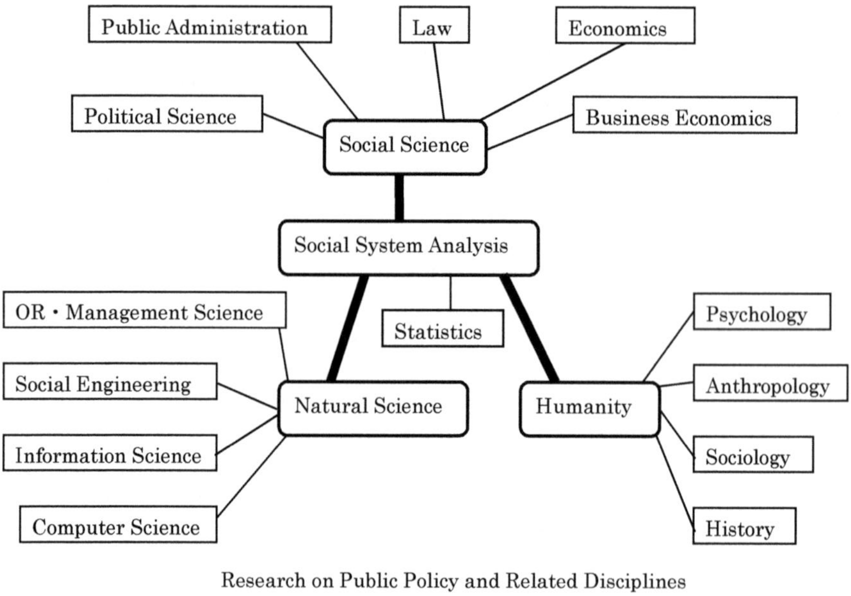

Fig. 1 Relationship between social systems analysis and various academic disciplines

major roles are as follows: (i) quantitative data analysis, (ii) mathematical modeling analysis, (iii) theory building analysis. In the following sections, we give several examples representing each type, respectively.

2 Quantitative Data Analysis

Quantitative data analysis is indispensable as a preliminary process for finding and defining problems. This process sometimes brings us new problems and new insights into the actual world. Thus, we can find an appropriate mathematical model.

We will give several examples of quantitative data analysis.

(1) Population data analysis

There are 61 cities in Japan with populations of over 300 thousand people. The graph of the descending order data is shown in Fig. 2.

Applying Jipp's law (George Kingsley Zipf, 1941) given by the following mathematical formula to the descending order data given in Fig. 2

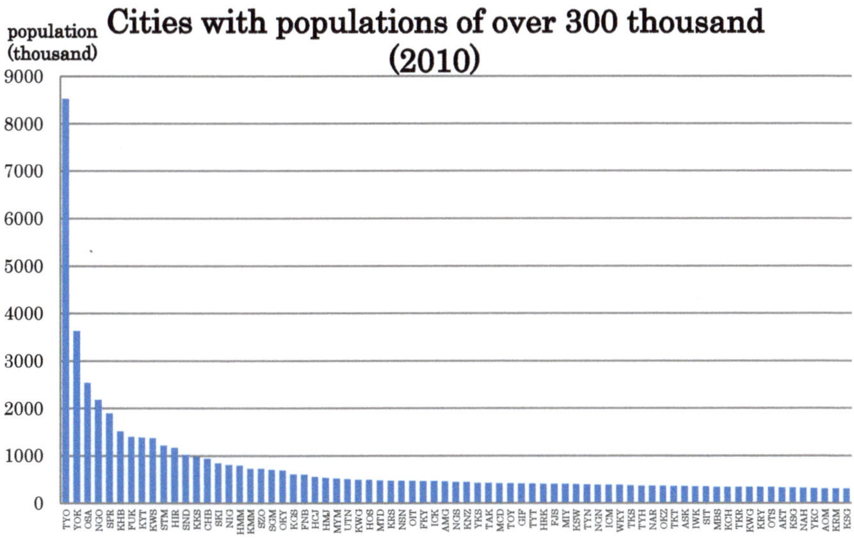

Fig. 2 Graph of the descending order population data in Japan

$$y = ax^b$$

where y: population, x: rank, a and b: parameters, and we can estimate the parameters a and b. We can interpret the estimate of parameter b as indicating the degree of concentration to the top populated cities in Japan or the degree of domination ("dominance power") by them. Thus, by investigating the historical trend, geographically regional characteristics, and international comparison, we can obtain new facts and insights with their reasons.

(2) Traffic and accident data analysis

Using survey data (Road Traffic Census) conducted by the Ministry of Land, Infrastructure, Transport and Tourism, we find the following: (i) regression models showing the relation between the factors given below were obtained. Figure 3 shows a diagram indicating the relation between various factors determining the traffic situation in Japan. (ii) Twenty-four-hour traffic data in the whole country follows the Weibull distribution. (iii) Each city's 24 h traffic data follows a logistic distribution. (iv) Twenty-four-hour travel speed data follows a Gamma distribution. (v) Investigating serious train accidents occurring in Japan over the last 35 years, we elucidated the causes for them and then have shown countermeasures to prevent these train accidents [1, 2]. (vi) Applying Bayes' theorem to actual traffic accident data, we concluded that wearing a seat belt while driving is almost four times more effective than not for avoiding deaths and serious injuries due to accidents.

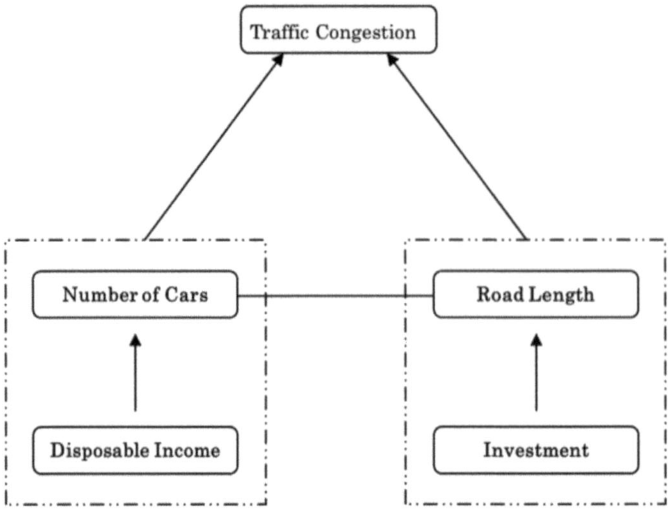

Fig. 3 Diagram indicating the relation between determining factors of the traffic situation in Japan

(3) Higher education policy data analysis

Japanese primary and secondary education systems have been evaluated highly and favorably in both domestically and internationally. However, higher education system focusing upon public and private universities in Japan has been argued quite frequently regarding how to reform and how to improve them in order to attain and maintain much higher quality. In [3], the Japanese government's subsidy policy for private universities has been investigated using actual historical data during the period from 1975 to 2004, and applying statistical approaches. We found the structural properties of the policy by applying statistical data analysis techniques including correlative rank approaches. Thus, we measured quantitatively the "dominance power" of the top-ranking subsidy-recipient private universities in Japan. Conclusively, we showed that the number of faculty members and the number of students were the most influential factors for general and special subsidies in Japan, respectively.

(4) Energy and environmental policy data analysis

The Kyoto Protocol, an international treaty linked to the United Nations Framework Convention on Climate Change (UNFCCC), was agreed in Kyoto, Japan, 1997, and entered into force in 2005. The CO_2 emission reduction targets set for OECD Annex I countries to the Kyoto Protocol have raised questions about its rationality. In [4], we tried to evaluate the emission reduction targets in the Kyoto Protocol by reviewing and investigating the energy supply and consumption structure for major developed and developing countries. The energy supply and consumption are focused upon these factors such as per capita energy consumption, per capita emission, and the share of clean energy in the energy supply structure for both 13 OECD and non-OECD Annex I and non-Annex I countries. Arguing that the emission reduction targets set in the Kyoto Protocol for various Annex I countries were neither rational nor consistent from the viewpoint of detailed energy data analysis, we recommended a more rational and simple approach, e.g., a common per capita emission norm in order to set an emission reduction target.

(5) Agricultural policy data analysis

In Japan, food self-sufficiency ratio has been measured and discussed commonly in calorie base for a long time as its sharp drop from 69% in 1960 to 39% in 2005, staying around since then until now, has become a very big political and economic issue. In [5], we investigate an internationally more common indicator, "weight-based food self-sufficiency ratio," which is more favorable for Japan than a calorie-based one, for a 45-year period from 1961 to 2005. We apply the factorial component analysis technique in order to measure the affecting and dynamically changing factors in the weight-based and calorie-based food self-sufficiency ratios. Thus, we found quantitatively the drivers of those changes. We conclude that Japanese experience of the drastic decline in its food self-sufficiency ratio during the above period was caused by the changes in the self-sufficiency ratio of the food groups/items rather than the quantity of the food supply. Presenting a list of major food groups and a number of characteristics for these changes, we made clear the causes of the above problem.

3 Mathematical Modeling Analysis

In the process of quantitative data analysis, various basic statistical theories can be applied. This process is necessary before we "build" a mathematical model. Once we find and define problems, necessary approaches (including mathematical modeling approaches) can be found rather "easily." When we try to build a mathematical model, it is necessary to investigate the quantitative data in detail as the first and preparatory stage. The common and usual process for mathematical modeling analysis can be expressed as follows: (i) define the problem → (ii) design the model → (iii) formulate the model → (iv) solve the model → (v) obtain an "optimal" solution. We need to say that the process for showing the justification and validity of the model is necessary before we try to apply our "optimal solution" to solve the problem occurring in the actual system. The diagram for indicating the common and usual mathematical modeling approach is shown in Fig. 4.

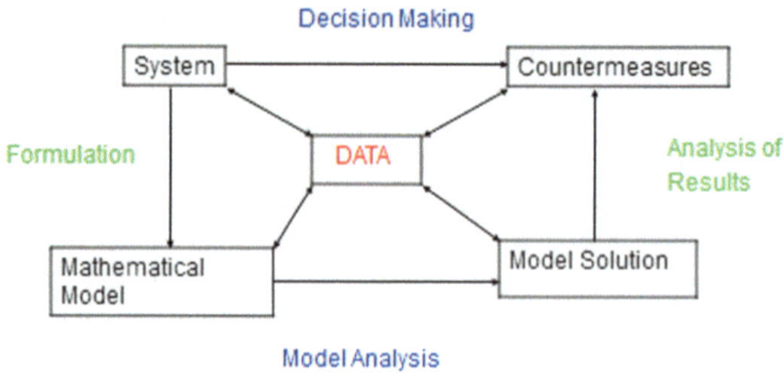

Fig. 4 Diagram for the mathematical modeling approach

Once the problem is defined such that we are interested in solving and finding an optimal solution, we first design our framework, formulate the problem, and build a mathematical model. This process is not so difficult when the problem is well and clearly defined and the quantitative data are available and investigated in detail.

In the public sector, decision making needs to be done when we make policies and we try to find or decide, e.g., which policy would be more desirable compared with others. Under such conditions in the public sector, criteria corresponding to the objective function in the optimization model might not be as clear as maximizing the profit, minimizing the cost, maximizing the efficiency, and/or minimizing the loss. Therefore, the criteria in the public sector cannot be applied to the whole population of the society while some other criteria like equality and fairness may be required also. We give examples of the mathematical modeling analyses used for making policies in the public sector.

(1) Optimal location model for the promotion branch offices in the local government [6]

We apply the mathematical modeling analysis technique in order to find the optimal integration of local promotion branch offices in Iwate Prefecture, a local government in Japan. Defining several indices for indicating the quality of services provided by local governmental offices, we try to obtain certain leveling for them. For the "demand" side, the indices we consider are the population, area, and the number of towns and villages, while for the "supply" side, the index we take is the number of staff in the offices. The research problems that we are interested in solving are as follows.

(i) What is the optimal integration of local promotion branch offices in Iwate Prefecture in Japan?

(ii) How would the optimal integration be changed by choosing different kinds of "gap indices," e.g., population gap, area gap?

(iii) Would the optimal integration be changed by choosing combined "gap indices"?

(iv) How do we evaluate the optimal integration of local promotion branch offices compared with the current situation?

(v) How can we derive a policy proposal obtained from the optimal integration by the model analysis?

In order to answer the questions above, we built a mixed-integer type optimization model to minimize the gap between the branch offices with respect to population and area. First, we define the diagram shown in Fig. 5 where the nodes correspond to the local branch offices and the edges correspond to the possibility that the "head" office can be integrated with the "tail" office.

Figure 6 shows how the objective function values can be different for each optimal solution under a different criterion for each fixed total number of local branch offices after integration, namely indicating that the optimal solution under the population gap minimization may be different from another optimal solution under the area gap minimization.

Then by linearly combining two criteria into the objective function, we try to investigate how the optimal solution under each criterion can be distant from others. Also, by adding more criteria such as the staff gap and city gap into consideration, we conclude that the best integration of local branch offices in Iwate Prefecture would be the one shown in Fig. 7.

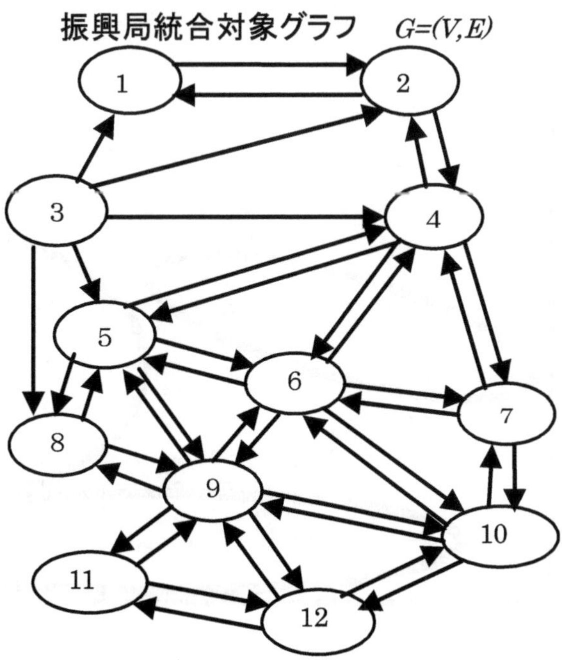

Fig. 5 Diagram of local promotion branch offices in Iwate and their integration possibilities

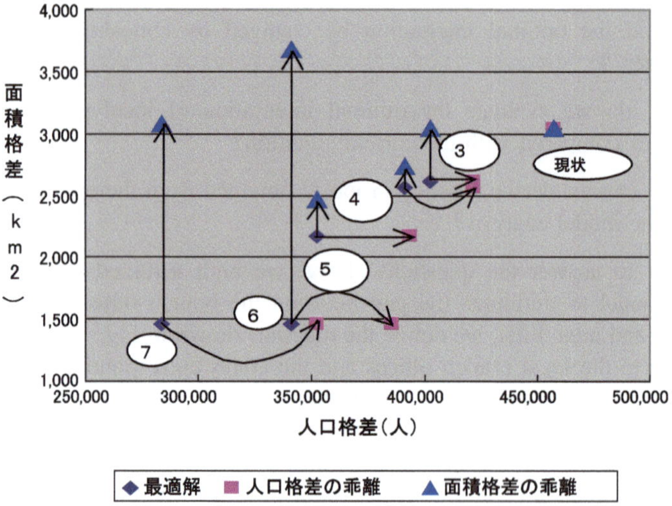

Fig. 6 Objective function gap under two criteria

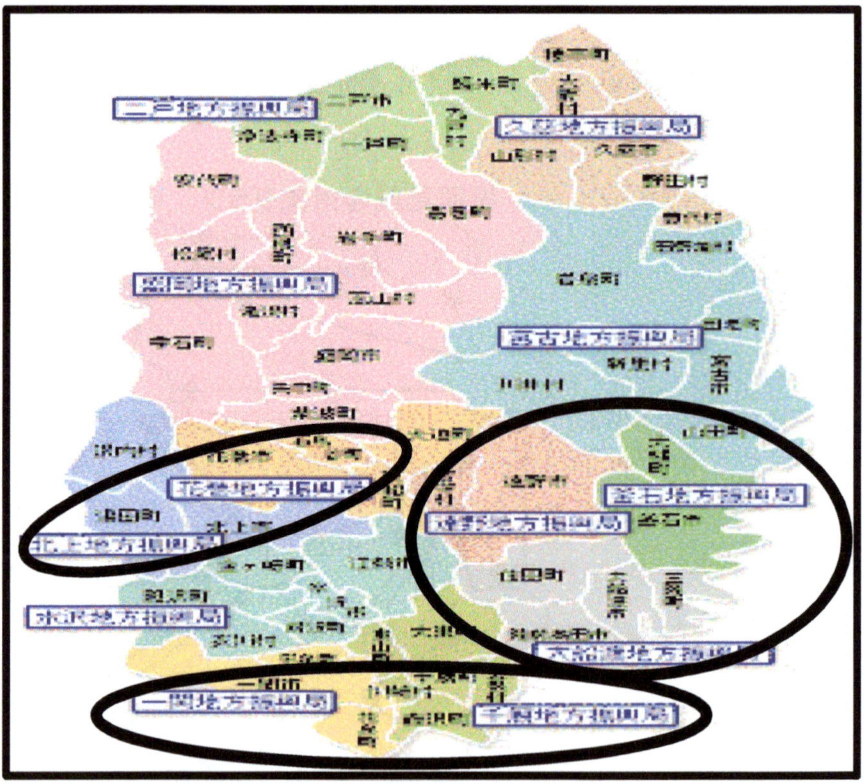

Fig. 7 Best integration of local branch offices in Iwate Prefecture

(2) Optimal railway track maintenance scheduling model [7, 8]

Japanese Railway Utilities have been evaluated highly due to their services for especially passengers maintaining safe, reliable, and punctual system as major transportation utilities in Japan for a long time. We believe that this fact has been based upon the reliable and sophisticated track maintenance work conducted by these railway utilities. As seen in [7, 8], we have been building and revising optimal track maintenance scheduling (OTMS) models to be used actually and practically in Japanese Railway Utilities in order to maintain the adequate track condition for achieving the efficient management of the railway service. Our optimization models have been already used by several major Japanese railway companies to develop their optimal track maintenance schedule.

Our OTMS optimization model tries to minimize the total maintenance cost by taking both the track tamping cost and the risk of derailment accidents into account. Our optimization model is formulated as an all-integer linear programming model to obtain the optimal railway track maintenance plan. Our model shows quantitatively the relation between the tamping cost and the quality level of track irregularities. In [7, 8], we show how to estimate the tamping cost by the quality level of the track irregularities and also how to estimate the risk of derailment accidents caused by large track irregularities considering both casualties and the probability of occurrence. In these papers, we confirm that our models are effective and useful enough to obtain an optimal track maintenance strategy.

However, once a train accident occurs in the real railway system, the losses caused by the accidents are huge and enormous not only for the railway company but also for the society itself. Thus, we can say that it is critically important to identify measures for quantitatively evaluating the risk of accidents to ensure safe and stable transportation. We propose a method for estimating the track maintenance cost and related cost associated with the risk of train derailment due to the railway track irregularities. We also try to show a desirable condition of railway tracks by applying the accident cost estimation method and the OTMS model to an actual railway section. Finally, we demonstrate how to apply these models to actual railway networks in general in addition to several special cases by validating the estimation results using actual numerical data.

4 Theory Building Analysis

OR theory and techniques have been developed and used to solve various types of problems occurring in the social system of the society or the real world. In other words, in order to solve various types of problems we are required to develop a new theory or a new technique. We give two examples such that the first one has been considered for a long time, e.g., hundreds of years so far, but is still unsolved even now, and the second one is a newly proposed problem that needs to be solved furthermore for the coming still unsolved problems.

(1) Apportionment problem [9, 10]

In the last 200 years, many apportionment methods have been proposed and various types of properties, which they should desirably satisfy, have been proposed and investigated. Given the total number of seats and the distribution of each constituency's population, the apportionment problem tries to allocate seats "fairly" among political constituencies. The apportionment problem can be formulated as follows: Given the set of N political constituencies as $S = \{1, 2, \ldots N\}$, the population of political constituency $i \in S$ as p_i, the total population as P, and the total number of seats as K, the "ideal" number of seats allocated to the constituency i, i.e., the "exact quota" q_i, is given by $q_i = \frac{p_i K}{P}$, $i \in S$ where $P = \sum_{i \in S} p_i$. Then the apportionment problem is to partition a given positive integer K into nonnegative integral parts $\{d_i | i \in S\}$ such that

$$\sum_{i \in S} d_i = K; \; d_i \geq 0, \text{ integer}, \; i \in S$$

and such that these parts are "as near as possible" proportional, respectively, to a set of nonnegative integers $\{p_1, p_2, \ldots, p_N\}$, i.e., $\{q_1, q_2, \ldots, q_N\}$.

Let $v(d)$ be a monotone increasing function defined for all integers $d \geq 0$ and also satisfying $d \leq v(d) \leq d + 1$. Then we define the rounding process for the divisor λ by

$$\left[\frac{p_i}{\lambda}\right]_r = d_i \quad i \in S$$

where

$$v(d_i - 1) < \frac{p_i}{\lambda} \leq v(d_i) \quad i \in S$$

Defining the rank function $r(p_i, d_i)$ as

$$r(p_i, d_i) = \frac{p_i}{v(d_i)} \quad i \in S$$

then we can write the above relation as

$$\max_{d_i \geq 0} r(p_i, d_i) \leq \min_{d_j > 0} r(p_j, d_j - 1)$$

Based upon different divisor functions, we can define an infinite number of different divisor methods. There are five traditional divisor methods as well as the parametric divisor method (*PDM*). The method of greatest divisors, which we denote by *GDM*, was also called the Jefferson method in Balinski and Young's papers. The method of major fractions, which we denote by *MFM*, was called the Webster method, the equal proportion method (*EPM*), the harmonic method (*HMM*), and the smallest divisor method (*SDM*) after the names of their advocates, i.e., the Hill method, the Dean method, and the Adams method, respectively.

Using a parameter t_0 such that $0 \leq t_0 \leq 1$, the divisor function of the parametric divisor method, which we denote by *PDM*, can be written as

$$\nu_{PD}(d_i, t_0) = d_i + t_0$$

We generalize the parametric divisor method as follows, which we call the general parametric divisor method (*GPDM*).

$$\nu(d_i; t_0, t_1) = (d_i + t_0)^{t_1}(d_i - t_0 + 1)^{1-t_1}$$

where we assume $0 \leq t_0, t_1 \leq 1$. Then traditional divisor methods including *PDM* and *GPDM* and their corresponding global optimization criteria can be summarized as in Table 1.

Unbiasedness is an important unsolved issue in the apportionment problem. It may be said that the *LFM* method would be the most unbiased as it satisfies the quota property, i.e., difference between allocated number of seats and the real quota being less than 1, while all other divisor methods do not. However, deciding which divisor method should be the most unbiased is very difficult to answer correctly as it is not completely solved yet. We may be able to say that *EPM* or *MFM* among the divisor methods will give closely similar allocations to *LFM,* and also *PDM* with parameter values approximately $0.46 \leq t \leq 0.48$ will be close to *LFM* as shown in [10]. We have been looking for the most unbiased divisor method.

(2) Shortest path counting problem and path counting problem [11, 12]

The shortest path counting problem (SPCP) asks how many shortest paths each edge of the network $N = (V, E)$ with the vertex set V and the edge set E is contained in. There are $n(n-1)$ shortest paths in the network with $|V| = n$, so among all these shortest paths, the SPCP requires us to count the number of shortest paths passing each edge. This implies that we can measure the importance of each edge in the network. When the network $N = (V, E)$ has certain special structures such as trees, grid type, circular type, polar type, we can obtain theoretical results for the SPCP in the form of the explicit expression of the number of shortest paths passing each

Table 1 Divisor methods and parameter values

GDM	*MFM*	*EPM*	*SDM*	*PDM*	*GPDM*
(1,0), (0,1)	(1/2,*)	(0,1/2), (1,1/2)	(0,0), (1,1)	(*,0)	(*,*)

*Arbitrary

edge in these networks. Then we consider the weight of the shortest paths for each edge in the network obtaining explicit expressions and finding their relation to the median and center location problem in the network. Furthermore, we consider the optimal connection problem of two networks from the viewpoint of minimizing the total weight of the shortest paths in the newly combined network. Finally, we can find the correspondence between these theoretical results of the SPCP and their implication in the location problems.

As an application of the SPCP, we can measure the "importance of traffic road segments in Tokyo." Figure 8 shows the road network of Tokyo.

The Tokyo metropolitan traffic road network in Fig. 8 [11, p. 560] has 529 vertices and 855 edges. Applying the SPCP to the network in Fig. 8 and calculating the frequency distributions on the weight of edges for each network of Tokyo, we find that the gap (ratio) between the maximum and the minimum of the weights of edges is extremely large as around 10^4. The huge gap results from the assumption that any shortest path between any two different vertices can be always available theoretically as each edge in the network has an infinite capacity. Figure 9 [11, p. 561] shows the cumulative distributions indicating the set of edges belonging to the corresponding weights interval between the smallest weight and above for each on the Tokyo metropolitan traffic road network given in Fig. 8. Thus, the first figure (1) in Fig. 9 corresponds to the original traffic road network given in Fig. 8. From these cumulative distributions networks, we can see that the actual traffic road segments appearing in Fig. 9 with the largest weights correspond to the most

Fig. 8 Traffic road network of Tokyo [11, p. 560]

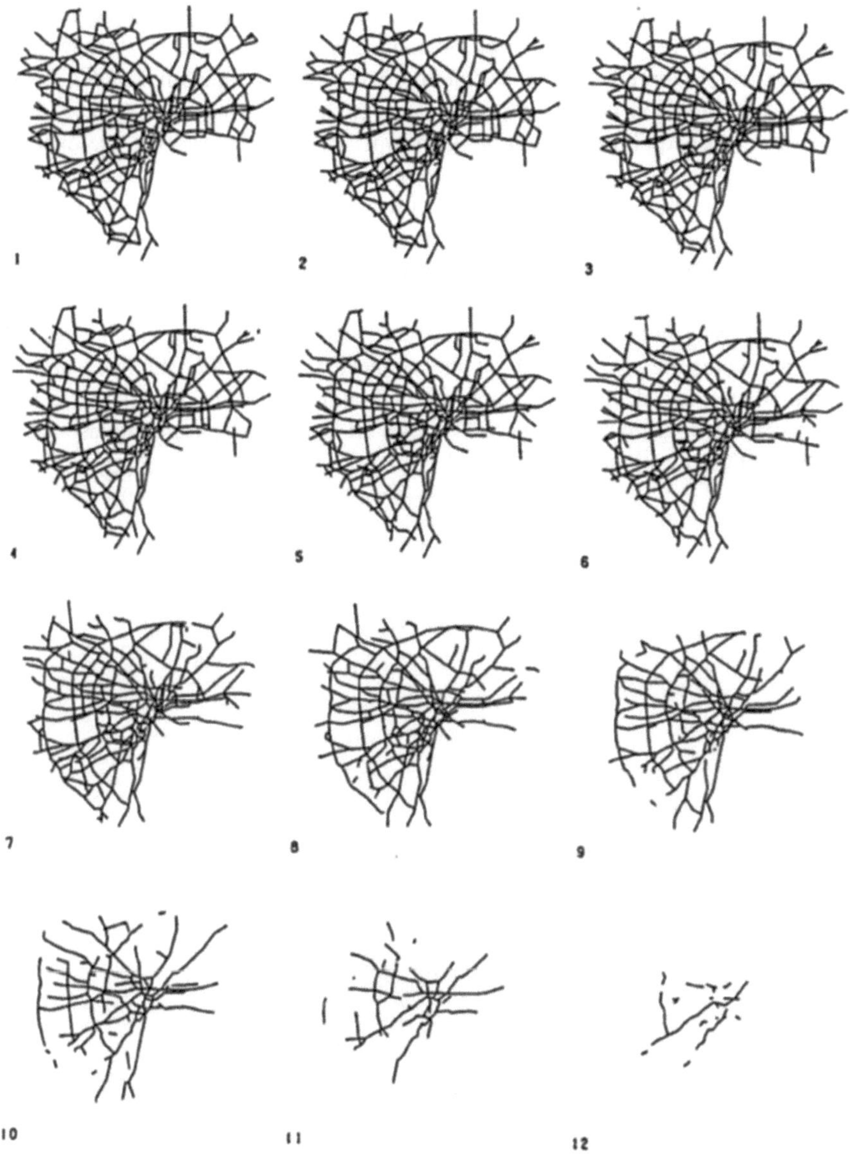

Fig. 9 Cumulative distribution of the weight of edges [11, p. 561]

congested traffic road segments in Tokyo. This implies that highly important traffic road segments remain in these graphs consecutively and continuously.

Firstly, we assume that a certain number of edges (nodes) are deleted from the originally given network. Then the path counting problem (PCP) tries to calculate how many paths exist between any two different nodes in a network after deletion.

By applying path counting methods for the PCP, we propose a method for measuring quantitatively the strength of the connectivity (robustness in other word) of the network-structured system. We define two types of function denoted by the connectivity function and the expected connectivity function, respectively, for the given network. Applying the Monte Carlo method, we can estimate the expected edge (node) deletion connectivity functions when an arbitrary number of edges (nodes) are deleted from the original network. Thus, we can approximate the expected edge deletion connectivity function by using an appropriate nonlinear function with two parameters, which we call the survivability function. We also show the numerical results of applying path counting methods in order to evaluate the connectivity quantitatively. The PCP has several different research objectives such as evaluating the strength of the connectedness of the "lifeline" network quantitatively, and developing a general methodology for the quantitative evaluation technique. Figure 10 [11, p. 567] shows all the Japanese traffic roads, and we apply the PCP to this network.

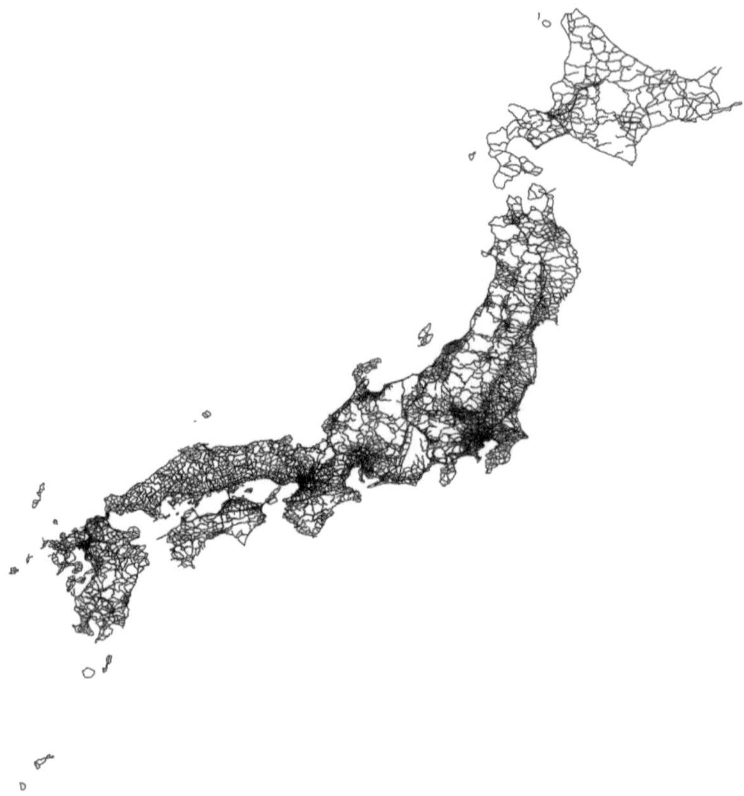

Fig. 10 Traffic road network in Japan [11, p. 567]

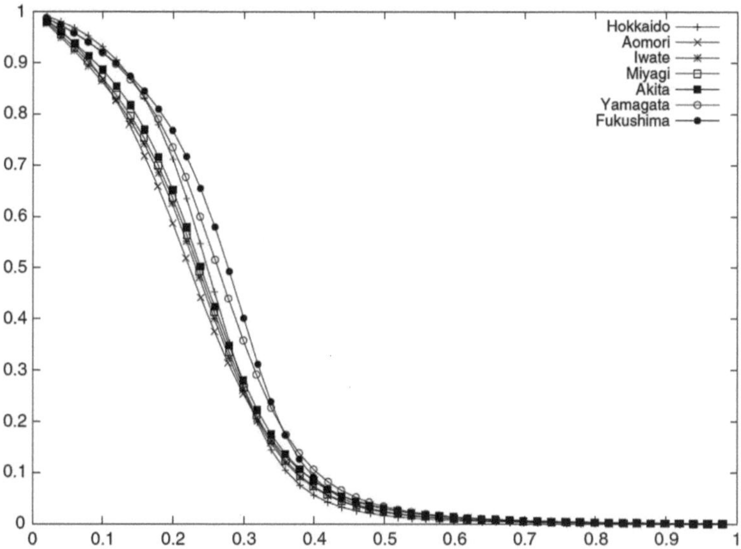

Fig. 11 Expected stable-connected functions for Hokkaido and Tohoku [11, p. 568]

Figure 11 [11, p. 568] illustrates the expected stable-connected functions mentioned above for all the prefectures in Hokkaido and the Tohoku regions. We see that Hokkaido and all six prefectures in Tohoku, which are located in the nothern part of Japan, show closely similar curves.

We attempt to approximate the expected stable-connected function using the following function with two parameters p and q which we call the survivability function.

$$f(x) = \frac{x^{pq}}{x^{pq} + (1 - x^p)^q}$$

Applying the survivability function to the expected stable-connected functions for all 47 prefectures' traffic road networks and estimating the parameters p and q, we obtain the distribution of the parameter estimates as shown in Fig. 12 [11, p. 572].

5 Summary and Future Perspectives

Dr. Brenda Dietrich, former President of INFORMS, stated[1] that we have been working in the "OR comfort zone" including transportation problems, scheduling problems, production systems, facility locations, the airline industry where we can

[1] *OR/MS Today*, April, 2007.

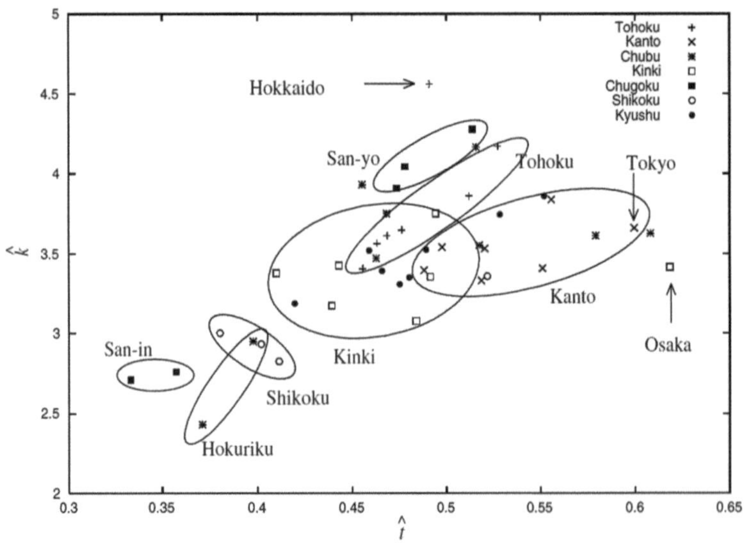

Fig. 12 Parameter estimates for all prefectures in Japan [11, p. 572]

find and obtain optimal solutions rather easily from our mathematical modeling formulation. However, from now on we should "venture outside the OR comfort zone." We believe that the application of OR in the public sector should be one of these areas we should venture into by challenging various types of "complex" problems as there remain many "unsolved" problems in the public sector including energy, environment, transportation, traffic, agriculture, social security, medical policy, education, risk management, natural disaster [13, 14].

Applying OR theory and techniques to solve societal problems, it is true that mathematical modeling analysis could be one of the strong tools [15]. However, we believe that the following would be necessary to be considered and paid attention to:

(i) The problems we are interested in solving and the solutions that are required need to be clearly defined and explained.
(ii) The objectives and reasons for using mathematical models need to be explained in detail so that people can understand them.
(iii) The decision variables, constraints, and criteria need to be explained by using "words" people understand.
(iv) The validity and justification of the mathematical models need to be explained and shown using actual data.

Additionally, when we try to apply mathematical modeling techniques to solve actual problems, we believe we should not aim at obtaining a model solution and an optimal solution only. We should rather have a good command of the models by manipulating them freely and easily. Namely, we should try to derive new insights and useful information from our model as much as possible. This process will bring us the validity of the model and then justify our results. In this sense, sensitivity analysis would be sometimes a more important and useful tool than finding a model solution and an optimal solution.

Regarding future prospects and future problems for applying OR theory and techniques, mathematical model analyses still contain various kinds of "unsolved" problems. Thus, the applicability of mathematical models in the future is highly expected as there exist so many fields to which the mathematical modeling approach has never been applied before.

On the future prospects for OR in the public sector, we still have unexplored potential applicability in the fields of public policy such as medical health care, public administration, higher education, science and technology policy, innovation, IoT and ICT, risk management, global management of natural resources, climate change, environmental problems, energy problems. These global problems arise in the areas of environment, energy resources, complex societal problems such as recycling, information technology industries, natural disasters, and other emergent risk situations; they all require appropriate policy decision making. They are worthy challenges to OR researchers. Policy evaluation has also become more necessary, important, and also very common with the emerging attention to the new public management (so-called New Public Management (NPM)). Also, much quantitative data has been gathered and prepared by all Japanese governmental organizations.

We believe that OR theory and techniques should take advantage of data analysis techniques more carefully and effectively as this approach should be the first step we need to check before we apply, e.g., modeling techniques, and moreover this could be easily explained and comfortably understood by the practitioners including administrative decision makers. We know that there are still so many policy issues and problems that have been neither considered seriously nor taken up as research issues by OR researchers in spite of the fact that their solutions are important and urgently needed in the near future. Finally, we conclude the paper by proposing several future research issues related to our "remaining problems." We believe it is desirably effective if we could contribute to improve and develop certain new ideas and techniques regarding the following items.

- Developing new data analytical or modeling techniques for analyzing complex societal problems.
- Developing a new theory and some computational techniques to deal with uncertainty, fuzziness, complexity, drastic and emergent changes, and so on.
- Developing more accurate and justifiable theoretical approaches regarding, e.g., the shortest path theory and the apportionment problem.

We hope that OR theory and techniques could contribute more in the future in challenging and solving urgent, complex, and important societal problems.

References

1. Miwa, M., Gozun, B., Oyama, T.: Statistical data analyses to elucidate the causes and improve the countermeasures for preventing train accidents in Japan. Int. Trans. Oper. Res. IFORS **13** (3), 229–251 (2006)
2. Oyama, T., Miwa, M.: Investigating serious train accidents and natural disaster data in Japan. Commun. Oper. Res. Oper. Res. Soc. Jpn. **53**(10), 11–17 (2008) (in Japanese)
3. Anwar, S., Oyama, T.: Statistical data analysis for investigating government subsidy policy for private universities. J. High. Educ. (Springer) **55**(4), 407–423 (2007). On line ISSN 1573-174X
4. Srivastava, D.C., Oyama, T.: Evaluating the emission reduction targets in UNFCCC Kyoto protocol by applying primary energy data analyses. J. Asian Public Policy **2**(1), 36–56 (2009)
5. Yoshii, K., Oyama, T.: A quantitative factorial component analysis to investigate the recent changes of Japan's weight-based food self-sufficiency ratio. Am. J. Oper. Res. **6**(1), 44–60 (2016)
6. Ojima, J., Oyama, T.: Index gap minimization model analyses for local promotion branches of Iwate Prefecture. Commun. Oper. Res. Oper. Res. Soc. Jpn. **48**(8), 567–573 (2003) (in Japanese)
7. Miwa, M., Oyama, T.: All-integer type linear programming model analyses for the optimal railway track maintenance scheduling. Oper. Res. Soc. India OPSEARCH **41**(3), 35–45 (2004)
8. Oyama, T., Miwa, M.: Mathematical modeling analyses for obtaining an optimal railway track maintenance schedule. Jpn. J. Ind. Appl. Math. **23**(2), 207–224 (2006)
9. Oyama, T.: On a parametric divisor method for the apportionment problem. J. Oper. Res. Soc. Jpn. **34**(2), 187–221 (1991)
10. Oyama, T., Ichimori, T.: On the unbiasedness of the parametric divisor method for the apportionment problem. J. Oper. Res. Soc. Jpn. **38**(3), 301–321 (1995)
11. Oyama, T., Morohoshi, H.: Applying the shortest path counting problem to evaluate the importance of city road segments and the connectedness of the network-structured system. Int. Trans. Oper. Res. **11**(5), 555–574 (2004)
12. Kobayashi, K., Morohosi, H., Oyama, T.: Applying path-counting methods for measuring the robustness of the network-structured system. Int. Trans. Oper. Res. IFORS **16**(3), 371–389 (2009)
13. Nemhauser, G.L., Rinnooy Kan, A.H.G.: Handbooks in Operations Research and Management Science. North-Holland, Amsterdam, Netherland (1994) (trans. Oyama, T.: Koukyou Seisaku OR, 776 pp. Asakura Shoten Publishing Co., 1998)
14. Oyama, T.: Saitekika moderu bunseki (Optimization Model Analyses), 372 pp. Nikka Giren Publishing Co. (1993) (in Japanese)
15. Oyama, T.: Mathematical programming model analyses and basic formulation techniques. Commun. Oper. Res. Oper. Res. Soc. Jpn. **43**(2), 71–75 (1998) (in Japanese)

Facility Location and Distribution Planning in a Disrupted Supply Chain

Himanshu Shrivastava, Pankaj Dutta, Mohan Krishnamoorthy and Pravin Suryawanshi

Abstract Most facility location models in the literature assume that facilities will never fail. In addition, models that focus on distribution planning assume that transportation routes are disruption-free. However, in reality, both the transportation routes and the facilities are subject to various sorts of disruptions. Further, not many supply chain models in the literature study perishable products. In this paper, we address issues of facility location and distribution planning in a supply chain network for perishable products under uncertain environments. We consider demand uncertainty along with random disruptions in the transportation routes and in the facilities. We formulate a mixed-integer optimisation model. Our model considers several capacitated manufacturers and several retailers with multiple transportation routes. We investigate optimal facility location and distribution strategies that minimise the total cost of the supply chain. We demonstrate the effectiveness of our model through an illustrative example and observe that a resilient supply chain needs to have a different design when compared to a disruption-free supply chain. The effects of various disruption uncertainties are also studied through statistical analysis.

Keywords Supply chain · Facility location · Distribution planning
Uncertainty · Perishable products

H. Shrivastava
IIT Bombay, IITB-Monash Research Academy, Powai, Mumbai 400076, Maharashtra, India

P. Dutta (✉) · P. Suryawanshi
Shailesh J. Mehta School of Management, Indian Institute of Technology Bombay, Mumbai 400076, Maharashtra, India
e-mail: pdutta@iitb.ac.in

M. Krishnamoorthy
Department of Mechanical and Aerospace Engineering, Monash University, Melbourne, VIC 3168, Australia

S. Kar et al. (eds.), *Operations Research and Optimization*, Springer Proceedings in Mathematics & Statistics 225, https://doi.org/10.1007/978-981-10-7814-9_19

1 Introduction

A supply chain (SC) must perform its planned operations effectively and efficiently and remain competitive in global markets. Thus, a SC has to consider multiple objectives such as (a) an increased level of responsiveness, (b) a reduction in the overall cost of the supply chain cost and (c) better distribution management [1]. Supply chain planning decisions are categorised on the basis of the time horizon into strategic, tactical and operational. According to Simchi et al. [2], strategic decisions will be long-term decisions and, typically, this includes the location of plants, the number and location of warehouses, the modes of transportation, the product that is to be manufactured or stored at various locations and the types of information systems that need to be employed. Tactical decisions reflect mid-term planning scenarios and deal with procurement contracts, production schedules and guidelines for meeting quality and safety standards. Operational decisions include planning that is related to machine/personnel/vehicle scheduling, sequencing, lot sizing, defining vehicle routes and so on.

The design of a supply chain network is an important aspect of supply chain management. This primarily involves the determining of facility location and distribution strategies in the supply chain. In the review paper, Melo et al. [3], the authors emphasise the role that facility location decisions play in the making of strategic supply chain decisions. Klibi et al. [4] describe a design of the supply chain network (SCN) by considering uncertain factors; they present a comprehensive review of the natural environmental factors that are responsible for SC disruptions. The paper covers aspects of SC modelling under uncertainty, robustness and resilience.

Qiang et al. [5] state, "supply chain disruption risk[s] are the most pressing issue[s] [that are] faced by firms in today's competitive global environment". Baghalian et al. [6] point out that disruptions are inevitable and are present in most business scenarios. One of the highlights of the World Economic Forum Report (2013) about global risk indicates that extreme weather events, major natural disaster and weapons of mass destruction can hinder the working of supply chains, which, in turn, results in financial losses to the industry. The report shows that there is, on an average, a 7% reduction in the share price of the affected companies [7]. This is alarming and indicates the pressing need for SC professionals to modify the existing working style and improve upon mitigation policies for the management of SC glitches. SC disruptions can be classified as "high-likelihood-low-impact, medium-likelihood-moderate-impact and low-likelihood-high-impact" [8, 9]. Further, Ray et al. [9] proposed an optimal ordering policy for sourcing decisions under disruption by maximising the expected profit and simultaneously minimising the buyers variance in a two-echelon SC structure. On the other hand, Ferrari [10] tries to ascertain the causes of major supply chain disruptions. The authors conclude by stating that "supply chain disruption remains a key executive level concern, and disruption takes on many dimensions, including lost business and industry competitive dimensions".

One of the early studies on facility location considering disruption was carried by Drezner et al. [11]. The authors used the reliability theory in order to capture the disruption effect. The authors had considered a predetermined probability of failure in facilities which were unreliable and could fail at any time. Ivanov et al. [12], in their recent study on SC disruption review, constructed a "risk-recovery" matrix that primarily includes the prominent risks that are responsible for the disruption in the SC and recovery strategies that are followed by various authors. The authors have also explained the various methodologies that are implemented to mitigate consequences due to SC disruptions.

Gupta et al. [13] proposed a stochastic programming-based approach in order to plan for manufacturing decisions that are termed as "here-and-now" decisions, which are made before the realisation of the demand. The logistics decisions, which are termed as "wait-and-see", are made when practitioners realise the uncertainty in the demand pattern. The authors have used the CPLEX optimisation solver, and the framework is illustrated using a real-life case study. Nasiri et al. [14] proposed two models for the designing of an optimal supply chain distribution network under demand uncertainty. Location and allocations decisions are made in the first model, while decisions that are related to production plan, such as production quantity, are made in the second model. The authors used a mixed-integer nonlinear model and solved it using the Lagrangian approach. Tang et al. [15] proposed robust strategies in order to handle disruption scenarios. Outlining the two properties for the strategies, the author stated, "(a) these strategies will manage the inherent fluctuations efficiently regardless of major disruptions, and (b) these strategies will make the SC more resilient in the face of major disruptions". Claypool et al. [16] designed a supply chain network for a new product in which the novelty lies in the combining of the effects of risk due to product development and due to SC. Sadghiani et al. [17] developed a retail SCN by considering operational and disruption risks. The authors validated the model by using illustrative examples and a real-life case study in retail SC. A review study by Snyder et al. [18] in the field of supply chain network design under disruptions describes various modelling approaches in the context of SC disruptions. The authors gathered 180 research articles under the four disruption-mitigating categories, namely "(a) mitigating disruption through inventories; (b) mitigating disruptions through sourcing and demand flexibility; (c) mitigating disruptions through facility location and (d) mitigating disruptions through interaction with external partners."

The impact of disruption in global SC is extravagant in its magnitude, because the impact in this case trickles between interlinked countries. Therefore, global enterprises should mitigate these disruptions and reduce vulnerabilities to disruption with resilient techniques. Mitigating disruptions has become an important research issue in the recent past [18–20].

The literature that has been discussed has been limited to “regular products”, for which perishability is not a major concern. India stands second in the production of fruits and vegetables in the world, after China.[1] Moreover, the SC challenges of perishable goods are unexceptional when compared to the regular products, because the value of the product deteriorates significantly over time. In addition to quality-level challenges, the production and distribution of perishable goods are non-administrable. Furthermore, economic chocks, government implication, product varieties and management issues are unavoidable in the overall working of an SC. Shankar et al. [21] and Nasiri et al. [14] have developed a production–distribution problem under demand uncertainty for regular products. However, the diminishing value of the product is not taken into consideration. The review article on agri-food SC that was proposed by Ahumada and Villalobos [22] sheds light on the mathematical models that are developed in order to address SC-related challenges for both non-perishable and fresh products. Authors Pathumnakul et al. [23] addressed the inventory problem of cultivated shrimp and attempted to ascertain the optimal harvest that could maximise a farmer’s bottom line by optimising the SC cost. Along similar lines, authors Lin et al. [24] discussed the optimal inventory levels, the price and the profit in the white shrimp SC industry in Taiwan. Negi et al. [25] studied the SC of the fruits and vegetables sector in India and addressed the following objectives: “(a) to identify the factors affecting [the] supply chain of fruits and vegetables sector in India and (b) to suggest mitigation strategies for the identified challenges in [the] supply chain of fruits and vegetables sector”. In India, the food and grocery industries account for approximately 31% of India’s consumption basket. This industry is currently valued at USD 39.71 billion and is expected to reach USD 65.4 billion with a Compounded Annual Growth Rate (CAGR) of 11% by 2018.[2]

The most cited examples (from the literature) of disruptions that severely affected the operations of SC in the past are Hurricane Katrina and Hurricane Rita in 2005 on the US Gulf Coast. These natural calamities had crippled the oil refineries and resulted in huge losses. The adverse effects were also palpable due to the destruction of large quantities of lumber and coffee produce, and the rerouting of bananas and other fresh produce [20]. The destruction highlighted the fact that, in the future, the designing of an SC network that is resilient to disruption is important. This research article is motivated by the need to quantify and mitigate the effects of disruptions in SC in the case of perishable products.

Shrivastava et al. [26] have studied the resilient supply chain network of the perishable product under random disruptions. The objective of their paper is “to address some practical issues of decision-making under uncertain environments; the focus is the designing of an optimal supply chain distribution network for perishable products under uncertain demand”. They considered the disruption in the transportation links and formulated a mixed-integer optimisation model. However, they have assumed that the facilities are disruption-free. In reality, the facilities are also prone to disruption risks. In such a case, it could be challenging to determine the location of the

[1] http://mofpi.nic.in/documents/reports/annual-report.

[2] http://www.ibef.org/industry/indian-food-industry.aspx.

facility when it is subjected to disruption risks. Another limitation of their study is that they have considered only single transportation routes between the supply chain entities. In reality, there could be multiple routes of transportation with the possibility of different risks of disruptions in each route.

The present paper extends the study of Shrivastava et al. [26] by considering multiple routes of transportation and disruption in facilities. The paper also examines the supply chain network for the perishable product under uncertain demand. The aim is to determine optimal facility locations and a distribution strategy in which the transportation routes and the facilities are subjected to disruption risks.

We have organised the rest of the sections of the study as follows: Sect. 2, which deals with the problem description and model formulation; Sect. 3, which presents an illustrative example of the developed model; Sect. 4, which presents the uncertainty analysis and Sect. 5, which concludes our study and suggests an area for future research.

2 Problem Description and Model Formulation

In this paper, we assume a two-echelon single-period supply chain system that consists of several manufacturers and retailers. The manufacturer produces a single product that is perishable. The demands are realised by the retailers, who anticipate their demand and order it from the manufacturer at the start of the period. The potential location and the capacities of the manufacturers are known in advance. There are multiple routes of transportation between each manufacturer and retailer. We assume that these transportation routes are subjected to disruptions, as a result of which some quantity of finished goods may be fully or partially lost in the transportation routes. Also, the manufacturing units are assumed to be prone to disruptions. If disruption occurs in the manufacturing units, the units may fully or partially lose their capacities. In order to ensure a full supply to the retailer, the manufacturers outsource the disrupted quantity from a third party manufacturer. It is assumed that the third party manufacturer has infinite capacity. We assume that demand and disruption are uncertain and follow a known probability distribution function. In this paper, we use the mixed-integer programming approach to formulate the mathematical model that determines the optimal supply chain structure under probabilistic disruptions. We also intend to determine a suitable distribution planning, while minimising the total supply chain's cost. We use the following notations to formulate mathematical model:

Indices:

$m \in M \longrightarrow$ The set of potential locations for manufacturers;
$r \in R \longrightarrow$ The set of retailers;
$f \in F \longrightarrow$ The set of transportation routes;

Decision variables:

$y_m \longrightarrow$ Binary variable equals 1 if manufacturer is open at candidate location m and 0 otherwise;
$x_{mrf} \longrightarrow$ Quantity of final product shipped from manufacturer m to retailer r through route f;
$x_{m\eta} \longrightarrow$ Quantity shipped from third party manufacturer to primary manufacturer m;

Parameters:

$F_m \longrightarrow$ Manufacturer's fixed opening cost at candidate location m;
$D_r \longrightarrow$ Demand at retailer r;
$E(D_r) \longrightarrow$ Expected demand at retailer r;
$F(D_r) \longrightarrow$ Cumulative distribution function of D_r;
$K_r \longrightarrow$ Handling cost per unit at retailer r which includes holding cost and processing/packaging cost;
$\phi_m \longrightarrow$ Capacity of manufacturer m;
$L_m \longrightarrow$ Sum of unit production and unit holding cost at manufacturer m;
$B \longrightarrow$ Budget limit of opening manufacturer's facilities;
$C_{mrf} \longrightarrow$ Unit cost of transportation from manufacturer m to retailer r through route f;
$C_{m\eta} \longrightarrow$ Unit cost of transportation from third party manufacturer to primary manufacturer m;
$\theta_m \longrightarrow$ Percentage of capacity disrupted at manufacturer m;
$\gamma_{mrf} \longrightarrow$ Fraction of supply disruption in the transportation route f between m and r;
$\upsilon \longrightarrow$ Unit penalty cost of disruption;
$\Omega \longrightarrow$ Desired level of fill rate;
$C_S \longrightarrow$ Unit shortage cost to retailer;
$C_E \longrightarrow$ Unit excess cost to retailer.

We assume that γ_{mrf} and θ_m follow a certain known distribution whose mean and standard deviation are known in advance. We deployed the same formulation style as used by Shrivastava et al. [26].

The total cost of the supply chain from manufacturer m to retailer r through route f:

$$F_m \cdot y_m + L_m \cdot x_{mrf} + C_{mrf} \cdot x_{mrf} + \gamma_{mrf} \cdot x_{mrf} \cdot \upsilon + y_m \cdot x_{m\eta} \cdot C_{m\eta} \quad (1)$$

The first term in Eq. (1) indicates the fixed opening cost of the manufacturer's facilities, and the second term denotes the production and holding costs at manufacturer m, while the third term indicates the transportation cost from manufacturer m to retailer r. The fourth term in the above equation denotes the penalty cost due to disruption in the transportation routes. If disruption occurred $\gamma_{mrf}\%$ of supply is assumed to be disrupted. Hence, the quantity arriving at the retailer r is $(1-\gamma_{mrf})\cdot x_{mrf}$. The last term in the above equation computes the transportation cost from third party manufacturer's location to primary manufacturer's locations when the disruption occurs at the primary manufacturing units.

The total cost at retailer r:

$$\sum_{m\in M}\sum_{f\in F}K_r\cdot(1-\gamma_{mrf})\cdot x_{mrf}+C_E\left(\sum_{m\in M}\sum_{f\in F}(1-\gamma_{mrf})\cdot x_{mrf}-D_r\right)^+ \\ +C_S\left(D_r-\sum_{m\in M}\sum_{f\in F}(1-\gamma_{mrf})\cdot x_{mrf}\right)^+ \quad (2)$$

where, $A^+=\max\{A,0\}$.

In retailer's total cost expression, first term denotes the handling cost (which is a combination of holding cost and processing/packaging cost) while the second term is the excess cost and the last term is the shortage cost. To capture the product perishability, we are using news vendor concept [27–30] for managing inventory at the retailer. Equation (2) is simplified by using the following equations:

$$\left(\sum_{m\in M}\sum_{f\in F}(1-\gamma_{mrf})\cdot x_{mrf}-D_r\right)^+=\int_0^{\sum_{m\in M}\sum_{f\in F}(1-\gamma_{mrf})\cdot x_{mrf}}F\left(D_r\right)dD_r \quad (3)$$

$$\left(D_r-\sum_{m\in M}\sum_{f\in F}(1-\gamma_{mrf})\cdot x_{mrf}\right)^+=\left(\sum_{m\in M}\sum_{f\in F}(1-\gamma_{mrf})\cdot x_{mrf}-D_r\right)^+ \\ -\sum_{m\in M}\sum_{f\in F}(1-\gamma_{mrf})\cdot x_{mrf}+E(D_r) \quad (4)$$

From Eqs. (3) and (4), the final expression of total cost at retailer r(T_r) is:

$$T_r=\sum_{m\in M}K_r\cdot(1-\gamma_{mrf})\cdot x_{mrf}+C_E\left(\int_0^{\sum_{m\in M}\sum_{f\in F}(1-\gamma_{mrf})\cdot x_{mrf}}F\left(D_r\right)dD_r\right) \\ +C_S\left(\int_0^{\sum_{m\in M}\sum_{f\in F}(1-\gamma_{mrf})\cdot x_{mrf}}F\left(D_r\right)dD_r-\sum_{m\in M}\sum_{f\in F}(1-\gamma_{mrf})\cdot x_{mrf}+E(D_r)\right) \quad (5)$$

The total cost of the supply chain is the sum of Eqs. (1) and (5), and on rearranging the resulting equation, we get the following mathematical model:

Objective function:

$$
\begin{aligned}
Min\ \ U = & \sum_{m\in M} F_m \cdot y_m + \sum_{m\in M}\sum_{r\in R} P_m \cdot x_{mrf} + \sum_{r\in R}\sum_{m\in M} C_{mrf} \cdot x_{mrf} \\
& + \sum_{r\in R}\sum_{m\in M} \gamma_{mrf} \cdot x_{mrf} \cdot \upsilon_{mrf} + \sum_{m\in M} y_m \cdot x_{m\eta} \cdot C_{m\eta} \\
& + \sum_{r\in R}\sum_{m\in M} K_r \cdot (1-\gamma_{mrf}) \cdot x_{mrf} \\
& + (C_E + C_s)\left[\sum_{r\in R}\left(\int_0^{\sum_{m\in M}(1-\gamma_{mrf})\cdot x_{mrf}} F(D_r)\, dD_r\right)\right] \\
& - C_S\left[\sum_{r\in R}\left(\sum_{m\in M}(1-\gamma_{mrf})\cdot x_{mrf} - E(D_r)\right)\right]
\end{aligned}
\tag{6}
$$

Subject to:

$$\sum_{r\in R}\sum_{f\in F} x_{mrf} \le x_{m\eta} + \left((1-\theta_m)y_m\right)\phi_m \ \ ; \ \ \forall\ m \in M\ ;\ \forall\ f \in F \tag{7}$$

$$\sum_{m\in M} F_m \cdot y_m \le B \tag{8}$$

$$\Omega \le \frac{\sum_{f\in F}\sum_{m\in M}(1-\gamma_{mrf})\cdot x_{mrf}}{E(D_r)} \ ; \qquad \forall\ \ r \in R \tag{9}$$

$$x_{mrf} \ge 0\ ; \qquad \forall\ \ m \in M,\ \ \forall\ \ r \in R,\ \ \forall\ \ f \in F \tag{10}$$

$$y_m \in \{0,1\} \qquad \forall\ \ m \in M \tag{11}$$

The objective function minimises the total cost of the supply chain network. Constraint Eq. (7) imposes disruption capacity constraint which ensures that the supply to the retailer should not be affected by the disruption at the manufacturing facilities. Equation (8) represents the budget limit. Constraint Eq. (9) ensures that service level should be greater or equal to $\Omega\%$. Constraint Eq. (10) and Eq. (11), respectively, impose the non-negativity and binary restrictions.

The decision variables address the optimal network structure. The decision variable in our model includes binary variables that represents the existence of manufacturers and the continuous variable that represent the material flow from manufacturers to retailers.

The mathematical model explained above is nonlinear due to its nonlinear objective function described in Eq. (6). The term responsible for nonlinearity is $y_m \cdot x_{m\eta}$.

To handle this nonlinearity, we define new variable, Ξ_m, such that $\Xi_m = y_m \cdot x_{m\eta}$, and add the following additional constraints to the model:

$$\Xi_m \le x_{m\eta} \;;\quad \forall \quad m \in M \tag{12}$$

$$\Xi_m \le N \cdot y_m \;;\quad \forall \quad m \in M \quad (\textit{where } N \textit{ is a large number}) \tag{13}$$

$$\Xi_{ms} \ge x_{m\eta} + N \cdot (y_m - 1) \;;\quad \forall \quad m \in M \tag{14}$$

We assume demand to be uniformly distributed. However, the model can be used for other distributions too. The uniform demand distribution, $F(D)$, in the interval $[a, b]$ is given as:

$$F(D) = \frac{D - a}{b - a} \qquad a \le D \le b \tag{15}$$

Substituting $F(D)$ in the objective function, the resulting expression is:

$$\begin{aligned}
Min\;\; U = & \sum_{m \in M} F_m \cdot y_m + \sum_{m \in M} \sum_{r \in R} \sum_{f \in F} L_m \cdot x_{mrf} + \sum_{r \in R} \sum_{m \in M} \sum_{f \in F} C_{mrf} \cdot x_{mrf} \\
& + \sum_{r \in R} \sum_{m \in M} \sum_{f \in F} \gamma_{mrf} \cdot x_{mrf} \cdot \upsilon + \sum_{m \in M} y_m \cdot x_{m\eta} \cdot C_{m\eta} \\
& + \sum_{r \in R} \sum_{m \in M} \sum_{f \in F} K_r \cdot (1 - \gamma_{mrf}) \cdot x_{mrf} \\
& + \left(C_E + C_s\right) \left[\sum_{r \in R} \left(\frac{\left(\sum_{m \in M} \sum_{f \in F} (1 - \gamma_{mrf}) \cdot x_{mrf} \right)^2}{2 \cdot (b_r - a_r)} \right. \right. \\
& \qquad \left. \left. - a_r \cdot \frac{\left(\sum_{m \in M} \sum_{f \in F} (1 - \gamma_{mrf}) \cdot x_{mrf} \right)}{b_r - a_r} \right) \right] \\
& - C_S \left[\sum_{r \in R} \left(\sum_{m \in M} \sum_{f \in F} (1 - \gamma_{mrf}) \cdot x_{mrf} - E(D_r) \right) \right]
\end{aligned} \tag{16}$$

subject to: Eqs. (7)–(11) and Eqs. (12)–(14).

The above expression is quadratic expression, and hence we have mixed-integer quadratic model.

3 Illustrative Example

In this section, we validate our model through a two-echelon supply chain design, which is subjected to disruption risks at facilities and transportation routes. We consider four manufacturing units and five retail zones and assume that there are two routes from each manufacturing unit to each retail zone. We solved our model by using the default settings of the CPLEX optimisation solver. The input parameters of the problem are shown in Tables 1 and 3. The unit excess cost and unit shortage cost are assumed to be 1 and 2, respectively. The disruption in the transportation routes and facilities are characterised by γ and θ, respectively. The disruption probability matrix (γ) for both the routes is shown in Table 2. θ for $m1$ is assumed to be 0.35, while it is 0.20, 0.05 and 0.15 for $m2$, $m3$ and $m4$, respectively.

The total cost of the supply chain is 44,980. It is observed that this design (which is the resilient design) selects all the four manufacturers. The quantity shipment decisions from the manufacturing units to the retailers are shown in Table 4. The quantity that needs to be outsourced from the third party manufacturer are 35, 30, 0 and 0 for manufacturer $m1$, $m2$, $m3$ and $m4$, respectively. We have also analysed the disruption-free design. In the disruption-free design, γ and θ are considered to be zero. The total cost of the supply chain for the disruption-free design is 41,330. This disruption-free design selects only three manufacturers, $m2$, $m3$ and $m4$. The quantity shipment decisions from the manufacturing units to the retailers are shown in Table 5. Here it should be noted that the supply chain network structure that is obtained for the

Table 1 Manufacturers and retailers input parameters

Manufacturers				Retailers	
Manufactuer	Capacity	Fixed opening cost	Per unit production cost	Retailer	Handling cost
m1	100	2000	30	r1	9
m2	250	3500	35	r2	12
m3	180	2000	40	r3	9.5
m4	200	2500	37	r4	10
				r5	11

Table 2 Disruption probabilities

Supply disruption probability in route f1						Supply disruption probability in route f2					
	r1	r2	r3	r4	r5		r1	r2	r3	r4	r5
m1	0.02	0.04	0.03	0.04	0.01	m1	0.3	0.25	0.1	0.17	0.15
m2	0.04	0.02	0.04	0.05	0.03	m2	0.22	0.16	0.12	0.26	0.27
m3	0.05	0.02	0.04	0.05	0.03	m3	0.15	0.1	0.8	0.05	0.07
m4	0.07	0.03	0.3	0.2	0.02	m4	0.07	0.03	0.3	0.2	0.02

Table 3 Transportation cost parameters

Unit transportation cost in route f1: manufacturers to retailers						Unit transportation cost in route f2: manufacturers to retailers					
	r1	r2	r3	r4	r5		r1	r2	r3	r4	r5
m1	8	15	7.5	8.5	9	m1	10	12	9	8	11
m2	12	11	7	8	10	m2	9	13	8	7	10.5
m3	8	9	7	8	9	m3	9	8	8.5	7.5	8.5
m4	9	10	9.5	12	8	m4	11	9	12	10	7

Table 4 Distribution decisions in resilient model

Quantity shipment through route f1:						Quantity shipment through route f2:					
	r1	r2	r3	r4	r5		r1	r2	r3	r4	r5
m1	100	0	0	0	0	m1	0	0	0	0	0
m2	0	0	103.1	25.7	0	m2	21.2	0	0	0	0
m3	0	113.5	0	0	0	m3	0	0	0	57.7	0
m4	0	0	0	0	0	m4	0	69	0	0	100

Table 5 Distribution decisions in disruption-free model

Quantity shipment through route f1:						Quantity shipment through route f2:					
	r1	r2	r3	r4	r5		r1	r2	r3	r4	r5
m1	0	0	0	0	0	m1	0	0	0	0	0
m2	0	0	70.8	0	0	m2	100	0	0	79.2	0
m3	0	0	28.2	0	0	m3	0	96	0	0	0
m4	21	0	0	0	0	m4	0	82.2	0	0	100

resilient supply chain is different from the disruption-free supply chain. The resilient supply chain selects all four manufacturers, while the disruption-free supply has only three manufacturers. Also, the total cost of the resilient supply chain is 8% higher than that of the disruption-free supply chain.

We further analyse the effect of fill-rate measures on the total cost of the supply chain. We observed that as the fill rate increases the total cost of the supply chain also increases. The variation of the total SC with the fill-rate measures is observed through the graph shown in Fig. 1.

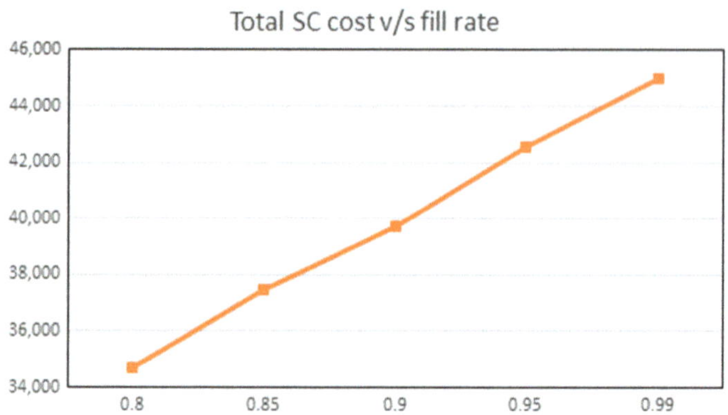

Fig. 1 Variation of the total SC cost with fill rate

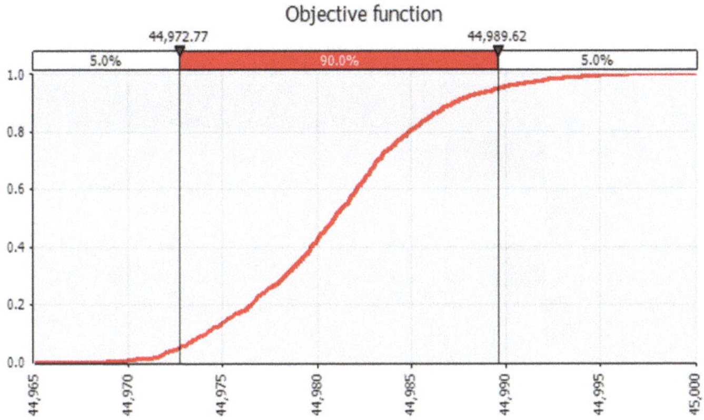

Fig. 2 Percentile analysis of the total SC's cost

4 Uncertainty Analysis

This section analyses the disruption effects that are present in the facilities and the transportation routes. We have assumed that these disruptions are uncertain and that they follow a normal distribution with known mean and variance. On executing a simulation of 1000 iterations by using @Risk,[3] the uncertainty effect (due to disruptions in the manufacturing facilities and in the transportation routes) is analysed through various graphs (Figs. 2 and 3).

The graph shown in Fig. 4 represents the overall nature of the objective function (the total cost of the supply chain). The Kolmogorov–Smirnov normality test is performed on the 1000 observed data of the total cost of the supply chain, and it is found that the outcome is also normal. Through simulation and t-test analysis by using @Risk, it is observed that the overall cost of the supply chain would lie between 44,972.7 and 44,989.6 with 90% confidence. The chance that the total cost exceeds 44,989.6 is only 5%. Figure 2 statistically summarises the objective function.

We now analyse the effect of the uncertain parameters, γ and θ on the supply chain by using the tornado graph. Figure 3 shows the tornado graph of top five most dominated uncertain parameters. In this figure, we calculated the variability on the total cost due to uncertainty in the parameters using the simulation output of the Pearson coefficient value. The γ in the transportation route, $f2$, which is between $m4$ and $r3$, is highly effective and causes a huge variation in the total cost of the supply chain. In other words, this is the most risky route. The route $f1$, between $m4$ and $r3$, causes approximately 63.5% variability in the total cost of the supply chain. However, route f1, between $m4$ and $r3$, is less risky than route f2, because the variation in the cost of its supply chain is lesser. This route is responsible for 11% variation in the total cost. This route is best for the risk-averse decision maker, while the risk-seeking decision maker could choose route $f2$. Similarly, the least variation

[3]http://www.palisade.com/risk/.

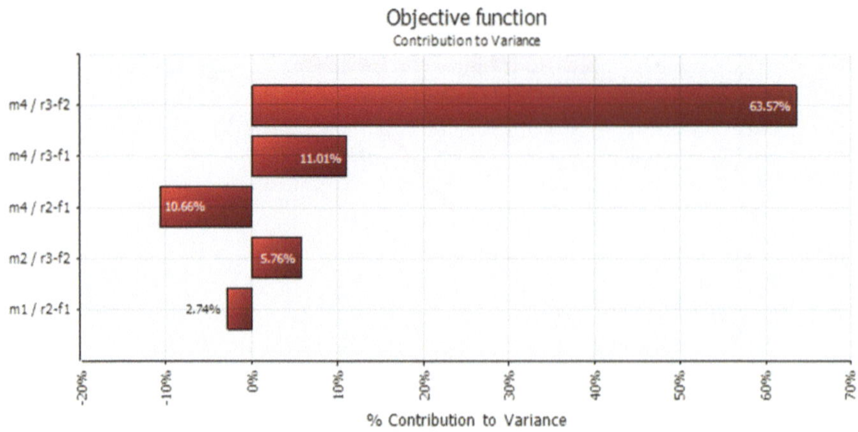

Fig. 3 Effect of uncertainty on supply chain's total cost

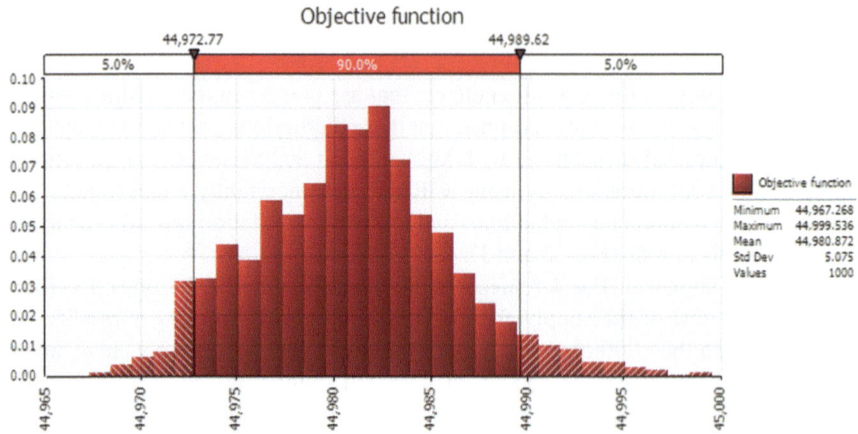

Fig. 4 Variation in supply chain's total cost due to uncertain disruptions

in the total cost of the supply chain is observed in route $f1$, between $m1$ and $r2$. The Pearson coefficient in this route is negative, and the variability due to this route is 2.7%. It should be noted that the disruption in facilities causes very less variation in the total cost, and hence, the uncertainty effect of θ is not dominant.

5 Conclusions and Future Work

In this paper, we have extended the model of Shrivastava et al. [26] and formulated the problem of facility location and the allocation of a two-echelon supply chain system under uncertainty as a mixed-integer quadratic model. The model addresses the decision variable, which corresponds to the location of the manufacturer and the quantity flow from the manufacturer to the retailer. We have considered disruption in the transportation routes and in the facilities, simultaneously. During disruptions, the manufacturing facilities may fully or partially lose their capacity. In order to ensure full supply to the retailer, we assumed that the manufacturer outsources its disrupted capacity from the third party manufacturer. We observed that the supply chain decisions in the resilient model and the disruption-free model are not same. We have also carried out an extensive analysis of the uncertain disruptions that are present in the transportation link between the manufacturers and the retailers, and in the manufacturing facilities. We have statistically studied the overall nature of the cost function. In the current parameter setting, we have found that the disruption parameter, γ, is highly effective in the link, $f2$, which is between $m4$ and $r3$; this parameter also causes a large amount of variation in the cost function.

Realising a more realistic supply chain that has a greater number of echelons could be a possible extension of this study. We have assumed a single-product and single-period model, which can be extended for multi-products and multi-periods. Along with demand and disruption uncertainties, cost parameter uncertainties can be considered as well.

References

1. Khalifehzadeh, S., Seifbarghy, M., Naderi, B.: A four-echelon supply chain network design with shortage: mathematical modeling and solution methods. J. Manuf. Syst. **35**, 164–175 (2015)
2. Simchi-Levi, D., Kaminsky, P., Simchi-Levi, E.: Managing the supply chain: the definitive guide for the business professional. McGraw-Hill Companies (2004)
3. Melo, T., Nickel, S., Saldanha-da Gama, F.: Network design decisions in supply chain planning. Fraunhofer-Institut für Techno-und Wirtschaftsmathematik, Fraunhofer (ITWM) (2008)
4. Klibi, W., Martel, A., Guitouni, A.: The design of robust value-creating supply chain networks: a critical review. Eur. J. Op. Res. **203**(2), 283–293 (2010)
5. Qiang, Q., Nagurney, A., Dong, J.: Modeling of supply chain risk under disruptions with performance measurement and robustness analysis. In: Managing Supply Chain Risk and Vulnerability, pp. 91–111. Springer (2009)
6. Baghalian, A., Rezapour, S., Farahani, R.Z.: Robust supply chain network design with service level against disruptions and demand uncertainties: a real-life case. Eur. J. Op. Res. **227**(1), 199–215 (2013)
7. Bhatia, G., Lane, C., Wain, A.: Building resilience in supply chains. An initiative of the risk response network in collaboration with accenture. In: World Economic Forum. Geneva, Switzerland (2013)
8. Oke, A., Gopalakrishnan, M.: Managing disruptions in supply chains: a case study of a retail supply chain. Int. J. Prod. Econ. **118**(1), 168–174 (2009)
9. Ray, P., Jenamani, M.: Mean-variance analysis of sourcing decision under disruption risk. Eur. J. Op. Res. **250**(2), 679–689 (2016)
10. Ferrari, B.: New survey data reflecting on causes of major supply chain disruption. http://www.theferrarigroup.com/supply-chain-matters/2012/11/07/new-survey-data-reflecting-on-causes-of-major-supply-chain-disruption/, Jan 2012
11. Drezner, Z., Hamacher, H.W.: Facility Location: Applications and Theory. Springer Science & Business Media (2001)

12. Ivanov, D., Dolgui, A., Sokolov, B., Ivanova, M.: Literature review on disruption recovery in the supply chain. Int. J. Prod. Res., 1–17 (2017)
13. Gupta, A., Maranas, C.D.: Managing demand uncertainty in supply chain planning. Comput. Chem. Eng. **27**(8), 1219–1227 (2003)
14. Nasiri, G.R., Zolfaghari, R., Davoudpour, H.: An integrated supply chain production-distribution planning with stochastic demands. Comput. Ind. Eng. **77**, 35–45 (2014)
15. Tang, C.S.: Robust strategies for mitigating supply chain disruptions. Int. J. Logist. Res. Appl. **9**(1), 33–45 (2006)
16. Claypool, E., Norman, B.A., Needy, K.L.: Modeling risk in a design for supply chain problem. Comput. Ind. Eng. **78**, 44–54 (2014)
17. Sadghiani, N.S., Torabi, S., Sahebjamnia, N.: Retail supply chain network design under operational and disruption risks
18. Snyder, L.V., Atan, Z., Peng, P., Rong, Y., Schmitt, A.J., Sinsoysal, B.: Or/ms models for supply chain disruptions: a review. IIE Trans. 1–21 (2015)
19. Azad, N., Davoudpour, H., Saharidis, G.K., Shiripour, M.: A new model to mitigating random disruption risks of facility and transportation in supply chain network design. Int. J. Adv. Manuf. Technol. **70**(9–12), 1757–1774 (2014)
20. Snyder, L.V., Scaparra, M.P., Daskin, M.S., Church, R.L.: Planning for disruptions in supply chain networks. Tutor. Op. Res. 234–257 (2006); Trans. Res. Part E Logist. Trans. Rev. **75**, 95–114 (2015)
21. Shankar, B.L., Basavarajappa, S., Chen, J.C., Kadadevaramath, R.S.: Location and allocation decisions for multi-echelon supply chain network-a multi-objective evolutionary approach. Expert Syst. Appl. **40**(2), 551–562 (2013)
22. Ahumada, O., Villalobos, J.R.: Application of planning models in the agri-food supply chain: a review. Eur. J. Op. Res. **196**(1), 1–20 (2009)
23. Pathumnakul, S., Piewthongngam, K., Khamjan, S.: Integrating a shrimp-growth function, farming skills information, and a supply allocation algorithm to manage the shrimp supply chain. Comput. Electr. Agri. **66**(1), 93–105 (2009)
24. Lin, D.-Y., Wu, M.-H.: Pricing and inventory problem in shrimp supply chain: a case study of taiwan's white shrimp industry. Aquaculture **456**, 24–35 (2016)
25. Negi, S., Anand, N.: Issues and challenges in the supply chain of fruits and vegetables sector in india: a review. Int. J. Manag. Value Supply Chains **6**(2), 2012 (2015)
26. Shrivastava, H., Dutta, P., Krishnamoorthy, M., Suryawanshi, P.: Designing a resilient supply chain network for perishable products with random disruptions. In: Proceedings of the International MultiConference of Engineers and Computer Scientists, vol. 2 (2017)
27. Khouja, M.: The single-period (news-vendor) problem: literature review and suggestions for future research. Omega **27**(5), 537–553 (1999)
28. Dutta, P., Chakraborty, D., Roy, A.R.: A single-period inventory model with fuzzy random variable demand. Math. Comput. Model. **41**(8), 915–922 (2005)
29. Dutta, P., Chakraborty, D.: Incorporating one-way substitution policy into the newsboy problem with imprecise customer demand. Eur. J. Op. Res. **200**(1), 99–110 (2010)
30. Qin, Z., Kar, S.: Single-period inventory problem under uncertain environment. Appl. Math. Comput. **219**(18), 9630–9638 (2013)

A Hybrid Heuristic for Restricted 4-Dimensional TSP (r-4DTSP)

Arindam Roy, Goutam Chakraborty, Indadul Khan, Samir Maity and Manoranjan Maiti

Abstract In this paper, we proposed a hybridized soft computing technique to solve a restricted 4-dimensional TSP (r-4DTSP) where different paths with various numbers of conveyances are available to travel between two cities. Here, some restrictions on paths and conveyances are imposed. The algorithm is a hybridization of genetic algorithm (GA) and swap operator-based particle swarm optimization (PSO). The initial solutions are produced by proposed GA which used as swarm in PSO. The said hybrid algorithm (GA-PSO) is tested against some test functions, and efficiency of the proposed algorithm is established. The r-4DTSPs are considered with crisp costs. The models are illustrated with some numerical data.

Keywords Hybrid algorithm · GA-PSO · r-4DTSP

1 Introduction

Optimization has been an active area of research for several decades. In optimization, the TSP is one of the most intensively studied problems. TSP is a well-known NP-hard combinatorial optimization problem [1]. Different types of TSPs have been solved by the researchers during last two decades. These are TSPs with time windows [2], stochastic TSP [3], double TSP [4], asymmetric TSP [5], TSP with precedence

A. Roy (✉)
Department of Computer Science, Prabhat Kumar College, Contai, Purba Medinipur 721401, W.B, India
e-mail: royarindamroy@yahoo.com

G. Chakraborty
Faculty of Software and Information Science, Iwate Prefectural University, Takizawa, Japan

I. Khan · S. Maity
Department of Computer Science, Vidyasagar University, Medinipur 721102, W.B, India

M. Maiti
Department of Applied Mathematics with Oceanology and Computer Programming, Vidyasagar University, Medinipur 721102, W.B, India

S. Kar et al. (eds.), *Operations Research and Optimization*, Springer Proceedings in Mathematics & Statistics 225, https://doi.org/10.1007/978-981-10-7814-9_20

constraints [6], etc. All these developed TSPs are two-dimensional. Recently, different types of classical 2D, 3D (solid) TSPs are studied by Maity et al. [7, 8] in different uncertain environments. Nowadays, real-world optimization problems are increasingly complex, and better algorithms are always needed to solve them. TSPs require a great powerful heuristics learning technique to find the near-optimal solutions. The research on the efficient algorithm for TSP is still a frontier subject.

In real life, for travelling from one city to another, different paths are available with a set of conveyances at each station. In that case, a salesperson has to design the tour for minimum cost maintaining the TSP conditions and using the appropriate paths with suitable conveyances at different cities. This problem is called 4-dimensional TSP (4DTSP), which involves 'paths' and 'vehicles' between two cities—from 'origin' to 'destination.' Again, it is not always possible to travel in a particular path or vehicles; i.e., the number of alternatives paths and vehicles is not same at different stations. For this reason, it is designed as a restricted 4DTSP (r-4DTSP).

The intelligent algorithm is another resolution for TSP. The motivation behind hybridizations of different algorithmic concepts is usually to obtain better-performing systems that exploit and unite advantages of the individual pure strategies; i.e., such hybrids are believed to be benefited from synergy. Recently, some intelligent algorithms such as anterior artificial neural network [9], particle swarm optimization (PSO), and the combinations of ACO with simulated annealing (SA) [10] have been applied for TSP. A survey of hybrid metaheuristics in combinatorial optimization by Bluma et al. [11] is done. Also, a novel imperialist competitive algorithm for generalized TSP has been proposed by Ardalan et al. [12]. Psychas et al. [13] are advised a hybrid evolutionary algorithm for the multi-objective TSP.

PSO is a population-based swarm intelligence algorithm that was proposed by Kennedy and Eberhart [14]. This algorithm simulates the ability of bird flocking and fish schooling, organizing a heuristic learning mechanism to achieve various goals [15–17]. The PSO algorithm involves a simple learning strategy that efficiently selects the best solution from particle positional values. Different kinds of discrete PSO are developed by researchers to solve many discrete optimization problems. Wang et al. [18] proposed swap operator-based discrete PSO to solve TSP. Again, Mahi et al. [19] developed a new hybrid method based on PSO-ACO-3-Opt for solving TSP. Marinkis et al. [20] proposed a hybrid PSO for vehicle routing problem.

Proposed hybrid heuristic algorithm which is the combination of GA and PSO to solve the r-4DTSP. Here initial solutions are given by GA, then PSO which uses swap sequence and swap operator on different nodes for update the solution. For the GA, classified the fitness of the chromosomes, then each chromosome has a linguistic value—Very Very Small (VVS), Very Small (VS), Small (S), Medium (M), High (H), Very High (VH), and Very Very High (VVH). According to the linguistic values, probability of crossover p_c is created of each chromosome. Comparison crossover [7] and a new generation-dependent random mutation are also implemented in the present algorithm. The proposed algorithm is tested with standard data set from TSPLIB [22] against the simple GA (SGA) which is the combination of Roulette Wheel Selection (RWS), cyclic crossover, and random mutation, and hence, the efficiency of the new hybrid algorithm is established.

2 Mathematical Preliminaries

Rough Variable [21]
Let $(\Lambda, \Delta, \kappa, \Pi)$ be a rough variable ξ which is a measurable function from the rough space $(\Lambda, \Delta, \kappa, \Pi)$ to the set of real numbers, i.e., for every Borel set(B) of $\Re$, $\{\lambda \in \Lambda | \xi(\lambda) \in B\} \in \kappa$.

Here, first time the trust measure for 7-point scale of the rough event $\hat{\xi} \geq r, Tr\{\hat{\xi} \geq r\}$ and its function curve is presented, where r is a crisp number and $\hat{\xi}$ is a rough variable given by $\hat{\xi} = ([\mathrm{a, b}], [\mathrm{c, d}]), 0 \leq c \leq e \leq f \leq a \leq b \leq g \leq h \leq d$.

$$Tr\{\hat{\xi} \geq r\} = \begin{cases} 0 & \text{for d} \leq \text{r} \\ \frac{(d-r)}{4(d-c)} & \text{for } h \leq r \leq d \\ \frac{(d-r)}{4(d-c)} + \frac{(h-r)}{4(h-e)} & \text{for } g \leq r \leq h \\ \frac{1}{4}(\frac{(d-r)}{(d-c)} + \frac{(h-r)}{(h-e)} + \frac{(g-r)}{(g-f)}) & \text{for } b \leq r \leq g \\ \frac{1}{4}(\frac{(d-r)}{(d-c)} + \frac{(h-r)}{(h-e)} + \frac{(g-r)}{(g-e)} + \frac{(b-r)}{(b-e)}) & \text{for } a \leq r \leq b \\ \frac{1}{4}(\frac{(d-r)}{(d-c)} + \frac{(h-r)}{(h-e)} + \frac{(g-r)}{(g-e)} + 1) & \text{for } f \leq r \leq a \\ \frac{1}{4}(\frac{(d-r)}{(d-c)} + \frac{(h-r)}{(h-e)} + 2) & \text{for } e \leq r \leq f \\ \frac{1}{4}(\frac{(d-r)}{(d-c)} + 3) & \text{for } c \leq r \leq e \\ 1 & \text{for } r \leq c. \end{cases} \quad (1)$$

3 Proposed Hybrid GA-PSO

We propose an evolutionary hybrid algorithm, GA-PSO using the swap sequence-based PSO with rough set-based rank selection (7-point), comparison crossover, and generation-dependent random mutation. The proposed GA-PSO and its procedures are presented below.

3.1 Representation

Here, a complete tour of N cities represents a solution of ants. So, an N-dimensional integer vector $X_i = (x_{i1}, x_{i2}, ..., x_{iN})$, $Y_i = (r_{i1}, r_{i2}, ..., r_{is})$, and $Z_i = (v_{i1}, v_{i2}, ..., v_{iP})$ are used as cities with route, and vehicles to represent a solution, where $x_{i1}, x_{i2}, ...,$ x_{iN} represent N consecutive cities in a tour. In the algorithm, initially GA is used to produce a set of paths (tours) for the salesman which is a set of potential solutions for the PSO part of the algorithm.

3.2 GA

3.2.1 Rough Set-Based Rank Classification

After finding the initial solution, we rank the fitness and then classify the rank depending on the minimum, average, and maximum rank. Since rank is represented by crisp values, we construct the common rough values from it,
Rough Rank $=([r_1 \times \text{avg rank}, r_2 \times \text{avg rank}], [r_3 \times \text{avg rank}, r_4 \times \text{avg rank}])$,
where $r_1 = \frac{Max - Avg}{Avg}$, $r_2 = \frac{Max + Min}{2}$, $r_3 = \frac{Max - Min}{2}$, $r_4 = \frac{Avg - Min}{Avg}$.

According to the rank of the chromosome, it belongs to any one of the common rough rank values and corresponding p_c's are created of each chromosome as VVL, VL, L, M, H, VH, VVH.

3.2.2 Comparison Crossover

(i) **Determination of Probability of Crossover ($\mathbf{p_c}$)**: For a pair of chromosomes (X_i, X_j), we construct the following rough set. At first, the states of X_i and X_j, i.e., (VVS, VS, S, M, H, VH, VVH), are determined by making trust measures of rough values w.r.t their ranks in common rough rank region given in Eq. (1). After the determination of states of rank intervals of the chromosomes, their crossover probabilities are determined as linguistic variables (VVL, VL, L, H, VH, VVH) using rough trust measures which are presented in Table 1 following Eq. (1).
(ii) **Crossover Mechanism**: For crossover, we choose two individuals (parents) to produce new individuals (child). To get optimal result of a TSP, we compare the cost between two corresponding cities and minimum cost route will be taken using the following algorithm (cf. [7]).

Table 1 Rough-extended trust-based linguistic

Gene	VVS	VS	S	M	H	VH	VVH
VVS	VVL	VL	VL	L	VL	VL	VVL
VS	VL	VL	L	M	L	VL	VL
S	VL	L	L	H	L	L	VL
M	L	M	H	VH	H	M	L
H	VL	L	L	H	L	L	VL
VH	VL	VL	L	M	L	VL	VL
VVH	VVL	VL	VL	L	VL	VL	VVL

3.2.3 Generation-Dependent Random Mutation

(a) **Generation-Oriented Mutation (Variable Method)**: Here, we model a new form of mutation mechanism where probability of mutation (p_m) is determined by

$$p_m = \frac{k}{\sqrt{1 + Current\, generation\, number}}, k \in [0, 1].$$

(b) **Selection for Mutation**: For each solution of P(t), generate a random number r from the range [0, 1]. If $r < p_m$, and then the solution is taken for mutation. Here, p_m decreases gradually as generation increases. After calculating the p_m, mutation operation follows the conventional random mutation.

3.3 *Particle Swarm Optimization (PSO)*

After finding the paths by above GA, we use the swap sequence for updating the paths. A PSO normally starts with a set of potential solution (called swarm) of the decision-making problem [14] .

For the TSP where swap sequence and swap operations are used to find velocity of a particle and its updating [18]. For swap sequence based PSO where different nodes /cities are used to update a solution. A sequence of swap operators known as swap sequence which to transform a solution to updated solution. We used the procedure of Wang et al. [18].

3.3.1 Discrete PSO Updating

Now the original PSO updated for TSP as follows:

$$\left\{ \begin{array}{l} V_i(t+1) = wV_i(t) \oplus c_1 r_1 (X_{pbest}(t) \ominus X_i(t)) \oplus c_2 r_2 (X_{gbest}(t) \ominus X_i(t)), \\ X_i(t+1) = X_i(t) \oplus V_i(t+1) \end{array} \right\} \tag{2}$$

as earlier given the parameters r_1, r_2, c_1, c_2 and w, now $c_1 r_1 (X_{pbest}(t) \ominus X_i(t))$ gives all swap operators in BSS. Similarly for the $c_2 r_2 (X_{gbest}(t) \ominus X_i(t))$ also.

Hybrid Algorithm (GA-PSO):

Input: Set $iter_{GA} = 0$, *maxiter* and Max_{gen} (S_0), Population Size (pop_size), Probability of Mutation (p_m), Problem Data (cost matrix, time matrix, route and vehicle set).

Output: The optimum and near optimum solutions.

Step 1. Start

Step 2. Set initial generation $iter_{GA} = 0$, $iter_{PSO} = 0$ and $Max_{gen}(S_0)$.

Step 3. Initialize // randomly generate the path

Step 4. Evaluate// determine the fitness

Step 5. For $(iter_{GA} \leq S_0)$
Step 6. Rank the chromosome.
Step 7. Sum the rank of all individual chromosomes.
Step 8. Clustered the rank.
Step 9. Develop the linguistic values as VVF, VLF, LF, MF, HF, VHF, VVHF
Step 10. Trust based p_c created.
Step 11. Crossover operation.
Step 12. Mutation operation.
Step 13. Update the chromosome.
Step 14. Update the rank.
Step 15. Find best optimum and near optimum solutions.
Step 16. $iter_{GA} = iter_{GA}+1$
Step 17. End for
Step 18. Set initial solution find from GA.
Step 19. For $(iter_{PSO} \leq maxiter_1)$
Step 20. Initialize the $X_i(t)$, $Y_i(t)$, $Z_i(t)$
Step 21. Determine X_{pbest}, $Xgbest$
Step 22. Updating by Eq. (2)
Step 23. $iter_{PSO} = iter_{PSO} + 1$
Step 24. End for
Step 25. Store the best solutions
Step 26. Store global and near optimum solutions.
Step 27. End

4 Proposed Restricted 4-Dimensional TSP (r-4DTSP)

4.1 Classical TSP with Time Constraints (2DTSP)

Let $c(i,j)$ and $t(i,j)$ be the cost and time, respectively, for travelling from ith city to jth city. Then, the problem can be mathematically formulated as:

$$\left.\begin{array}{l} \text{Minimize} \quad Z = \sum_{i \neq j} c(i,j)x_{ij} \\ \text{subject to} \quad \sum_{i=1}^{N} x_{ij} = 1 \ \text{ for } j = 1, 2, \ldots, N \\ \qquad \sum_{j=1}^{N} x_{ij} = 1 \ \text{ for } i = 1, 2, \ldots, N \\ \qquad \sum_{i \in S}^{N} \sum_{j \in S}^{N} x_{ij} \leq |S| - 1, \forall S \subset Q \\ \qquad \sum_{i=1}^{N} \sum_{j=1}^{N} t(i,j)x_{ij} \leq t_{max} \\ where \ \ x_{ij} \in \{0, 1\}, i, j = 1, 2, \ldots, N. \end{array}\right\} \tag{3}$$

where Q = {1, 2, 3, …, N} set of nodes, x_{ij} is the decision variable, and $x_{ij} = 1$ if the salesman travels from city-i to city-j, otherwise $x_{ij} = 0$ and t_{max} is the maximum time that should be maintained for the entire tour to avoid unwanted delay. Then, the above CTSP reduces to

$$\left.\begin{array}{c} \text{determine a complete tour} \quad (x_1, x_2, \ldots, x_N, x_1) \\ \text{to minimize} \quad Z = \sum_{i=1}^{N-1} c(x_i, x_{i+1}) + c(x_N, x_1) \\ \text{subject to} \quad \sum_{i=1}^{N-1} t(x_i, x_{i+1}) + t(x_N, x_1) \leq t_{max} \\ where \;\; x_i \neq x_j, i, j = 1, 2, \ldots, N. \end{array}\right\} \tag{4}$$

along with sub-tour elimination criteria

$$\left.\begin{array}{c} \sum_{i \in S}^{N} \sum_{j \in S}^{N} x_{ij} \leq |S| - 1, \forall S \subset Q \\ where \;\; x_{ij} \in \{0, 1\}, i, j = 1, 2, \ldots, N. \end{array}\right\} \tag{5}$$

4.2 STSP with Time Constraints (3DTSP)

Let $c(i, j, k)$ and $t(i, j, k)$ be the cost and time, respectively, for travelling from ith city to jth city using kth type conveyance. Then, the salesman has to determine a complete tour $(x_1, x_2, \ldots, x_N, x_1)$ and corresponding conveyance types $(v_1, v_2, \ldots, v_P)$ to be used for the tour, where $x_i \in \{1, 2, \ldots, N\}$ for $i = 1, 2, \ldots, N$, $v_i \in \{1, 2, \ldots P\}$ for $i = 1, 2, \ldots, N$ and all x_i's are distinct. Then, the problem can be mathematically formulated as:

$$\left.\begin{array}{c} \text{minimize} \quad Z = \sum_{i=1}^{N-1} c(x_i, x_{i+1}, v_i) + c(x_N, x_1, v_l), \\ \text{subject to} \quad \sum_{i=1}^{N-1} t(x_i, x_{i+1}, v_i) + t(x_N, x_1, v_l) \leq t_{max}, \\ where \;\; x_i \neq x_j, i, j = 1, 2 \ldots, N, \quad v_i, v_l \in \{1, 2, \ldots or\, P\} \end{array}\right\} \tag{6}$$

along with Eq. (5).

4.3 4DTSP with Time Constraints (4DTSP)

Let $c(i, j, r, k)$ and $t(i, j, r, k)$ be the cost and time, respectively, for travelling from ith city to jth city by the rth route using kth type conveyance. Then, the salesman has to determine a complete tour $(x_1, x_2, ..., x_N, x_1)$ and corresponding available route types $(r_1, r_2, ..., r_S)$ with conveyance types $(v_1, v_2, ..., v_P)$ to be used for the tour, where $x_i \in \{1, 2, \ldots, N\}$ for $i = 1, 2, ..., N$, $r_i \in \{1, 2, \ldots S\}$ and $v_i \in \{1, 2, \ldots P\}$ for $i = 1, 2, ..., N$ and all x_i's are distinct. Then, the problem can be mathematically formulated as:

$$\left.\begin{array}{r}\text{minimize}\quad Z = \sum\limits_{i=1}^{N-1} c(x_i, x_{i+1}, r_i, v_i) + c(x_N, x_1, r_l, v_l),\\ \text{subject to}\quad \sum\limits_{i=1}^{N-1} t(x_i, x_{i+1}, r_i, v_i) + t(x_N, x_1, r_l, v_l) \le t_{max},\\ where\ \ x_i \neq x_j, i, j = 1, 2 \ldots N,\quad r_i, r_l \in \{1, 2, \ldots, or\, S\},\quad v_i, v_l \in \{1, 2, \ldots, or\, P\}\end{array}\right\} \tag{7}$$

along with Eq. (5).

4.4 4DTSP in Restricted Routes with Time Constraints (r-4DTSP)

In real life, it is seen that in all stations, all types of routes may not be available due to the geographical position of the station, weather conditions, etc. So, it is more realistic that restricted routes are considered to travel different stations. Let $c(i, j, r, k)$ and $t(i, j, r, k)$ be the cost and time, respectively, for travelling from ith city to jth city by the rth route using kth type conveyance. Then, the salesman has to determine a complete tour $(x_1, x_2, ..., x_N, x_1)$ and corresponding available route types $(r_{m1}, r_{m2}, ..., r_{ms})$ with conveyance types $(v_{q1}, v_{q2}, ..., v_{qp})$ providing maximum available $s_1(\le S)$ and $p_1(\le P)$ types of routes and conveyances to be used for the tour, where $x_i \in \{1, 2, \ldots, N\}$ for $i = 1, 2, ..., N$, $r_{mi} \in \{1, 2, \ldots, s_1\}$ and $v_{qi} \in \{1, 2, \ldots, p_1\}$ for $i = 1, 2, ..., N$ and all x_i's are distinct. Then, the problem can be mathematically formulated as:

$$\left.\begin{array}{r}\text{minimize}\quad Z = \sum\limits_{i=1}^{N-1} c(x_i, x_{i+1}, r_{mi}, v_{qi}) + c(x_N, x_1, r_{ml}, v_{ql}),\\ \text{subject to}\quad \sum\limits_{i=1}^{N-1} t(x_i, x_{i+1}, r_{mi}, v_{qi}) + t(x_N, x_1, r_{ml}, v_{ql}) \le t_{max},\\ where\ \ x_i \neq x_j, i, j = 1, 2, \ldots, N, m = 1, 2, \ldots s_1,\ q = 1, 2, \ldots, p_1,\\ r_{mi}, r_{ml} \in \{1, 2, \ldots, or\ s_1\}, v_{qi}, v_{ql} \in \{1, 2, \ldots, or\ p_1\},\end{array}\right\} \tag{8}$$

along with Eq. (5).

5 Numerical Experiments

5.1 *Testing for Hybrid GA-PSO*

The performance of the proposed GA-PSO algorithm is established solving 15 standards benchmarked from TSPLIB [22]. Table 2 gives the results of hybrid GA-PSO with the standard GA and ACO with their hybridization ACO-GA. We compare the problems in terms of total cost. The results are under 20 independent runs, the average results, best found solution. The results are taken for four algorithms as proposed hybrid algorithm (HA) GA-PSO, ACO-GA, known ACO, and simple GA.

The parameters for the hybrid GA-PSO are set as those in Table 3 for different nodes of the TSP. As increases of the size of the TSP as increases of the pop_{size}, Maxgen, ant numbers for convergence of the optimal solution (Table 4).

5.2 *r-4DTSP with Time Constraint in Crisp Environment*

For r-4DTSP, here we consider three types of conveyances and maximum three types of route are consider as Eq. (8). The cost and time matrices for the r-4DTSP are represented below: Here, we consider a deterministic 2DTSP given by Eq. (4). The problem is solved by proposed hybrid GA-PSO, and the results are presented in Table 5.

Here, we have taken maximum generation $= 1000$, and we see that as time factor decreases the corresponding cost increases as real-life demand. Here, we consider a deterministic 3DTSP given by Eq.(6). The problem is solved by GA-PSO, and the results are presented in Table 6.

Here, we consider a deterministic 4DTSP given by Eq. (7). The problem is solved by GA-PSO, and the results are presented in Table 7.

Again, we consider a deterministic restricted 4DTSP given by Eq. (8). The problem is solved by GA-PSO, and the results are presented in Table 8.

6 Conclusion

In this paper, a new hybrid heuristic algorithm GA-PSO is proposed and illustrated in r-4DTSP formulated in different environments. In the proposed algorithm, where initial solution are generated by GA then swap operator based discrete PSO used for optimized the TSP path. Here, GA is applied with a new rough 7-point selection of rank of each chromosome and comparison crossover is used along with virgin generation-dependent random mutation. Restricted 4DTSP is first time introduced in the area of TSPs and regarded as highly NP-hard combinatorial

Table 2 Results for standard TSP problem (TSPLIB)

Instances	Average	Result			Best	Found	Result	
	GA-PSO	ACO-GA	ACO	GA	GA-PSO	ACO-GA	ACO	GA
fri26	938.23	938.51	939.63	939.64	937	937	937	937
bays29	2023.89	2024.23	2022.78	2022.56	2020	2020	2020	2020
bayg29	1611.34	1610.34	1611.02	1610.97	1610	1610	1610	1610
dantzig42	702.97	703.34	703.51	700.07	699	699	703	699
eil51	428.43	429.8	432.98	429.31	426	426	430	426
berlin52	7565.12	7648.9	7936.35	7654.87	7542	7542	7883	7623
st70	684.25	679.34	699.51	682.17	675	675	687	675
eil76	545.67	549.65	567.27	545.86	538	538	547	547
pr76	108221.45	108265.76	108634.71	108572.32	108159	108159	108346	108258
rat99	1215.21	1217.52	1236.46	1218.71	1211	1211	1223	1211
kroa100	21324.23	21421.78	21567.82	21431.75	21282	21282	21427	21378
kroc100	20879.53	20934.87	20956.23	20971.75	20750	20750	20802	20831
kroa150	26754.67	26805.76	26952.34	26743.89	26524	26524	26871	26701
krob200	29456.97	29550.7	30887.34	29965.27	29413	29413	29944	29789
pr299	49352.67	49765.6	52945.78	50831.43	48743	48743	49765	49391

Table 3 Parameters for GA-PSO, ACO, and simple GA

Size (N)	Maxgen	$Iter_{PSO}$	$Iter_{ACO}$	$Iter_{GA}$	Maxiter	Ant number (n)	Popsize	p_c	p_m	δ_1
$N \leq 50$	200	30	80	120	100	30	50	0.35	0.1	0.2
$50 < N \leq 100$	300	40	120	180	200	50	100	0.3	0.15	0.2
$100 < N \leq 150$	400	40	200	300	300	80	100	0.35	0.2	0.3
$150 < N \leq 200$	500	50	200	400	400	100	130	0.4	0.2	0.3
$200 < N \leq 250$	600	60	250	450	400	100	150	0.45	0.2	0.3
$250 < N \leq 300$	900	80	400	500	500	100	150	0.45	0.25	0.3

Table 4 Input data: crisp r-4DTSP

Crisp cost matrix (10 × 10) with three route and conveyances

i/j	1	2	3	4	5
1	∞	(35, 36, 27) (24, 34, 25) (17, 23, 26)	(18, 39, 30) (19, 24, 26) (30, 24, 31)	(20, 33, 34) (23, 27, 22) (23, 22, 28)	(30, 21, 62) (32,14, 18) (31, 43, 32)
2	(35, 26, 17) (33, 34, 28) (22, 27, 29)	∞	(40, 21, 32) (57, 28, 39) (13, 27, 19)	(18, 29, 10) (18, 39, 20) (15, 21, 32)	(35, 26, 37) (27, 36, 30) (31, 54, 23)
3	(38, 30, 29) (23, 45, 18) (17, 28, 35)	(17, 58, 34) (23, 24, 27) (37, 27, 19)	∞	(12, 25, 14) (44, 38, 37) (39, 23, 43)	(42, 25, 46) (29, 30, 46) (43, 33, 54)
4	(28, 20, 11) (18, 19, 16) (56, 23, 19)	(10, 22, 14) (18, 28, 32) (333, 46, 28)	(17, 8,29) (37, 11, 44) (48, 29, 10)	∞	(30, 19, 24) (30, 17, 11) (41, 37, 21)
5	(17, 15, 9) (34, 29, 11) (17, 29, 10)	(42, 23, 34) (45, 19, 20) (15, 29, 30)	(35, 36, 37) (29, 10, 28) (37, 25, 18)	(20, 31, 43) (36, 29, 13) (52, 19, 38)	∞
6	(22, 25, 17) (17, 27, 15) (23, 24, 27)	(17, 15, 9) (11, 34, 13) (43, 25, 28)	(32, 37, 33) (45, 48, 10) (23, 24, 27)	(43, 25, 28) (54, 38, 20) (28, 29, 17)	(23, 24, 27) (55, 38, 43) (45, 56, 57)
7	(30, 21, 62) (30, 21, 62)	(35, 26, 17) (43, 25, 28) (43, 25, 28)	(32, 37, 33) (24, 34, 25) (48, 29, 10)	(17, 27, 15) (53, 67, 18) (18, 15, 13)	(23, 24, 27) (18, 15, 13) (18, 28, 29)
8	(43, 25, 28) (11, 34, 13) (43, 25, 28)	(53, 67, 18) (18, 15, 13) (30, 21, 62)	(18, 15, 13) (18, 28, 29) (45, 56, 27)	(34, 56, 15) (45, 56, 27) (35, 26, 17)	(23, 24, 27) (28, 25, 26) (17, 27, 15)
9	(18, 15, 13) (18, 15, 13) (19, 18, 17)	(17, 15, 9) (11, 34, 13) (17, 27, 15)	(45, 56, 27) (35, 26, 17) (23, 24, 27)	(54, 37, 29) (24, 34, 25) (18, 15, 13)	(23, 24, 27) (18, 28, 29) (45, 56, 27)
10	(21, 34, 13) (30, 21, 62) (43, 25, 28)	(43, 25, 28) (11, 34, 13) (23, 24, 27)	(12, 33, 13) (16, 34, 13) (23, 24, 27)	(11, 34, 23) (23, 24, 27) (18, 15, 13)	(17, 27, 15) (24, 34, 25) (17, 27, 15)

(continued)

Table 4 (continued)

Crisp cost matrix (10 × 10) with three route and conveyances

i/j	6	7	8	9	10
1	(23, 24, 27) (28, 36, 29) (57, 28, 39)	(41, 37, 21) (31, 45, 62) (24, 11, 28)	(17, 15, 9) (67, 38, 29) (11, 34, 13)	(35, 36,37) (45, 38, 29) (19, 28, 17)	(23, 45, 18) (47, 39, 20) (17, 29, 10)
2	(17, 27, 15) (45, 25, 16) (43, 25, 28)	(18, 23, 16) (23, 26, 22) (19, 28, 38)	(21, 24, 15) (41, 39, 20) (23, 25, 27)	(18, 28, 19) (17, 28, 19) (32, 37, 33)	(35, 36, 37) (27, 26, 29) (23, 27, 28)
3	(19, 27, 35) (34, 27, 18) (21, 26, 16)	(29, 19, 24) (27, 28, 17) (15, 17, 19)	(17, 17, 19) (18, 27, 16) (21, 27, 28)	(17, 16, 19) (24, 22, 29) (21, 26, 28)	(15, 18, 19) (17, 18, 19) (17, 22, 28)
4	(31, 32, 18) (17, 27, 15) (32, 37, 33)	(17, 43, 23) (11, 34, 13) (30, 21, 62)	(23, 27, 29) (35, 26, 17) (36, 28, 22)	(35, 36, 37) (28, 36, 29) (17, 10, 19)	(21, 28, 29) (33, 21, 38) (67, 26, 38)
5	(32, 37, 33) (28, 36, 29) (35, 26, 17)	(28, 36, 29) (32, 15, 33) (17, 34, 23)	(17, 19, 10) (17, 18, 14) (29, 27, 27)	(21, 22, 29) (22, 29, 30) (35, 36, 37)	(28, 28, 19) (34, 33, 37) (43, 36, 23)
6	∞	(22, 26, 17) (28, 36, 29) (47, 46, 35)	(17, 16, 19) (17, 54, 29) (35, 28, 47)	(22, 17, 16) (28, 39, 10) (24, 34, 25)	(31, 28, 29) (39, 40, 29) (48, 29, 10)
7	(48, 29, 10) (33, 27, 26) (28, 25, 29)	∞	(30, 38, 40) (23, 24, 27) (35, 28, 19)	(56, 53, 61) (28, 39, 28) (53, 67, 18)	(17, 28, 19) (18, 15, 13) (18, 28, 29)
8	(17, 27, 15) (17, 27, 15) (45, 56, 27)	(17, 15, 9) (17, 10, 11) (17, 12, 11)	∞	(17, 27, 15) (23, 24, 27) (23, 17, 19)	(45, 56, 27) (32, 18, 19) (24, 27, 20)
9	(48, 29, 10) (17, 27, 15) (19, 18, 17)	(19, 18, 17) (20, 26, 19)	(12, 34, 13) (17, 19, 10) (28, 36, 29)	∞	(37, 45, 28) (54, 37, 29) (22, 32, 16)
10	(48, 29, 10) (53, 67, 18) (35, 36, 37)	(17, 27, 15) (18, 28, 29) (18, 28, 29)	(54, 37, 29) (45, 56, 27) (28, 36, 29)	(54, 37, 29) (19, 18, 17) (17, 27, 15)	∞

(continued)

Table 4 (continued)

Crisp time matrix(10×10) with three route and conveyances respectively

i/j	1	2	3	4	5
1	∞	(0.69, 0.68, 0.75) (0.32, 0.45, 0.71) (0.16, 0.18, 0.19)	(0.84, 0.63, 0.7) (0.24, 0.62, 0.44) (0.18, 0.19, 0.31)	(0.82, 0.7, 0.71) (0.36, 0.64, 0.72) (0.25, 0.28, 0.29)	(0.72, 0.8, 0.42) (0.32, 0.42, 0.26) (0.27, 0.28, 0.29)
2	(0.7, 0.66, 0.61) (0.8, 0.75, 0.71) (0.68, 0.7, 0.61)	∞	(0.76, 0.71, 0.69) (0.68, 0.61, 0.59) (0.6, 0.61, 0.4)	(0.67, 0.62, 0.6) (0.9, 0.85, 0.82) (0.29, 0.65, 0.32)	(0.75, 0.68, 0.65) (0.6, 0.58, 0.5) (0.56, 0.48, 0.35)
3	(0.55, 0.51, 0.48) (0.6, 0.56, 0.53) (0.61, 0.58, 0.56)	(0.72, 0.69, 0.62) (0.38, 0.31, 0.26) (0.6, 0.58, 0.51)	∞	(0.81, 0.76, 0.7) (0.71, 0.68, 0.66) (0.8, 0.76, 0.71)	(0.51, 0.46, 0.4) (0.7, 0.64, 0.61) (0.48, 0.44, 0.4)
4	(0.69, 0.64, 0.62) (0.78, 0.75, 0.71) (0.85, 0.83, 0.8)	(0.86, 0.81, 0.79) (0.76, 0.71, 0.69) (0.81, 0.78, 0.74)	(0.79, 0.75, 0.72) (0.9, 0.85, 0.82) (0.7, 0.64, 0.6)	∞	(0.65, 0.63, 0.6) (0.76, 0.72, 0.7) (0.78, 0.71, 0.69)
5	(0.8, 0.76, 0.71) (0.81, 0.79, 0.75) (0.88, 0.81, 0.79)	(0.55, 0.52, 0.49) (0.75, 0.74, 0.72) (0.61, 0.58, 0.54)	(0.6, 0.58, 0.4) (0.58, 0.55, 0.5) (0.59, 0.58, 0.54)	(0.78, 0.75, 0.71) (0.65, 0.62, 0.61) (0.55, 0.51, 0.48)	∞
6	(0.8, 0.75, 0.71) (0.81, 0.79, 0.76) (0.88, 0.85, 0.81)	(0.65, 0.63, 0.6) (0.75, 0.72, 0.7) (0.66, 0.61, 0.59)	(0.85, 0.82, 0.78) (0.7, 0.68, 0.62) (0.65, 0.62, 0.6)	(0.88, 0.84, 0.79) (0.87, 0.84, 0.8) (0.85, 0.81, 0.78)	(0.7, 0.67, 0.63) (0.6, 0.58, 0.55) (0.58, 0.54, 0.49)
7	(0.58, 0.54, 0.49) (0.56, 0.52, 0.48) (0.65, 0.62, 0.58)	(0.65, 0.63, 0.6) (0.44, 0.38, 0.33) (0.71, 0.65, 0.6)	(0.64, 0.6, 0.58) (0.6, 0.58, 0.55) (0.67, 0.64, 0.6)	(0.7, 0.68, 0.65) (0.55, 0.51, 0.45) (0.71, 0.68, 0.64)	(0.56, 0.54, 0.51) (0.38, 0.32, 0.28) (0.55, 0.53, 0.51)
8	(0.56, 0.52, 0.49) (0.54, 0.52, 0.51) (0.5, 0.43, 0.4)	(0.7, 0.68, 0.65) (0.9, 0.88, 0.84) (0.8, 0.81, 0.78)	(0.64, 0.6, 0.58) (0.41, 0.38, 0.37) (0.51, 0.45, 0.4)	(0.56, 0.52, 0.5) (0.76, 0.74, 0.7) (0.56, 0.52, 0.49)	(0.62, 0.58, 0.53) (0.62, 0.57, 0.55) (0.52, 0.48, 0.45)
9	(0.56, 0.51, 0.48) (0.88, 0.85, 0.81) (0.68, 0.65, 0.51)	(0.58, 0.52, 0.5) (0.59, 0.57, 0.56) (0.58, 0.55, 0.53)	(0.9, 0.85, 0.82) (0.62, 0.61, 0.58) (0.6, 0.54, 0.5)	(0.7, 0.68, 0.64) (0.74, 0.7, 0.67) (0.68, 0.52, 0.58)	(0.78, 0.75, 0.71) (0.65, 0.61, 0.58) (0.74, 0.7, 0.68)
10	(0.78, 0.71, 0.69) (0.7, 0.67, 0.64) (0.69, 0.64, 0.6)	(0.66, 0.61, 0.58) (0.77, 0.74, 0.7) (0.78, 0.76, 0.71)	(0.69, 0.65, 0.62) (0.8, 0.76, 0.74) (0.68, 0.65, 0.63)	(0.74, 0.7, 0.68) (0.65, 0.6, 0.57) (0.76, 0.71, 0.68)	(0.83, 0.78, 0.75) (0.62, 0.58, 0.56) (0.75, 0.71, 0.66)

(continued)

Table 4 (continued)

Crisp time matrix(10×10) with three route and conveyances respectively

i/j	6	7	8	9	10
1	(0.45, 0.34, 0.28) (0.45, 0.56, 0.73) (0.23, 0.25, 0.32)	(0.33, 0.42, 0.45) (0.23, 0.45, 0.36) (0.31, 0.33, 0.34)	(0.22, 0.32, 0.42) (0.21, 0.52, 0.33) (0.41, 0.43, 0.45)	(0.42, 0.62, 0.45) (0.24, 0.26, 0.27) (0.32, 0.34, 0.36)	(0.43, 0.53, 0.52) (0.32, 0.28, 0.35) (0.43, 0.46, 0.47)
2	(0.68, 0.64, 0.61) (0.7, 0.65, 0.62) (0.17, 0.35, 0.52)	(0.69, 0.63, 0.6) (0.31, 0.26, 0.2) (0.41, 0.56, 0.22)	(0.51, 0.45, 0.4) (0.32, 0.34, 0.19) (0.42, 0.44, 0.12)	(0.6, 0.57, 0.53) (0.7, 0.69, 0.62) (0.37, 0.29, 0.52)	(0.8, 0.76, 0.71) (0.81, 0.76, 0.7) (0.61, 0.46, 0.73)
3	(0.59, 0.55, 0.52) (0.61, 0.58, 0.56) (0.62, 0.6, 0.57)	(0.8, 0.75, 0.71) (0.9, 0.86, 0.81) (0.89, 0.86, 0.81)	(0.65, 0.6, 0.59) (0.64, 0.6, 0.58) (0.68, 0.65, 0.61)	(0.58, 0.55, 0.51) (0.8, 0.76, 0.71) (0.55, 0.5, 0.48)	(0.67, 0.61, 0.58) (0.76, 0.71, 0.68) (0.64, 0.6, 0.57)
4	(0.69, 0.65, 0.62) (0.78, 0.75, 0.71) (0.68, 0.67, 0.65)	(0.78, 0.74, 0.71) (0.68, 0.65, 0.61) (0.6, 0.54, 0.5)	(0.6, 0.56, 0.52) (0.59, 0.58, 0.56) (0.79, 0.76, 0.72)	(0.85, 0.82, 0.8) (0.78, 0.74, 0.71) (0.71, 0.69, 0.64)	(0.68, 0.63, 0.59) (0.5, 0.45, 0.41) (0.6, 0.54, 0.5)
5	(0.62, 0.58, 0.55) (0.81, 0.75, 0.72) (0.55, 0.51, 0.45)	(0.51, 0.45, 0.41) (0.81, 0.78, 0.75) (0.71, 0.68, 0.66)	(0.67, 0.62, 0.59) (0.66, 0.61, 0.58) (0.82, 0.79, 0.75)	(0.8, 0.76, 0.7) (0.88, 0.81, 0.78) (0.9, 0.87, 0.81)	(0.69, 0.66, 0.62) (0.7, 0.68, 0.65) (0.9, 0.87, 0.83)
6	∞	(0.64, 0.6, 0.58) (0.55, 0.51, 0.46) (0.7, 0.68, 0.65)	(0.55, 0.52, 0.48) (0.65, 0.63, 0.6) (0.76, 0.71, 0.68)	(0.68, 0.61, 0.58) (0.73, 0.7, 0.68) (0.62, 0.58, 0.55)	(0.65, 0.61, 0.58) (0.55, 0.52, 0.48) (0.65, 0.62, 0.6)
7	(0.55, 0.51, 0.46) (0.75, 0.71, 0.68) (0.52, 0.47, 0.4)	∞	(0.85, 0.81, 0.78) (0.55, 0.54, 0.51) (0.75, 0.76, 0.72)	(0.65, 0.61, 0.59) (0.58, 0.54, 0.5) (0.65, 0.61, 0.58)	(0.78, 0.74, 0.69) (0.71, 0.68, 0.64) (0.65, 0.62, 0.58)
8	(0.55, 0.52, 0.48) (0.8, 0.77, 0.7) (0.88, 0.83, 0.8)	(0.55, 0.54, 0.51) (0.78, 0.72, 0.7) (0.54, 0.53, 0.5)	∞	(0.78, 0.76, 0.73) (0.43, 0.4, 0.36) (0.73, 0.7, 0.68)	(0.58, 0.56, 0.51) (0.6, 0.54, 0.5) (0.58, 0.54, 0.49)
9	(0.74, 0.7, 0.68) (0.64, 0.61, 0.59) (0.67, 0.64, 0.6)	(0.85, 0.81, 0.8) (0.62, 0.6, 0.57) (0.58, 0.54, 0.49)	(0.62, 0.6, 0.58) (0.65, 0.61, 0.6) (0.79, 0.75, 0.72)	∞	(0.69, 0.65, 0.63) (0.78, 0.73, 0.7) (0.72, 0.7, 0.68)
10	(0.65, 0.61, 0.58) (0.87, 0.83, 0.78) (0.68, 0.64, 0.59)	(0.59, 0.54, 0.5) (0.68, 0.64, 0.61) (0.59, 0.55, 0.51)	(0.55, 0.52, 0.47) (0.52, 0.48, 0.54) (0.64, 0.6, 0.58)	(0.64, 0.59, 0.58) (0.45, 0.41, 0.37) (0.61, 0.59, 0.58)	∞

Table 5 Results of 2DTSP in crisp environment

Algorithm	Path	Value	T_{max}
GA-PSO	2-7-3-1-5-9-10-4-6-8	143	Without T_{max}
	2-7-3-1-5-9-10-4-6-8	147	8.54
	4-7-8-1-5-9-10-3-6-2	157	8.51
	2-6-3-1-9-5-4-7-8-10	167	8.42
	4-6-2-8-5-9-10-7-3-1	176	8.25
	5-8-2-1-5-9-10-3-6-7	184	8.02
ACO	6-3-9-7-5-2-1-10-8-4	193	8.7
GA	2-8-5-7-6-10-4-3-9-1	197	8.7
ACO-GA	4-8-9-1-3-7-2-10-5-6	204	8.00
GA-PSO	5-7-3-2-4-6-8-10-9-1	193	
	8-7-3-2-4-6-5-10-9-1	206	
ACO	3-8-5-7-6-10-4-2-9-1	227	
GA	8-2-1-3-4-10-7-9-6-5	221	
GA-PSO	8-2-7-9-4-3-5-6-10-1	216	7.5
GA-PSO	4-8-9-1-3-7-2-10-5-6	204	
ACO	5-6-2-7-8-10-3-9-4-1	392	
GA	10-6-2-7-8-5-3-9-4-1	398	

Table 6 Results of 3DTSP in crisp environment

Algorithm	Path (Vehicle)	Cost	Time	T_{max}
GA-PSO	9(1)-7(2)-8(3)-4(1)-3(1)-2(2)-5(1)-1(1)-10(2)-6(2)	170	8.75	8.75
	2(2)-1(3)-10(1)-3(1)-6(2)-7(1)-4(2)-5(2)-10(1)-9(2)	193	8.62	
	6(1)-9(2)-10(1)-7(2)-3(1)-8(2)-5(1)-4(1)-2(1)-1(3)	205	8.59	
	6(1)-10(2)-5(1)-7(1)-4(2)-3(3)-1(2)-10(3)-9(1)-2(1)	213	8.54	
	6(1)-7(2)-9(2)-8(1)-4(1)-5(2)-1(2)-2(2)-3(2)-10(1)	228	8.46	
ACO	3(2)-10(1)-8(1)-2(3)-3(3)-1(3)-5(1)-4(2)-6(2)-8(1)	242	8.7	
GA	3(2)-5(1)-8(2)-4(1)-2(1)-10(3)-5(1)-4(2)-6(2)-7(1)	247	8.7	
	3(2)-7(1)-4(1)-3(1)-1(1)-5(2)-10(2)-8(1)-6(1)-2(3)	282	7.95	8.00
GA-PSO	7(2)-9(1)-8(1)-10(2)-1(2)-3(2)-6(2)-5(1)-4(3)-2(1)	315	7.71	7.75
	10(1)-7(2)-6(1)-5(3)-4(2)-2(3)-3(1)-1(2)-8(2)-9(1)	376	7.58	

optimization problems. Such r-4DTSPs are here formulated in crisp costs and time boundary and solved by the proposed GA-PSO. Here, development of GA-PSO is in general form and it can be applied in other discrete problems such as network optimization, graph theory, solid transportation problems, vehicle routing, VLSI chip design.

Table 7 Results of 4DTSP in crisp environment

Algorithm	Path (Route, Vehicle)	Cost	Time	T_{max}
GA-PSO	10(2, 1)-7(3, 2)-8(1, 3)-4(2, 1)-3(1, 1)-2(1, 2)-5(2, 1)-1(3, 1)-9(1, 2)-6(2, 2)	183	8.75	8.75
	2(1, 2)-10(2, 3)-1(1, 1)-4(1, 2)-6(1, 2)-7(3, 1)-3(2, 2)-5(1, 2)-10(2, 1)-9(2, 2)	187	8.67	
	6(1, 3)-9(2, 1)-10(1, 1)-7(1, 2)-3(1, 3)-8(2, 2)-5(3, 1)-4(2, 1)-2(1, 1)-1(2, 3)	216	8.53	
	6(2, 1)-10(2, 2)-5(1, 1)-7(2, 1)-4(2, 3)-3(3, 1)-1(2, 1)-10(3, 1)-9(2, 1)-2(3, 1)	219	8.42	
	6(1, 3)-7(2, 1)-9(2, 1)-8(1, 1)-4(2, 1)-5(2, 2)-1(1, 2)-2(3, 2)-3(1, 2)-10(3, 3)	245	8.34	
ACO	3(1, 2)-10(2, 1)-8(3, 1)-2(2, 3)-3(3, 1)-1(1, 1)-5(2, 1)-4(1, 2)-6(2, 2)-8(1, 2)	253	8.73	
GA	4(3, 3)-5(1, 2)-8(3, 1)-3(2, 3)-2(1, 1)-10(2, 3)-5(2, 1)-4(1, 2)-6(1, 2)-7(2, 2)	262	8.7	
	3(2, 3)-7(1, 2)-4(3, 1)-3(2, 1)-1(1, 1)-5(2, 1)-10(2, 2)-8(1, 3)-6(1, 1)-2(3, 2)	303	7.91	8.00
GA-PSO	8(3, 2)-7(2, 1)-9(3, 1)-10(2, 3)-1(2, 2)-3(2, 1)-6(2, 1)-5(1, 2)-4(3, 3)-2(1, 2)	338	7.66	7.75
	10(1, 2)-7(, 12)-6(3, 1)-5(3, 2)-4(2, 2)-2(1, 3)-3(2, 1)-1(3, 2)-8(2, 2)-9(2, 1)	381	7.48	

Table 8 Results of r-4DTSP in crisp environment

Algorithm	Path (Route, Vehicle)	Cost	Time	T_{max}
GA-PSO	10(1, 1)-7(3, 1)-8(1, 3)-4(2, 1)-3(1, 1)-2(1, 2)-5(2, 1)-1(3, 1)-9(1, 2)-6(2, 2)	192	8.75	8.75
	2(1, 2)-10(2, 2)-1(1, 1)-4(1, 2)-6(2, 2)-7(3, 1)-3(2, 1)-5(1, 2)-10(2, 1)-9(2, 2)	201	8.67	
	6(1, 3)-9(2, 1)-10(1, 1)-7(1, 2)-3(1, 3)-8(2, 2)-5(3, 1)-4(2, 1)-2(1, 1)-1(2, 3)	229	8.53	
	6(2, 1)-10(2, 2)-5(1, 1)-7(2, 1)-4(2, 3)-3(3, 1)-1(2, 1)-10(3, 1)-9(2, 1)-2(3, 1)	236	8.42	
	6(1, 3)-7(2, 1)-9(2, 1)-8(1, 1)-4(2, 1)-5(2, 2)-1(1, 2)-2(3, 2)-3(1, 2)-10(3, 3)	278	8.34	
ACO	3(1, 2)-10(2, 1)-8(3, 1)-2(2, 3)-3(3, 1)-1(1, 1)-5(2, 1)-4(1, 2)-6(2, 2)-8(1, 2)	253	8.73	
GA	4(3, 3)-5(1, 2)-8(3, 1)-3(2, 3)-2(1, 1)-10(2, 3)-5(2, 1)-4(1, 2)-6(1, 2)-7(2, 2)	281	8.7	
	3(2, 3)-7(1, 2)-4(3, 1)-3(2, 1)-1(1, 1)-5(2, 1)-10(2, 2)-8(1, 3)-6(1, 1)-2(3, 2)	303	7.91	8.00
GA-PSO	8(3, 2)-7(2, 1)-9(3, 1)-10(2, 3)-1(2, 2)-3(2, 1)-6(2, 1)-5(1, 2)-4(3, 3)-2(1, 2)	338	7.66	7.75
	10(1, 2)-7(, 12)-6(3, 1)-5(3, 2)-4(2, 2)-2(1, 3)-3(2, 1)-1(3, 2)-8(2, 2)-9(2, 1)	381	7.48	

Acknowledgements This research article is supported by University Grant Commission of India by grant number PSW-150/14-15 (ERO).

References

1. Lawler, E.L., Lenstra, J.K., Rinnooy Kan, A.H.G., Shmoys, D.B.: The traveling salesman problem: G. E. Re Guided tour of combinatorial optimization. Wiley, New York (1985)
2. Focacci, F., Lodi, A., Milano, M.: A hybrid exact algorithm for the TSPTW. Inform. J. Comput. **14**(4), 403–417 (2002)
3. Chang, T., Wan, Y., Tooi, W.: A stochastic dynamic travelling salesman problem with hard time windows. Eur. J. Op. Res. **198**(3), 748–759 (2009)
4. Petersen, H.L., Madsen, O.B.G.: The double travelling salesman problem with multiple stack—formulation heuristic solution approaches. Eur. J. Op. Res. **198**, 339–347 (2009)
5. Majumder, A.K., Bhunia, A.K.: Genetic algorithm for asymmetric traveling salesman problem with imprecise travel times. J. Comput. Appl. Math. **235**(9), 3063–3078 (2011)
6. Moon, C., Ki, J., Choi, G., Seo, Y.: An efficient genetic algorithm for the TSP with precedence constraints. EJOR **140**, 606–617 (2002)
7. Maity, S., Roy, A., Maiti, M., Modified, A.: Genetic algorithm for solving uncertain constrained solid travelling salesman problems. Comput. Ind. Eng. **83**, 273–296 (2015)
8. Maity, S., Roy, A., Maiti, M.: An imprecise multi-objective genetic algorithm for uncertain constrained multi-objective solid travelling salesman problem. Expert Syst. Appl. **46**, 196–223 (2016)
9. Ghaziri, H., Osman, I.H.: A neural network algorithm for TSP with backhauls. Comput. Ind. Eng. **44**(2), 267–281 (2003)
10. Chen, S.M., Chien, C.Y.: Solving the TSP based on the genetic simulated annealing ant colony system with PSO techniques. Expert Syst. Appl. **38**(12), 14439–14450 (2011)
11. Bluma, C., Puchingerb, J., Raidlc, G.R., Roli, A.: Hybrid metaheuristics in combinatorial optimization: a survey. Appl. Soft Comput. **11**, 4135–4151 (2011)
12. Ardalan, Z., Karimi, S., Poursabzi, O., Nader, B.: A novel imperialist competitive algorithm for generalized TSP. Appl. Soft Comput. **xxx**, XX–XX (2014)
13. Psychas, I.-D., Delimpasi, E., Marinakis, Y.: Hybrid evolutionary algorithms for the multiobjective traveling salesman problem. Expert Syst. Appl. https://doi.org/10.1016/j.eswa.2015.07.051
14. Kennedy, J., Eberhart, R.: Particle swarm optimization. IEEE Int. Conf. Neural Netw. **4**, 1942–1948 (1995)
15. Chen, Y.L.: Particle swarm optimization survey for traveling salesman problem. Master Thesis. Department of Industrial Engineering and Management National Taipei University of Technology, July 2006
16. Chen, C.-Y., Feng, H.-M.: Hybrid intelligent vision-based car-like vehicle backing systems design. Expert Syst. Appl. **36**, 7500–7509 (2009)
17. Feng, H.M.: Self-generation RBFNs using evolutional PSO learning. Neurocomputing **70**, 241–251 (2006)
18. Wang, K., Hunang, L., Zhou, C.-G., Pang, W.: Particle swarm optimization for TSP, 2nd Intr. In: IEEE Conference on Machine Learning and Cybernetics, pp. 1583–1585 (2003)
19. Mahi, M., Baykan, O.K., Kodaz, H.: A new hybrid method based on PSO, ACO and 3-Opt algorithm for TSP. Appl. Soft Comput. **30**, 484–490 (2015)
20. Marinakis, Y., Marinaki, M., Dounias, G.: A hybrid PSO algorithm for vehicle routing problem. Eng. Appl. Artif. Intel. **23**, 464–472 (2010)
21. Liu, B.: Theory and Practice of Uncertain Programming. Physica-Verlag, Heidelberg (2002)
22. TSPLIB http://comopt.ifi.uniidelberg.de/software/TSPLIB95/

An Integrated Imperfect Production-Inventory Model with Lot-Size-Dependent Lead-Time and Quality Control

Oshmita Dey and Anindita Mukherjee

Abstract In this article, an integrated single-vendor single-buyer production-inventory model with stochastic demand and imperfect production process is investigated. The lead-time is assumed to be dependent on the lot-size and a fixed delay due to non-productive times. A methodology is developed to derive the optimal vendor investment required to reduce the defect rate and thereby minimize the total cost of the integrated system. Under the n-shipment policy, an algorithm is proposed so as to minimize the expected integrated total cost and determine the optimal values of the number of shipments, lot-size, safety stock factor, and percentage of defectives. Numerical results are used to illustrate the effect of various parameters on the system.

Keywords Economic order quantity · Integrated model · Imperfect production
Process quality · Variable lead-time

1 Introduction

The integrated single-vendor single-buyer production-inventory problem is inspired by the expanding focus on supply chain management which has been proved to be an adequate means by which both the buyer's and the vendor's interest can be benefited simultaneously [8]. A significant amount of literature [1, 9, 11–13, 17, 19] is available in this regard. In the existing literature, it is mostly found that the demand is deterministic and that shortages are not allowed. This was first extended by Ben-Daya and Hariga [3] where the authors assumed the annual customer demand to be stochastic, thus allowing shortages. Since then various researchers [4, 6, 7, 14] and the references therein have extended the stochastic models under various assumptions. However, in most of these works, the production process quality is presumed to be perfect. Even in models with imperfect production, the production process quality is not taken to be a control parameter [2, 10, 15–17, 21–23]. Ouyang et al. [18]

O. Dey (✉) · A. Mukherjee
Department of Mathematics, Techno India University, Kolkata 700091, West Bengal, India
e-mail: oshmi_kgp@yahoo.co.uk

S. Kar et al. (eds.), *Operations Research and Optimization*, Springer Proceedings in Mathematics & Statistics 225, https://doi.org/10.1007/978-981-10-7814-9_21

did consider process quality improvement but neglected the duration of screening. Dey and Giri [5] extended this existing literature by assuming optimal vendor investment in a stochastic single-vendor single-buyer imperfect production-inventory model with non-negligible screening time. But, they assumed the lead-time to be constant. However, in reality, lead-time is usually not a constant and assuming it to be so is an unreal restriction imposed on the model. Recently, Glock [6] developed a model with variable lead-time extending the Ben-Daya and Hariga's model [3] and permitting batch shipments increasing by a fixed factor. Glock [7] further extended this model by studying the alternative methods for reducing the lead-time and its effect on the expected total costs. Ben-Daya and Hariga [3] assumed the lead-time is taken to be proportional to the lot-size produced by the vendor in addition to a fixed delay due to transportation, non-productive time, etc. This makes sense intuitively since, from a practical point of view, lead-time should be considered as a function of the production lot-size [3]. Keeping this argument in mind, a linear relationship between lead-time and lot-size, including non-productive time, is taken into consideration. Thus, in order to make the model more attuned to reality, the present paper extends Dey and Giri's model [5] by assuming the lead-time to be linearly dependent on the production lot-size and non-productive times.

2 The Model

2.1 Notations

- D expected demand rate in units per time for non-defective items
- P production rate, $p = \frac{1}{P}$
- A buyer's ordering cost per order
- F transportation cost per delivery
- B vendor's setup cost
- L lead-time
- h_v vendor's holding cost per item per year
- h_{b1} buyer's holding cost for defective items per item per year
- h_{b2} vendor's holding cost for non-defective items per item per year
- s buyer's unit screening cost
- x buyer's screening rate
- w vendor's unit warranty cost for defective items
- y percentage of defective items produced
- k safety stock factor
- π buyer's shortage cost per item per year
- η fractional opportunity cost
- δ percentage decrease in defective items per dollar increase in investment

2.2 Assumptions

- Items of a single product are ordered from a single vendor by a single buyer.
- Demand per unit time is normally distributed with mean D and standard deviation σ.
- An order of nQ (non-defective) items is placed by the buyer to the vendor. These items are produced and, on average, transferred to the buyer in n equal sized shipments by the vendor, n being a positive integer.
- The buyer follows the classical (Q, r) continuous review inventory policy.
- It is assumed that the lead-time depends on the lot-size as per the form $L = pQ + b$, where b is the fixed delay due to transportation, non-productive times, etc. The lead-time demand is defined as the demand during the lead-time period. The lead-time demand is normally distributed with mean $D(pQ + b)$ and standard deviation $\sigma\sqrt{pQ + b}$.
- The re-order point r = expected demand during lead-time + safety stock (SS), i.e., $r = D(pQ + b) + k\sigma\sqrt{pQ + b}$, where k is the safety stock factor.
- Shortages are allowed and completely backlogged.
- y_0 $(0 \leq y_0 \leq 1)$ is the percentage of defective items produced in each batch of size Q.
- The vendor's rate of production of non-defective items is greater than the demand rate, i.e., $P(1 - y_0) > D$.
- The screening rate x is fixed and is greater than the demand rate i.e., $x > D$.
- The vendor incurs a warranty cost for each defective item produced.
- The vendor invests money to improve the production process quality in terms of buying new equipment, improving machine maintenance and repair, worker training, etc. We consider the following logarithmic investment function $I(y)$ [20]:

$$I(y) = \frac{1}{\delta} \ln\left(\frac{y_0}{y}\right)$$

  where δ is the percentage decrease in y per dollar (or any other suitable currency) increase in investment and y_0 is the original percentage of defective items produced prior to investment.

It is assumed that the vendor accepts an order of size nQ for non-defective items from the buyer. The vendor then produces these nQ items all at once, and then, n batches of Q items are delivered each at a regular interval of $Q(1 - y)/D$ units of time on average. Hence, we can say that each ordering cycle is of length $Q(1 - y)/D$, and the complete production cycle is of length $nQ(1 - y)/D$.

2.3 Buyer's Perspective

The buyer is assumed to follow the classical (Q, r) continuous review inventory system. That is, the buyer places an order of Q items to the vendor once the inventory level falls to the re-order point r. The vendor delivers these items after a lead-time $L = pQ + b$. Here, the safety stock factor k is taken to be a decision variable instead of the re-order point r. On receiving the order from the vendor, the buyer inspects the items at a fixed non-negligible screening rate x. The defective items are discovered in each lot, kept in hold separately and returned to the vendor when the next lot of items arrive. Therefore, the buyer incurs two types of holding cost—one for defective items and one for non-defective items [5]. The average inventory level for non-defective items for the buyer (including those defective items which have not yet been detected before the end of the screening time Q/x) is given by Eq. (1) (Fig. 1).

$$\frac{nQ(1-y)}{D}\left[k\sigma\sqrt{pQ+b}+\frac{Q(1-y)}{2}+\frac{DQy}{2x(1-y)}\right] \tag{1}$$

Equivalently, the average inventory level for defective items is given as below:

$$nQ^2y\left[\frac{1-y}{D}-\frac{1}{2x}\right] \tag{2}$$

Thus, the annual expected total cost for the buyer including the ordering cost, shipment cost, holding cost, shortage cost, and screening cost is given as

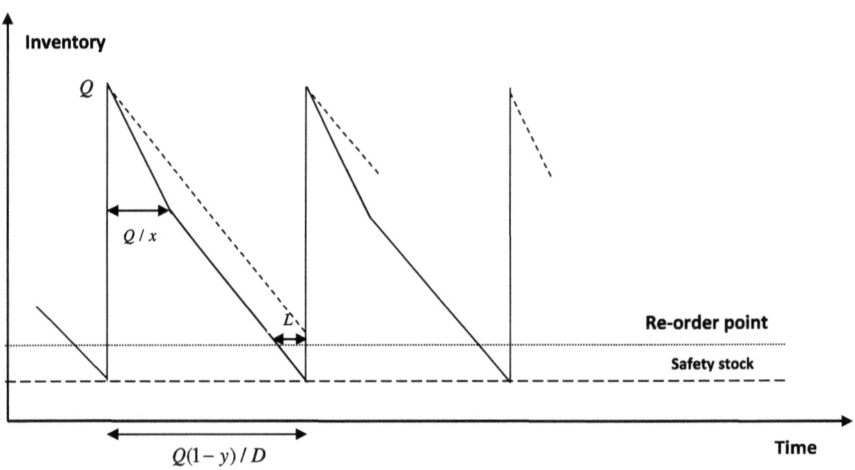

Fig. 1 Inventory of the buyer

$$ETCB(Q,k,n) = \frac{D(A+nF)}{nQ(1-y)} + h_{b1}\left[Qy - \frac{DQy}{2x(1-y)}\right] + h_{b2}\left[k\sigma\sqrt{pQ+b} + \frac{Q(1-y)}{2} + \frac{DQy}{2x(1-y)}\right] + \frac{\pi D\sigma\sqrt{pQ+b}\psi(k)}{Q(1-y)} + \frac{sD}{1-y} \quad (3)$$

where $\psi(k) = \int_k^{\infty}(z-k)\phi(z)dz$, $\phi(z)$ being the standard normal density function.

2.4 Vendor's Perspective

In the course of the production process, Q items are produced by the vendor in the first instance and then, these items are delivered to the buyer. Thenceforth, a quantity of Q items is delivered by the vendor to the buyer after an interval of every T units of time, where $T = Q(1-y)/D$. This process of delivering the items to the buyer is carried on till the vendor's production run is completed (Fig. 2).

Now, the average inventory holding cost for the vendor [15] is calculated as given below in Eq. (4):

$$EHCV = h_v\frac{Q}{2}\left[n(1 - \frac{Dp}{1-y}) - 1 + \frac{2Dp}{1-y}\right] \quad (4)$$

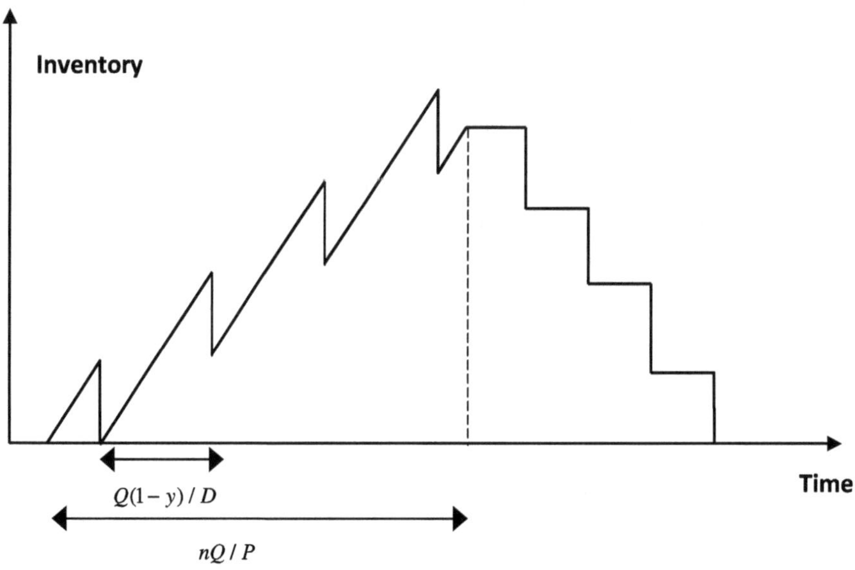

Fig. 2 Inventory of the vendor

Thus, the total cost incurred by the vendor is the sum of the setup cost, holding cost, warranty cost, and investment for reducing the percentage defective items [5] and it is given as

$$ETCV(Q, y, n) = \frac{BD}{nQ(1-y)} + h_v\frac{Q}{2}\left[n(1 - \frac{Dp}{1-y}) - 1 + \frac{2Dp}{1-y}\right] + \frac{wDy}{1-y} + \frac{\eta}{\delta}\ln\left(\frac{y_0}{y}\right) \quad (5)$$

where η is the fractional opportunity cost. It should be taken into account here that the logarithmic investment function considered above is convex in y.

2.5 Integrated System

The total expected annual cost of the integrated system can therefore be expressed as the sum of the buyer's and the vendor's total expected annual costs which is given as below:

$$ETC(Q, y, k, n) = \frac{D(A + B + nF)}{nQ(1-y)} + h_{b1}\left[Qy - \frac{DQy}{2x(1-y)}\right] + h_v\frac{Q}{2}\left[n(1 - \frac{Dp}{1-y}) - 1 + \frac{2Dp}{1-y}\right] + h_{b2}\left[k\sigma\sqrt{pQ+b} + \frac{Q(1-y)}{2} + \frac{DQy}{2x(1-y)}\right] + \frac{\pi D\sigma\sqrt{pQ+b}\psi(k)}{Q(1-y)} + \frac{(s+wy)D}{1-y} + \frac{\eta}{\delta}\ln\left(\frac{y_0}{y}\right) \quad (6)$$

Here, the control parameters are the lot-size Q, the percentage of defectives produced y, the safety stock factor k, and the number of shipments n.

Showing analytically that the expected total cost function, ETC, is convex in all the decision variables Q, y, k and n is not always possible. Nevertheless, the same can be demonstrated numerically. For given fixed values of n (where n is a positive integer) and y ($0 \leq y \leq y_0 \leq 1$), the convexity of total cost function ETC w.r.t Q and k can be easily shown by means of a 3D-graph (Fig. 4). Keeping this potential non-convexity in mind, an iterative algorithm is proposed, in the subsequent section, to derive the optimal values of Q, y, k, and n for which the expected annual total cost for the integrated system ETC is minimized.

3 Solution Procedure

Taking the second-order partial derivative of the total cost function ETC with respect to n, we find,

$$\frac{\partial^2 ETC}{\partial n^2} = \frac{2D(A+B)}{n^3 Q(1-y)} > 0 \qquad \forall \qquad n \geq 1 \tag{7}$$

Thus, from the above equation, we can conclude that *ETC* is convex in n.

Again, taking the second-order partial derivative of *ETC* with respect to *k* and *Q*, we get,

$$\frac{\partial^2 ETC}{\partial k^2} = \frac{D\sqrt{pQ+b}\,\sigma\pi\phi(k)}{Q(1-y)} > 0 \tag{8}$$

$$\frac{\partial^2 ETC}{\partial Q^2} = \frac{2DG(n)}{Q^3(1-y)} - \frac{h_{b2}k\sigma p^2}{4(pQ+b)^{\frac{3}{2}}} + \frac{\pi D\sigma\psi(k)}{(1-y)}\left[\frac{2\sqrt{pQ+b}}{Q^3} - \frac{p}{Q^2\sqrt{pQ+b}} - \frac{p^2}{4Q(pQ+b)^{\frac{3}{2}}}\right] > 0 \tag{9}$$

where $G(n) = \frac{A+B+nF}{n}$.

Hence, from Eqs. (8) and (9), *ETC* is seen to be convex in *k* and *Q* for fixed values of *n* and *y* $(0 \leq y \leq y_0 \leq 1)$. Although *y* is bounded, it is not possible to prove conclusively that *ETC* is convex in *y*. So in order to arrive at an optimal solution, the following procedure is followed:

For fixed value of *n*, the first derivative of *ETC* w.r.t *k* is set to zero. That is,

$$\frac{\partial ETC}{\partial k} = h_{b2} + \frac{\pi D}{Q(1-y)}(F(k)-1) = 0 \tag{10}$$

where $F(\cdot)$ is the cumulative distribution function.

Thus, we have,

$$\overline{F}(k) = \frac{h_{b2}Q(1-y)}{\pi D} \tag{11}$$

where $\overline{F}(\cdot) = 1 - F(\cdot)$.

Next, taking the first derivatives of *ETC* with respect to *Q* and *y* and setting those equal to zero, we get

$$\frac{\partial ETC}{\partial Q} = -\frac{DG(n)}{Q^2(1-y)} + yh_{b1}\{1 - \frac{D}{2x(1-y)}\} + h_{b2}\{\frac{1-y}{2} + \frac{Dy}{2x(1-y)}\} + \frac{h_{b2}k\sigma p}{2\sqrt{pQ+b}} + \frac{h_v}{2}\{-1 + n(1 - \frac{Dp}{1-y}) + \frac{2Dp}{1-y}\} - \frac{\pi D\sigma\psi(k)}{(1-y)}\left[-\frac{\sqrt{pQ+b}}{Q^2} + \frac{p}{2Q\sqrt{pQ+b}}\right] = 0 \tag{12}$$

and

$$\frac{\partial ETC}{\partial y} = \frac{Dw}{1-y} + \frac{D(s+wy)}{(1-y)^2} - \frac{\eta}{y\delta} + \frac{DG(n)}{Q(1-y)^2} + Qh_{b1}\{1 - \frac{D}{2x(1-y)}\}$$
$$-\frac{DQyh_{b1}}{2x(1-y)^2} + h_{b2}\{-\frac{Q}{2} + \frac{DQ}{2x(1-y)} + \frac{DQy}{(1-y)^2}\}$$
$$+\frac{Qh_v}{2}\{\frac{2Dp}{(1-y)^2} - \frac{Dnp}{(1-y)^2}\} - \frac{\pi D\sigma\sqrt{pQ + b\psi(k)}}{Q(1-y)^2} = 0, \quad (13)$$

respectively.

The algorithm presented by Dey and Giri [5] is modified and used here to derive the optimal solution. It is given as below:

The Algorithm

Step 1: Set $ETC^* = \infty, n = 1$
Step 2: Set $y = y_0$ and $k = 0$ and compute $\psi(k)$ and then compute $Q = Q_0$ using the values of $y_0, k, \psi(k)$ in equation (12)
Step 3: Compute k from (11) using Q_0, y and $\psi(k) = \int_k^\infty (z-k)\phi(z)dz$
Step 4: Compute y from (13) using the values k, Q_0 obtained in the previous step. If $y \geq y_0$, then set $y = y_0$.
Step 5: Compute Q from (12) using the updated values of k, y.
If $|Q - Q_0| \leq \epsilon$, then compute $ETC(Q, k, y, n)$ and go to Step 6.
Else set $Q_0 = Q$ and go back to Step 3.
Step 6: If $ETC^* \geq ETC$, we set $ETC^* = ETC, Q^* = Q, y^* = y, k^* = k, n = n + 1$ and go back to Step 2. Else put $n^* = n - 1$ and stop.
The corresponding values of the control parameters for $n^* = n - 1$ give the optimal solution.

It is to be noted here that we only get a local optimum by adopting the solution procedure mentioned. Since proving analytically that the objective function ETC is convex in all control parameters is not possible, we cannot say that the solution obtained above is a global optimum. In order to showcase the effects of the original process quality, the investment option and other model parameters on the optimal decisions, numerical studies are carried out in the following section.

4 Numerical Results and Discussions

For numerical studies, the following data set is considered:
$D = 1000, P = 3200, A = 50, F = 35, K = 400, L = 10/365, h_v = 4, h_{b1} = 6, h_{b2} = 10, s = 0.25, x = 2152, w = 20, \pi = 100, b = 0.01, \sigma = 5,\ \ y = 0.22, \eta = 0.2, \delta = 0.0002$

For fixed values of Q, k, n, it is shown that the total expected cost function ETC is convex in y $(0 \leq y \leq y_0)$ (Fig. 3). For fixed values of n, y, the convexity of ETC w.r.t. Q, k is shown in Fig. 4.

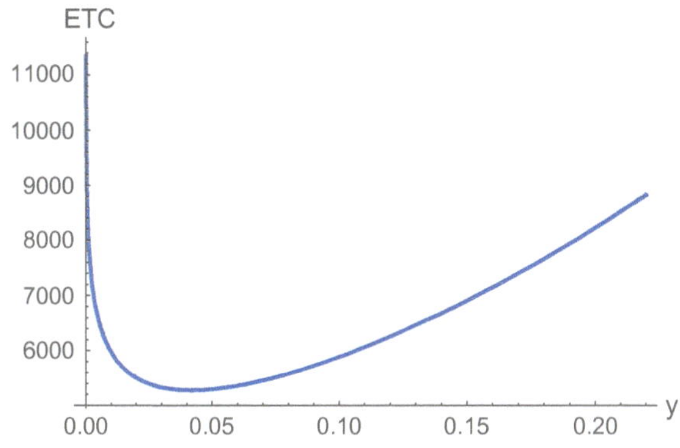

Fig. 3 ETC w.r.ty

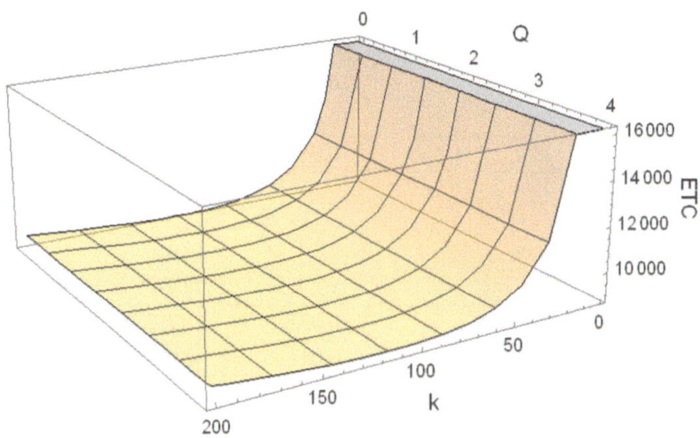

Fig. 4 ETC w.r.t Q, k

Table 1 depicts that the increase in the warranty cost w paid by the vendor results in an increase in the optimal total cost incurred by the supply chain. Also, with an increase in warranty cost, we find a decrease in the optimal value of the percentage of defective items. This is intuitively correct since if a higher warranty cost is to be paid by the vendor as a penalty for producing defective items, it would reasonably be beneficial for him if the number of defective items produced reduces considerably. Following the same logic, an increase in the value of b should imply an increase in the total cost incurred as also shown in Table 1.

A significant conclusion that can be reached from Table 2 is that the investment which is made in order to improve the production process quality is not independent of the original quality. That is, the necessity of an investment and the extent of it being beneficial is decided by the original production process quality. This is evident from Table 2 which clearly shows that investment to improve the production process quality is not needed when the original percentage of defectives is very low.

Table 1 Effect of parameters w and b

		Q^*	n^*	y^*	ETC^*
w	20	86.42	7	0.043	5213.31
	24	86.10	7	0.037	5378.61
	30	95.05	6	0.030	5584.26
b	0.005	86.38	7	0.043	5211.48
	0.010	86.42	7	0.043	5213.31
	0.100	96.01	6	0.043	5235.53

Table 2 Effect of y_0

y_0	Q^*	n^*	y^*	ETC^*	$I(y^*)$
0.010	86.41	7	0.043	2122.27	0.00
0.040	86.41	7	0.043	3508.57	0.00
0.100	86.42	7	0.043	4424.86	843.63
0.220	86.42	7	0.043	5213.31	1632.09
0.418	86.42	7	0.043	5855.17	2273.93
0.680	86.42	7	0.043	6341.78	2760.05

Table 3 Effect of demand rate

d	Q^*	n^*	y^*	ETC^*	$I(y^*)$
800	84.31	6	0.052	4752.10	1438.87
900	90.09	6	0.047	4993.20	1541.46
1000	86.42	7	0.043	5213.31	1632.09
1100	91.53	7	0.039	5413.49	1716.48
1200	96.60	7	0.037	5598.41	1794.05

However, with an increase in the value of y_0, the amount of investment required to optimize the supply chain also increases noticeably.

Table 3 shows that the production lot-size increases with an increase in demand rate, which is very obvious since the buyer would need to place an order of a larger quantity to satisfy the increase in demand. Also, an increase in the lot-size implies that there is an increase in number of both the defective and non-defective items produced, and consequently, the amount of investment needed to optimize the total cost will also increase. So, an increase in demand causes an increase in the optimal lot-size, the total expected cost incurred, and also the optimal vendor investment amount. All these intuitively correct effects are illustrated numerically.

It is seen from Table 4 that for very small values of y_0, the optimal value of ETC obtained for the two cases—with investment and without investment—differs by a small amount. However, as the percentage of defectives increases in the system, there

Table 4 Effect of investment

y_0	ETC* (with investment)	ETC* (without investment)
0.100	4424.86	5069.14
0.220	5213.31	8873.63
0.418	5855.17	18296.6
0.680	6341.78	48540.7

is a significant increase in the value of *ETC* without investment compared to that of *ETC* with investment. Therefore, it can reasonably be concluded that making an investment turns out to be significantly profitable for the supply chain as a whole, especially when the percentage of defectives produced is high.

5 Concluding Remarks

An attempt is made in this paper to analyze the problem of variable lead-time for an integrated single-vendor single-buyer imperfect production-inventory model under optimal vendor investment. It is shown that, as in the case of constant lead-time, for the variable lead-time model also, the investment by the vendor helps in reducing the production yield rate of non-defective items. Further, in case of the vendor making such an investment, the integrated system is better optimized in terms of minimizing the joint expected annual total cost. As a scope of future research, the variable lead-time may be assumed to be controllable. Also, setup cost reduction, inspection errors, variable shipment size, multiple buyers, etc., may also be included.

References

1. Banerjee, A.: A joint economic-lot-size model for purchaser and vendor. Decis. Sci. **17**, 292–311 (1986)
2. Ben-Daya, M., Hariga, M.: Economic lot scheduling problem with imperfect production process. J. Oper. Res. Soc. **51**, 875–881 (2000)
3. Ben-Daya, M., Hariga, M.: Integrated single vendor single buyer model with stochastic demand and variable lead-time. Int. J. Prod. Econ. **92**, 75–80 (2004)
4. Ben-Daya, M., Darwish, M., Ertogral, K.: The joint economic lot sizing problem: review and extensions. Eur. J. Oper. Res. **185**, 726–742 (2008)
5. Dey, O., Giri, B.C.: Optimal vendor investment for reducing defect rate in a vendor-buyer integrated system with imperfect production process. Int. J. Prod. Econ. **155**, 222–228 (2014)
6. Glock, C.H.: A comment: "integrated single vendor-single buyer model with stochastic demand and variable lead time". Int. J. Prod. Econ. **122**, 790–792 (2009)
7. Glock, C.H.: Lead timer eduction strategies in a single-vendor single-buyer integrated inventory model with lot size-dependent lead times and stochastic demand. Int. J. Prod. Econ. **136**, 37–44 (2012)

8. Goyal, S.K.: An integrated inventory model for a single supplier-single customer problem. Int. J. Prod. Res. **15**, 107–111 (1976)
9. Goyal, S.K.: A joint economic-lot-size model for purchaser and vendor: a comment. Decis. Sci. **19**, 236–241 (1988)
10. Goyal, S.K., Cárdenas-Barrón, L.E.: Note on: economic production quantity model for items with imperfect quality—a practical approach. Int. J. Prod. Econ. **77**, 85–87 (2002)
11. Ha, D., Kim, S.L.: Implementation of JIT purchasing: an integrated approach. Prod. Plan. Control **8**, 152–157 (1997)
12. Hill, R.M.: The single-vendor single-buyer integrated production-inventory model with a generalized policy. Eur. J. Oper. Res. **97**, 493–499 (1997)
13. Hill, R.M.: The optimal production and shipment policy for the single-vendor single-buyer integrated production-inventory problem. Int. J. Prod. Res. **37**, 2463–2475 (1999)
14. Hsiao, Y.-C.: Integrated logistic and inventory model for a two-stage supply chain controlled by the reorder and shipping points with sharing information. Int. J. Prod. Econ. **115**, 229–235 (2008)
15. Huang, C.K.: An optimal policy for a single-vendor single-buyer integrated production-inventory problem with process unreliability consideration. Int. J. Prod. Econ. **91**, 91–98 (2004)
16. Lin, H.J.: An integrated supply chain inventory model. Yugoslav J. Oper. Res. **22** (2012). https://doi.org/10.2298/YJOR110506019L
17. Ouyang, L.Y., Wu, K.S., Ho, C.H.: Analysis of optimal vendor-buyer integrated inventory policy involving defective items. Int. J. Adv. Manuf. Technol. **29**, 1232–1245 (2006)
18. Ouyang, L.Y., Wu, K.S., Ho, C.H.: Integrated vendor-buyer inventory model with quality improvement and lead-time reduction. Int. J. Prod. Econ. **108**, 349–358 (2007)
19. Pan, J.C.-H., Yang, J.S.: A study of an integrated inventory with controllable lead time. Int. J. Prod. Res. **40**, 1263–1273 (2002)
20. Porteus, E.L.: Optimal lot sizing, process quality improvement and setup cost reduction. Oper. Res. **36**, 137–144 (1986)
21. Salameh, M.K., Jaber, M.Y.: Economic production quantity model with for items with imperfect quality. Int. J. Prod. Econ. **64**, 59–64 (2000)
22. Shu, H., Zhou, X.: An optimal policy for a single-vendor and a single-buyer integrated system with setup cost reduction and process-quality improvement. Int. J. Syst. Sci. (2013). https://doi.org/10.1080/00207721.2013.786155
23. Zhang, X., Gerchak, Y.: Joint lot sizing and inspection policy in an EOQ model with random yield. IIE Trans. **22**, 41–47 (1990)

Fixed Charge Bulk Transportation Problem

Bindu Kaushal and Shalini Arora

Abstract This paper discusses an exact method to solve fixed charge bulk transportation problem (FCBTP). The fixed charge bulk transportation problem is a variant of the classical transportation problem in which a fixed cost is incurred in addition to the bulk transportation cost. This paper comprises of two sections. In Sect. 2, an algorithm based on lexi-search approach is proposed to solve FCBTP which gives the optimal solution in a finite number of iterations. Section 3 reports and corrects the errors which occurred in the paper entitled 'Solving the fixed charge problem by ranking the extreme point' by Murty (Oper. Res. 16(2): 268–279, 1968) [24]. Towards the end, some Concluding Remarks are given.

Keywords Fixed charge bulk transportation problem (FCBTP) · Bulk transportation problem (BTP) · Lexi-search

1 Introduction

In classical transportation problems, the aim is to find that scheduled flow of the homogeneous material from a number of sources to a number of destinations which is least expensive. In BTP and FCTP, also the cost is independent of the quantity shipped; hence in the FCBTP, the total cost of transportation is also independent of the transported quantity due to the addition of the step function which results in the objective being a step function. This could be effective in shipping bulk cargo, in handling bulk material such as cereals, milk, petroleum/crude oil, ores, coal, sand, wood chips, crushed rocks and stone in loose bulk form. Various authors have studied the

B. Kaushal (✉) · S. Arora
Indira Gandhi Delhi Technical University For Women, Kashmere Gate, New Delhi 110006, India
e-mail: bindukaushal27@gmail.com
URL: http://www.springer.com/lncs

S. Arora
e-mail: shaliniarora@igdtuw.ac.in
URL: http://www.springer.com/lncs

S. Kar et al. (eds.), *Operations Research and Optimization*, Springer Proceedings in Mathematics & Statistics 225, https://doi.org/10.1007/978-981-10-7814-9_22

fixed charge transportation problem and bulk transportation problem separately but the FCBTP includes both of these. The fixed charge transportation problem (FCTP) is an extension of classical transportation problem in which a fixed cost is associated with each route in addition to the usual cost coefficient. Unlike the usual cost, a fixed cost which affects the total cost is independent of the transported quantity. This might be the expense of leasing a vehicle, arrival charges at airplane terminal, setup cost required for assembling the item, and so forth.

Mathematically FCTP can be Stated As:

$$\min \sum_{i=1}^{m} \sum_{j=1}^{n} (c_{ij}x_{ij} + f_{ij}y_{ij})$$

subject to

$$\sum_{j=1}^{n} x_{ij} = a_i \quad \text{for } i = 1, 2, 3, \ldots, m$$

$$\sum_{i=1}^{m} x_{ij} = b_j \quad \text{for } j = 1, 2, 3, \ldots, n$$

$$x_{ij} \geq 0 \qquad \forall (i, j)$$

$$y_{ij} = \begin{cases} 1 & x_{ij} > 0 \\ 0 & x_{ij} = 0 \end{cases}$$

$$\sum_{i=1}^{m} a_i = \sum_{j=1}^{n} b_j \qquad a_i, b_j, c_{ij}, f_{ij} \geq 0$$

Here,
$I = (1, 2, \ldots, m)$: number of sources
$J = (1, 2, \ldots, n)$: number of destinations
a_i = availability at each source, b_j = requirement at each destination
x_{ij} = the quantity transported from ith source to jth destination
c_{ij} = cost of transportation from ith source to jth destination
$\sum_{i=1}^{m} a_i = \sum_{j=1}^{n} b_j$; this shows the case of balanced transportation problem
f_{ij} = fixed cost of transportation from ith source to jth destination.

The fixed charge problem was initially formulated in early 1950 s by Warren and George [33]. It was observed that the optimum will exist in one of the extreme points and a local minimum need not be the global minimum. Later on, Warren Hirsch and Hoffman [34] found the sufficient condition for attainment of infimum of concave function. Balinski [9] attempted to provide an approximate solution of FCTP and formulated it as an integer program. Murty [24] devised an exact algorithm for solution of the fixed charge transportation problem by ranking the extreme points which works efficiently when fixed charges are quite small as compared to the transportation cost. Section 3 reports some errors of [24] which affect the ranking limit to a large extent; however, the optimal solution remains the same perhaps because it appeared at an early stage of ranking limits and is explained in Sect. 3. An alternate approach to this algorithm was suggested by Gray [16], and the method is suitable when fixed charge dominates the variable cost. A vertex ranking algorithm based on Murty's extreme point ranking scheme for the fixed charge transportation problem was developed by Sadagopan and Ravindran [28]. Steinberg [30] developed an exact method based on branch and bound technique with some of the additional features that the computer storage remains constant throughout for any size of the problem. However, some adjacent extreme point techniques were also developed to solve FCTP [11, 13, 15, 26, 27]. Cooper [12] developed a 'simplex -like' algorithm which replaces several vectors in the basis at a time. Various approximation techniques [1, 2, 4, 20] have been developed to solve FCTP. A paradox in a fixed charge problem was discussed by Arora and Ahuja [6]. Thirwani [31] studied an enhanced flow in FCTP. Various discussions have been made to find either the exact or approximate solution for FCTP [3, 19, 21, 22]. Aguado [5] proposed a method to solve FCTP by the intensive use of Lagrangean relaxation techniques. Several real-world problems have been solved using branch and bound method [10, 25, 32]. Some authors [17, 18] have also used spanning tree-based genetic algorithm for solution of FCTP.

A zero-one transportation problem is also called a bulk transportation problem in which a homogeneous material is supplied in bulk from sources (with a fixed maximum capacity) to destinations (of known demands). In cost minimizing bulk transportation problem objective is to find that schedule flow of material which gives the minimum cost under the constraint that each destination is served by a single source but a source can serve to any number of destinations subject to its capacity. Here, transportation cost is independent of the quantity being transported.

Mathematically it can be stated as:

$$\min \sum_{i=1}^{m} \sum_{j=1}^{n} c_{ij} x_{ij}$$

subject to

$$\sum_{j=1}^{n} g_j x_{ij} \leq d_i \qquad i \in (1, 2, 3, \ldots, m)$$

$$\sum_{i=1}^{m} x_{ij} = 1 \qquad j \in (1, 2, 3, \ldots, n)$$

$$x_{ij} = \begin{cases} 1 & \text{if } j\text{th destination is supplied by } i\text{th source} \\ 0 & \text{otherwise} \end{cases}$$

where d_i is the availability at the ith source, g_j is the requirement at jth destination, c_{ij} is the cost of transportation from ith source to jth destination and is independent of the quantity shipped.

An additive algorithm for zero-one transportation problems was first proposed by Balas et al. [8] which is also called the filter method of Balas's as the variables can be grouped into subsets such that only one variable from each subset takes the value one while the remaining take it as zero. De Maio and Roveda [14] developed an implicit enumeration technique for this special class of transportation problem which is also called as the sequel of Balas filter method. Later on, Srinivasan and Thompson [29] presented a branch and bound algorithm for the same problem and showed that optimum solution to this special transportation problem is the basic feasible solution of some standard transportation problem. Murthy [23] solved bulk transportation problem using the lexi-search approach with some additional features that each destination is served by a single source but a source can serve to any number of destinations depending upon its capacity. A variant of time minimization assignment problem using lexi-search was developed by Arora and Puri [7]. A lexi-search approach is an implicit enumeration technique in which instead of enumerating all the solutions only manageable solutions are enumerated. In this, each solution is defined as a word and a partial word is used to define the block of words. The technique simplifies the search by eliminating the partial words which do not provide the better solution thereby reducing the search to a greater extent.

The algorithm using lexi-search technique is developed based on [7, 23] which gives the optimal feasible solution in a finite number of iterations. Initial upper bound on the objective function value is calculated using a heuristic and in many problems is very close to the optimal feasible solution. This paper consists of two sections; Sect. 2 deals with the solution of FCBTP. The mathematical model of the problem is explained in Sect. 2.1, and some related definitions and results are established in Sect. 2.2. Based upon these results, an algorithm is proposed in the Sect. 2.3. A numerical illustration to explain the process is presented in Sect. 2.4, and computational details are given in Sect. 2.5. Section 3 reports and corrects some errors which were observed in a paper entitled 'Solving the fixed charge by ranking the extreme points' by Murty [24]. Some conclusions based on the study are reported towards the end in Concluding Remarks.

2 Problem Description

2.1 Fixed Charge Bulk Transportation Problem (FCBTP)

The fixed charge bulk transportation problem is different from classical transportation problem due to the addition of fixed cost to the bulk cost. There is a set $I = \{1, 2, \dots, m\}$ of m sources and $J = \{1, 2, \dots, n\}$ of n destinations. The availability at each source $i \in I$ is d_i, and the requirement at each destination $j \in J$ is g_j. The bulk transportation cost from source i to destination j is denoted by c_{ij}, and fixed cost from source i to destination j is denoted by f_{ij} $(i \in I, j \in J)$. The total fixed charge bulk cost is denoted by Z. There is a restriction that each destination is served by a single source but a source can serve to any number of destinations depending upon its capacity. The objective is to find that feasible solution which minimizes the total cost. As the transportation is done in bulk, therefore variable x_{ij} is defined as

$$x_{ij} = \begin{cases} 1, & \text{if } i\text{th source serves } j\text{th destination} \\ 0, & \text{otherwise} \end{cases}$$

Mathematical Model:

Problem P_1:-

$$\min Z = \sum_{i=1}^{m} \sum_{j=1}^{n} (c_{ij}x_{ij} + f_{ij}y_{ij})$$

subject to

$$\sum_{j \in J} g_j x_{ij} \le d_i, \quad i \in I \qquad (i)$$

$$\sum_{i \in I} x_{ij} = 1, \quad j \in J \qquad (ii)$$

$$x_{ij} = 0 \text{ or } 1 \quad \forall (i,j) \in I X J \qquad (iii)$$

$$y_{ij} = \begin{cases} 1 & x_{ij} > 0 \\ 0 & x_{ij} = 0 \end{cases} \qquad (iv)$$

Feasibility check:-

$$\sum_{j \in J} g_j \le \sum_{i \in I} d_i \qquad (v)$$

A solution which satisfies (*i*), (*ii*), (*iii*), (*iv*) is called feasible solution. A feasible solution which minimizes the total cost is called optimal feasible solution. Equation (*v*) is necessary for existence of feasible solution of the problem.

2.2 *Theoretical Development*

Some Definitions and Results

Notation:-
Z_u:- Initial upper bound on the value of objective function
I_u:- Index of unassigned sources

$\uplus$:- Augmentation
$\cap$:- Negation of augmentation.

Alphabet Matrix:- It is a matrix which is formed by the position of the elements of bulk cost matrix when they are organized in non-decreasing order of their values. It is denoted by AB, and any jth column of this consists of the position of the entries of the jth column of bulk cost matrix when they are arranged in non-decreasing order of their values. Here, $ab(y,j)$ indicates yth entry in the jth column of matrix AB. So for any $y < z$ the corresponding bulk cost is $c_{ab(y,j)j} \leq c_{ab(z,j)j}$. Thus, any jth column of Alphabet Matrix AB consists of $[ab(1,j), ab(2,j), \ldots, ab(m,j)]'$ s.t. $c_{ab(1,j)j} \leq c_{ab(2,j)j} \leq \cdots \leq c_{ab(m,j)j}$.

Partial Word:- A partial word of length r is represented as $Pw = \left(ab(y_1, 1), ab(y_2, 2), \ldots, ab(y_r, r)\right), (i_1, \ldots, i_r), r \leq n$.

The partial solution corresponding to the partial word is denoted by X^{Pw}, and it consists of transportation to the jth destination by the source $ab(y_j, j), j = 1, 2, \ldots, r$ whereas the destinations $(r + 1, r + 2, \ldots, n)$ are still to be served. Each partial word Pw defines a block of words and is also called the leader of the block of words. Partial words are generated systematically by considering rows of AB and in decreasing order of their contribution to the objective function. If at any stage let a partial word of length r is under study, $r \leq n$, i.e. $Pw = \left(ab(y_1, 1), ab(y_2, 2), \ldots, ab(y_r, r)\right)$ then it means that all the partial word which start with $ab(y, 1)$ where $1 \leq y \leq y_1 - 1$ have not generated value better than Z_u. Contribution of Pw to the objective function is denoted by $Z(X^{Pw})$.

Theorem 1 *Let* $Pw = \left(ab(y_1, 1), ab(y_2, 2), \ldots, ab(y_r, r)\right), r \leq n$ *be a partial word for which* $Z(X^{Pw}) \geq Z_u, \forall y_r = 1, 2, \ldots, m$, *where* Z_u *is the upper bound on the optimal value of the objective function Z. Then the partial word* $\widetilde{Pw}$ *defined as* $\widetilde{Pw} = \left(ab(y_1, 1), ab(y_2, 2), \ldots, ab(y_{r-1}, r - 1)\right)$ *cannot generate a word with the corresponding value of* $Z(X^{\widetilde{Pw}})$ *less than* Z_u.

Proof $Z(X^{Pw}) \geq Z_u, \forall y_r = 1, 2, \ldots, m$. This imply that the value of the objective function for Pw is greater than upper bound and cannot generate a word with better value of the objective function, i.e. Z_u, when rth destination was checked for being served by all the m sources. This also means that supply to $(r - 1)$ destinations respectively by $ab(y_1, 1), ab(y_2, 2), \ldots, ab(y_{r-1}, r - 1)$ sources have not generated a partial word corresponding to which $Z(X^{Pw})$ is less than Z_u. This implies that $\widetilde{Pw}$ can not be augmented further to generate a word corresponding to which $Z\left(X^{Pw}\right)$ is less than Z_u.

Remark 1 If for the feasible partial word Pw, we find $Z(X^{Pw}) < Z_u$. Then the partial word $Pw(Z(X^{Pw}) < Z_u)$ may contain a word with the corresponding value $Z\left(X^{Pw}\right)$ less than Z_u, $\forall y_r = 1, 2, \ldots, m$. And if $Z(X^{Pw}) \geq Z_u$ where $Pw = \left(ab(y_1, 1), ab(y_2, 2), \ldots, ab(y_r, r)\right), r \leq n$, then the partial word is rejected and supply to some or all the first $(r - 1)$ destinations respectively by the sources $ab(y_1, 1), ab(y_2, 2), \ldots, ab(y_{r-1}, r - 1)$ must be rearranged, means to find the possibility of supply to $(r - 1)$th destinations by $ab(y, r - 1), y_{r-1} < y \leq m$. If $y_{r-1} = m$

then find the possibility of supply to $(r-2)$th destination by $ab(y, r-2)$ s.t. $y_{r-2} \leq y \leq m$, and so on. If $y_1 = y_2 = \ldots = y_r = m$ then partial word $\widetilde{Pw}$ can not be altered.

Theorem 2 *Let $Pw = ab(y_1, 1)$ be a partial word and $Z(X^{Pw}) \geq Z_u$, where Z_u be the upper bound on the value of the objective function. Then Z_u is the optimal value of the objective function.*

Proof As we have $Z(X^{Pw}) \geq Z_u$, then from the above Theorem 1 the partial word $ab(y_1, 1)$ cannot generate a feasible word corresponding to which $Z(X^{Pw})$ less than Z_u. Also as $\forall\ z > y_1,\ c_{ab(z,1)1} \geq c_{ab(y_1,1)1}$, it means that for every partial word $\widetilde{Pw} = (ab(z, 1)), z \geq y_1, Z(X^{\widetilde{Pw}}) \geq Z_u$. Therefore, no partial word $\widetilde{Pw} = ab(z, 1), z \geq y_1$ can generate a word for which $Z\left(X^{Pw}\right)$ less than Z_u. Further, as the words are generated systematically in decreasing order of their contribution to the objective function, it means that any word $ab(y, 1), 1 \leq y \leq y_1$, must have the corresponding value not less than Z_u. And hence Z_u is the optimal value of the objective function. Therefore, for every partial word for which its first entry $ab(y_1, 1)$ is such that $Z(X^{Pw}) \geq Z_u$, then for all $ab(z, 1), z \geq y_1$ it cannot contain a word for which $Z\left(X^{Pw}\right)$ less than Z_u. This implies that Z_u is the optimal value of the objective function.

Theorem 3 *Let us consider a partial word $Pw = (ab(m, 1), ab(m, 2), \ldots, ab(m, r))$, $r \leq n$ such that $Z(X^{Pw}) < Z_u$, Z_u be the upper bound on the value of the objective function. Let $\overline{Pw}$ be a partial word defined as $\overline{Pw} = Pw \uplus (ab(y_{r+1}, r+1)$ which is derived from Pw such that $Z(X^{\overline{Pw}}) \geq Z_u, \forall y_{r+1} = 1, 2, \ldots, m$ then Z_u is the optimal value of the objective function.*

Proof As $Z(X^{\overline{Pw}}) \geq Z_u, \forall y_{r+1} = 1, 2, \ldots, m$. Then by virtue of Theorem 1, it follows that Pw cannot contain a word for which the corresponding objective value is less than Z_u. Therefore by Remark 1, supply to r destinations respectively by the sources $ab(m, 1), \ldots, ab(m, r)$ must be altered. But these transportations are corresponding to last entries of r columns of Alphabet Matrix. It follows that Pw cannot be altered as all the partial word generated systematically in decreasing order of their contribution to the objective function. It means that before Pw, all the partial words, i.e. $\left(ab(y_1, 1), ab(y_2, 2), \ldots, ab(y_r, r)\right)$ for all possible permutation $\left(y_1, y_2, \ldots, y_r\right)$ where $r = 1, 2, \ldots, m$, have not generated value better than Z_u except for the case $y_1 = y_2 = \ldots = y_r = m$. Hence we cannot have a word with $Z(X^{Pw}) < Z_u$. So Z_u is the optimal value.

Remark 2 (i) If the partial word Pw is s.t, $\sum_{i \in I} d_i^u < \sum_{j=r+1}^{n} g_j$, where d_i^u is the updated availability at source i for $i \in I$ then it cannot contain a feasible word. And if $\sum_{i \in I} d_i^u \geq \sum_{j=r+1}^{n} g_j$ then partial word is called feasible partial word. A partial word $Pw = (ab(m, 1), ab(m, 2), \ldots, ab(m, r)), r \leq n$ for which $\sum_{j=r+1}^{n} g_j > \sum_{i \in I} d_i^u$, is rejected because of infeasibility.

(ii) If $d^u_{ab(y,r)}$ is the updated availability at any source $ab(y, r) \in \mathrm{I}$ is less than $g_r, \forall y = 1, 2, \ldots, m$, the partial word $\widetilde{Pw} = (ab(y_1, 1), \ldots, ab(y_m, m))$ is rejected as it can not contain a feasible word.

(iii) A partial word $Pw = (i_1, i_2, \ldots, i_r)$, $r \le n$ is also rejected if $\exists$ an $i \in I_u$ s.t, $d_i < g_j$, $\forall j \in \{r+1, \ldots, n\}$ because further augmentation of Pw cannot generate a word, i.e. feasible word.

(iv) If $n - r < |I_u|$, the partial word $Pw = \{i_1, i_2, \ldots, i_r\}$, $r < n$ is rejected.

Remark 3 **Method to find initial upper bound Z_u**

Step 0 Initially check the following conditions for all sources I and destination $J_n = J$

(a) If $\exists$ atleast one $j \in J$ s.t $g_j > d_i \;\; \forall i$, then go to Step (vi).
Or

(b) if $\sum_{i\in I} d_i < \sum_{j\in J} g_j$, then go to Step (vi).
Or

(c) if $\exists\, i \in I$ s.t., $d_i < g_j \; \forall\, j \in J$, then update $I_u = I - \{i\}$ as this source would not be able to serve any destination.
Now if $\sum_{i\in I} d_i < \sum_{j\in J} g_j$ then go to Step (vi).
else
Set $I = I_u, i = 1$, update $m = m - 1$ and go to Step (i).
If any of the above conditions are not satisfied, then set i=1 and go to Step (i).

Step (i) For $i \in I$ find $j \in J$ s.t $c_{ij_i} = \min_{j\in J, d_i \ge g_j} c_{ij}$. Set $x_{ij_i} = 1$ & $x_{sj_i} = 0 \, \forall\, s \ne i,\, s \in I$.
Update $d_i^u = d_i - g_j$, $d_i = d_i^u$, J = J − {j}.
Now check

(a) If $\exists$ atleast one $j \in J$ s.t $g_j > d_i \forall i$, then go to Step (vi).
Or

(b) if $\sum_{i\in I} d_i < \sum_{j\in J} g_j$, then go to Step (vi).
Or

(c) if $\exists\, i \in I$ s.t $d_i < g_j \; \forall\, j \in J$, then update $I_u = I - \{i\}$ as it would not serve to any destination.
Now if $\sum_{i\in I} d_i < \sum_{j\in J} g_j$, then go to Step (vi).
else
Set $I = I_u$, Find current value of i, update $m = m - 1$ and go to Step (ii).

If any of the above conditions are not satisfied, then go to Step (ii).

Step (ii) Set $i = i + 1$.

(a) If $i \leq m$, go to Step (i).
(b) If $i > m$, go to Step (iii).

Step (iii) Set $i = i - 1$

(a) If $i = m = n$ and $J = \phi$, then all m sources supplied uniquely to the destinations in the set $\{j_1, j_2, \ldots, j_m\}$. Thus, each destination is served uniquely by m sources and go to Step (vii).
(b) If $i = m < n$, go to Step (iv).

Step (iv) For the remaining n-m destinations proceed as follows:-
For $k = 1$, find
$\min_{i \in I_{m+k}} \left(\sum_{j \in J_n - J_{m+k}} (c_{ij} : x_{ij} = 1) + \min_{j \in J_{m+k}} (c_{ij}) \right) = \sum_{j \in J_n - J_{m+k}} (c_{i_{m+k}j} : x_{i_{m+k}j} = 1) + c_{i_{m+k}j_{m+k}}$
where $I_{m+k} = \{i \in I : d_i \geq g_j, j \in J_{m+k}\}$.
Thus, j_{m+k} destination served by the source i_{m+k} & $d^u_{i_{m+k}} = d^u_{i_{m+k}} - g_{j_{m+k}}$.
Set $x_{i_{m+k}j_{m+k}} = 1$ & $x_{ij_{m+k}} = 0 \; \forall \, i \in I, i \neq i_{m+k}$,
$\mathrm{J} = \mathrm{J} - \{\mathrm{j}_{m+k}\}$, $d_{i_{m+k}} = d^u_{i_{m+k}}$
If $J \neq \phi$, then check the following conditions, otherwise go to Step (vii).

(a) If $\exists$ atleast one $j \in J$ s.t $g_j > d_{i_{m+k}} \forall i$, then go to Step (vi).
Or
(b) if $\sum_{i \in I} d_{i_{m+k}} < \sum_{j \in J} g_j$, then go to Step (vi).
Or
(c) if $\exists \, i \in I$ s.t $d_{i_{m+k}} < g_j \; \forall \, j \in J$, then update $I_u = I - \{i_{m+k}\}$ as it would not supply to any destination.
Now if $\sum_{i \in I} d_{i_{m+k}} < \sum_{j \in J} g_j$, then go to Step (vi).
else
Set $I = I_u$, Find current value of k, update $m = m - 1$ and go to Step (v).

If any of the above conditions are not satisfied, then go to Step (v).

Step (v) Set k=k+1.

(a) If $k \leq n - m$, go to Step (iv).
(b) If $k > n - m$, go to Step (vii).

Step (vi) The given problem is infeasible.

Step (vii) Hence the upper bound Z_u can be calculated as
$Z_u = \sum_{(i \in I, j \in J)} (c_{ij} + f_{ij} : x_{ij} = 1)$

2.3 Algorithm

In the algorithm, partial word will be updated in the following three ways.

A-I (Augmentation) If $Pw = \big(ab(y_1, 1), ab(y_2, 2), \ldots, ab(y_r, r)\big)$ be the partial word of length r for which $Z(X)^{Pw} < Z_u$, then it is augmented with $(r + 1)$ element say $ab(y_{r+1}, r + 1)$ as $Pw = Pw \uplus ab(y_{r+1}, r + 1)$, $d^u_{ab(y_{r+1},r+1)} = d^u_{ab(y_{r+1},r+1)} - g_{r+1}$. If $d^u_i = 0$, update $I_u = I_u \cap \{i\}$, set $x_{ab(y_{r+1},r+1)r+1} = 1$

A-II (Negation) If $Pw = \big(ab(y_1, 1), ab(y_2, 2), \ldots, ab(y_r, r)\big)$ be a partial word of length r for which either $Z(X^{Pw}) \geq Z_u$ or the above Remark 2 holds, then update the partial word as follows :

$$Pw = Pw \cap ab(y_r, r), \quad d^u_{ab(y_r,r)} = d^u_{ab(y_r,r)} + b_r$$

$$I_u = I_u \uplus ab(y_r, r), \text{if } ab(y_r, r) \neq ab(y_t, t), 1 \leq t \leq r - 1; \text{set} x_{ab(y_r,r)r} = 0$$

A-III (Negation of more than one element) If Pw be a partial word of length r and more than one element say last $(r - s)$ elements are to be removed, $r \leq n$. If $Pw = (ab(y_1, 1) \ldots ab(y_r, r)), r \leq n$, then it is updated as follows:

$$Pw = Pw \cap (ab(y_{s+1}, s + 1) \ldots ab(y_r, r))$$

In this I_u and the availabilities are updated for each source $\{ab(y_{s+1}, s + 1) \ldots ab(y_r, r)\}$ as in (A-II) & set$x_{ab(y_t,t)t} = 0 \quad \forall t \in \{s + 1, \ldots, r\}$.

The algorithm runs in the following steps, and some notations are used throughout the process, i.e. J is used to show the position of a column in the Alphabet Matrix AB, K is used to show the position of an entry in the column of AB which is under investigation.

Step 0 (Initialization) Initialize $J = 1, I = I_u, K = 1, Pw = \phi$, compute AB, Z_u (Ref. Sect. 2.2, Remark 3), & d_i, g_j are available and go to Step II.

Step I If $K < m$, then update as in (A-II). Set $K = K + 1$ and go to Step II. If $K = m$, then update as in (A-II). Set $J = J - 1$ and for this J go to Step III.

Step II If $d^u_{ab(K,J)} < g_j$, then go to Step IV.

If $\sum_{i=1}^{m} d_i < \sum_{j \in J}^{n} g_j$ then go to Step I (Ref. Sect. 2.2, Remark 2(i))

If $I_u = \phi$ and $\exists$ an $i \in I_u$ for which $d_i < g_j \quad \forall j \in \{J, J + 1 \ldots n\}$ or $n - r < |I_u|$ then go to Step I (Ref. Sect. 2.2, Remark 2(iii)).
If none of the above holds, then update as (A-I). Find $Z(X^{Pw})$.
If $Z(X^{Pw}) < Z_u$, then go to Step V (Ref. Sect. 2.2, Remark 1).

$Z(X^{Pw}) \geq Z_u$ and $J > 1$, go to Step I (Ref. Sect. 2.2, Remark 1).
$Z(X^{Pw}) \geq Z_u$ and $J = 1$, go to Step VI (Ref. Sect. 2.2, Theorem 2).

Step III If $J \geq 1$ and each $y_j = m$ in $ab(y_j, j)$, then go to step VI. Otherwise set $K = y_j$ and go to Step I (Ref. Sect. 2.2, Theorem 3).

Step IV If $K < m$, then set $K = K + 1$, go to Step II. If $K = m$ and

(i) $J > 1$, Set $J = J - 1, K = y_j$ and go to Step I.
(ii) $J = 1$, go to Step III.

Step V If $J < n$, then $J = J + 1$ and $K = 1$, go to Step II.
If $J = n$, we get word w for which $Z(X^w) < Z_u$. Set $w = Pw$, $Z = Z(X^w)$, go to Step III.

Step VI Stop. Z_u is the optimal fixed charge bulk transportation cost.

2.4 Numerical Illustration

Consider the following FCBTP having four sources with respective availabilities $d(i), i = 1, 2, 3, 4$ and five destinations with demands $g(j), j = 1, 2, 3, 4, 5$ respectively as shown in Table 1.

Step 0 Initialize $J = 1, I = I_u = \{1, 2, 3, 4\}, Pw = \phi, Z_u = 21$ (Ref. Sect. 2.2, Remark 3) is the upper bound, AB (Table 2), $d(i)$, $g(j)$ are available; $K = 1$, go to Step II.

Table 1 In each cell, the entry in the lower right corner shows the bulk cost and upper left corner shows the fixed cost of transportation

3 2	3 3	2 4	3 7	1 1	5
1 4	1 1	5 1	4 8	4 8	4
4 1	3 7	4 11	1 1	3 6	3
4 ∞	4 ∞	3 10	3 3	1 5	2
3	3	2	2	1	

Table 2 Alphabet matrix

3	2	2	3	1
1	1	1	4	4
2	3	4	1	3
4	4	3	2	2

Step II Find $ab(K,J) = ab(1,1) = 3$. If $d^u_{ab(K,J)} = d^u_3 = 3 = g_1$ then update as in (A - I). $Pw = Pw \uplus ab(1,1) = \{3\}$, $d^u_{ab(1,1)} = d^u_{ab(1,1)} - g_1 = 0$. As $ab(1,1) = 3 \in I_u$ then update $I_u \cap ab(1,1) = I_u \cap \{3\} = \{1,2,4\}$, set $x_{ab(1,1),1} = 1$, $Z(X^{Pw}) = 5 < Z_u = 21$ then go to Step V.

Step V $J = 1 < n$ then set $J = 2$ and $K = 1$ and go to Step II.

Step II Find $ab(K,J) = ab(1,2) = 2$, $d^u_2 > g_2$ then update as in (A-I). $Pw = Pw \uplus ab(1,2) = \{3,2\}$, $d^u_2 - g_2 = 1$, set $x_{2,2} = 1$, $Z(X^{Pw}) = 7 < Z_u, I_u = \{1,2,4\}$ then go to Step V.

Step V $J = 2 < n$ then set $J = 3$ and $K = 1$ and go to Step II.

Step II Find $ab(K,J) = ab(1,3) = 2$, $d^u_2 < g_3$ then go to Step IV.

Step IV $K = 1 < m$ then set $K = 2$ and go to Step II.

Step II Find $ab(K,J) = ab(2,3) = 1$, $d^u_1 = 5 \geq g_3$ then update as in (A-I). $Pw = Pw \uplus ab(2,3) = \{3,2,1\}, d^u_1 = 5 - 2 = 3, I_u = \{1,2,4\}$, set $x_{1,3} = 1, Z(X^{Pw}) = 13 < 21$ then go to Step V.

Step V $J = 3 < n$, then set $J = 4$ and $K = 1$ and go to Step II.

Step II Find $ab(K,J) = ab(1,4) = 2$, $d^u_3 < g_4$ then go to Step IV.

Step II Find $ab(K,J) = ab(2,4) = 4$, $d^u_4 = 2 \geq g_4$ then update as in (A-I). $Pw = Pw \uplus ab(2,4) = \{3,2,1,4\}$, $d^u_2 = 2 - 2 = 0$, $I_u = \{1,2,4\} \cap \{4\} = \{1,2\}$, set $x_{4,4} = 1$, $Z(X^{Pw}) = 19 < 21$ then go to Step V.

Step V $J = 4 < n$, then set $J = 5$ and $K = 1$ and go to Step II.

Step II Find $ab(K,J) = ab(1,5) = 1$, $d^u_1 = 3 \geq g_5$ then update as in (A-I). $Pw = Pw \uplus ab(1,5) = \{3,2,1,4,1\}$, $d^u_1 = 3 - 1 = 2$, $I_u = \{1,2\}$, set $x_{1,5} = 1$, $Z(X^{Pw}) = 21 \geq 21$ then go to Step I.

Step I $K = 1 < 5 = m$ then update as in (A-II), set $K = K + 1 = 2$ and go to Step II.

And continuing like this, we obtain a word $\{3,2,1,4,1\}$ corresponding to which the value of the objective is 21. This is the optimal value of fixed charge bulk transportation problem, and the run-time is also calculated for the same using the test code in MATLAB which is equal to 0.0825 second.

Table 3 Average run-time of FCBTP for randomly generated problem of different sizes (taken over about 1000 instances) using MATLAB

Source	Destination	Run-time(sec)
10	11	0.2099
10	15	0.0699
10	20	0.0862
20	21	0.1280
20	25	0.2015
20	30	0.2379
20	40	0.3474
30	31	0.3528
30	35	0.4197
30	40	0.4818
40	41	0.6461
40	45	0.8564
40	50	0.9730
50	51	1.1975
50	55	1.4085
50	60	1.5546

2.5 Computational Details

The algorithm has been coded in MATLAB and runs efficiently for various randomly generated FCBTP of different sizes. The problems are tested using Intel Processor i5 with 2.40 GHZ, 4 GB RAM on 64-bit Windows operating system. Some of them are reported in the above Table 3.

3 Observations in the Paper [24]

Some observations are made from a paper entitled 'Solving the fixed charge by ranking the extreme points' by Murty [24]. It is noticed that some errors were there, which majorly affect the ranking limit of the solution procedure. In the paper [24], the extreme points were ranked in increasing order of their objective value. The mathematical model of [24] is defined as

$$Minimize\ \xi(x) = \sum_j d_j(1 - \delta_{0,x_j}) + \sum_j c_j x_j$$

Subject *to*

Table 4 Solution S_2

4	**9**	18	**14**	5	21	11
9	31	25	6		17	**26**
6	89	7	16	**8**	**24**	9
72	8	**35**	6	17	**31**	**9**
16	3	9	**40**	86	29	6
$Z_2 = 2230$		$D_2 = 59(\mathbf{46})$			$\xi = 2289$	

Table 5 Solution S_5

4	**9**	24	**14**	13	27	17
3	25	25	0	**8**	17	**18**
	83	7	10	2	**38**	9
66	2	**35**	**6**	17	**17**	**17**
22	3	15	**34**	92	35	12
$Z_5 = 2250$		$D_5 = 105(\mathbf{80})$			$\xi = 2355$	

$$Ax = b,\ x \geq 0$$

$$\delta_{0,xj} = \begin{cases} 1 & x_j > 0 \\ 0 & x_j = 0 \end{cases}$$

In this paper [24], we found the solution of fixed charge transportation problem by ranking the extreme points with respect to Z. The same numerical example which was discussed in the paper [24] is solved on same lines, and only the tables which contain the errors are shown above in Tables 4 and 5. The ranking of the basic feasible solution in the algorithm took place using pivoting which was same as done in simplex method, and each ranking stage was labelled as $\{S_1, S_2, \ldots, \ldots\}$. Therefore, the next element in the ranking sequence can easily be obtained with each iteration using pivoting. When ranking is done using pivoting at stage 2 and stage 5, i.e. corresponding to the solution S_2 and S_5, the total cost which is the sum of variable cost and fixed cost was wrong. Because some errors were found in calculating the fixed cost corresponding to the solutions S_2 and S_5. The reported fixed cost values at stages 2 and 5 corresponding to the solutions S_2 and S_5 were 46 and 80 which are actually 59 and 105. Hence, the solution which was found after stage 6 corresponding to the solution S_6, as done in the paper [24], will have to be carried out up to stage 13, i.e. S_{13}, as a number of basic feasible solutions which must exist in the ranking limit were missed by the author. So, due to the errors in solutions S_2 and S_5, the whole solution got affected, however the final solution is same as it appeared in early stage of ranking corresponding to S_2.

The errors which occurred in the paper are shown in Tables 4 and 5, and the wrong values reported in the paper marked bold, i.e. (D_2&D_5). Each cell contains the relative cost whereas each bold entry inside it shows the basic feasible solutions.

At each stage, Z shows minimum cost of transportation, D shows fixed cost incurred with each transportation. ξ shows the total cost of transportation, i.e. the sum of minimum cost and fixed cost of transportation. Hence, the ranking limit which is reported by the author after stage 6, i.e. for $Z \leq 2260$ was wrong and the corrected limit upto which the ranking has to be carried out is stage 13, i.e. for $Z \leq 2273$.

4 Concluding Remarks

1. An exact method to solve fixed charge bulk transportation problem is developed using lexi-search approach. The algorithm converges to the optimality in a finite number of steps because:
   (a) Maximum number of generated words are n^m.
   (b) Every new word in the process gives a tighter upper bound on the optimal value.
   (c) All the partial words that yield a value greater than upper bound are rejected.
   (d) Infeasible partial words when encountered in the process are also rejected.
2. It is also noted that as the partial word under investigation needs to be stored in the active memory hence almost negligible active memory is required. The time complexity of the algorithm is found to be $O(n^2)$. The algorithm has been coded in MATLAB and verified for variety of test problem. The algorithm runs successfully for randomly generated problems of different sizes for different input data. All the instances, i.e. availabilities, demands, bulk cost, fixed cost, sources and destinations are generated randomly in MATLAB following the uniform distribution as reported in Sect. 2.5 computational details.
3. Consider a problem P_2 which is defined as

   Problem P_2

$$\min Z = \sum_{i=1}^{m} \sum_{j=1}^{n} (c_{ij} + f_{ij}) x_{ij}$$

subject to

$$\sum_{j \in J} g_j x_{ij} \leq d_i, \qquad i \in I \qquad (i)$$

$$\sum_{i \in I} x_{ij} = 1, \qquad j \in J \qquad (ii)$$

$$x_{ij} = 0 \text{ or } 1 \qquad \forall (i,j) \in IXJ \qquad (iii)$$

Feasibility check:-

$$\sum_{j \in J} g_j \leq \sum_{i \in I} d_i \qquad (iv)$$

It can be seen that an optimal solution of problem P_1 is an optimal solution of problem P_2. From here it is concluded that the fixed charge bulk transportation problem is equivalent to a cost bulk transportation problem which can be solved using [14, 23, 29] although a method to solve FCBTP is developed using lexi-search in this paper for the problem P_1.

4. The proposed study can also be extended for more generalized case when $n'(m < n' < n)$ destinations are to be served or when each source has to serve atleast a specified number of destinations.

Acknowledgements The authors are thankful to the honourable reviewers for their significant comments which have enhanced the quality of our manuscript 'Fixed Charge Bulk Transportation Problem' to a great extent.

References

1. Adlakha, V., Kowalski, K.: On the fixed-charge transportation problem. Omega **27**(27), 381–388 (1999)
2. Adlakha, V., Kowalski, K.: A simple heuristic for solving small fixed-charge transportation problems. Omega **31**(3), 205–211 (2003)
3. Adlakha, V., Kowalski, K., Vemuganti, R.R., Lev, B.: More-for-less algorithm for fixed charge transportation problems. OMEGA Int. J. Manag. Sci. **35**(1), 116–127 (2007)
4. Adlakha, V., Kowalski, K., Wang, S., Lev, B., Shen, W.: On approximation of the fixed charge transportation problem. Omega **43**, 64–70 (2014)
5. Aguado, J.: Fixed charge transportation problems: a new heuristic approach based on lagrangean relaxation and the solving of core problems. Ann. Oper. Res. **172**, 45–69 (2009)
6. Arora, S.R., Ahuja, A.: A Paradox in a fixed charge transportation problem. Indian J. Pure Appl. Math. **31**(7), 809–822 (2000)
7. Arora, S., Puri, M.C.: A variant of time minimizing assignment problem. Eur. J. Oper. Res. **110**, 314–325 (1998)
8. Balas, E., Glover, F., Zionts, S.: An additive algorithm for solving linear programs with zero one variables. Oper. Res. INFORMS **13**(4), 517–549 (1965)
9. Balinski, M.L.: Fixed-cost transportation problems. Nav. Res. Logist. Q. **8**(01), 41–54 (1961)
10. Caramia, M., Guerreio, F.: A heuristic approach to long-haul freight transportation with multiple objective functions. OMEGA Int. J. Manag. Sci. **37**, 600–614 (2009)
11. Cooper, L., Drebes, C.: An approximate solution method for the fixed charge problem. Nav. Res. Logist. Q. **14**(1) (1967)
12. Cooper, L.: The fixed charge problem-I: a new heuristic method. Comput. Math. Appl. **1**(1), 89–95 (1975)
13. Cooper, L., Olson, A.M.: Random Perturbations and MI-MII Heuristics for the Fixed Charge Problem
14. De Maio, A., Roveda, C.: An all zero-one algorithm for a certain class of transportation problems. INFORMS **19**(6), 1406–1418 (1969)
15. Denzler, D.R.: An approximative algorithm for the fixed charge problem. Nav. Res. Logist. Q. **16**(3) (1969)

16. Gray, P.: Exact solution of the fixed-charge transportation problem. Oper. Res. **19**(6), 1529–1538 (1971)
17. Hajiaghaei-Keshteli, M., Molla-Alizadeh-Zavardehi, S., Tavakkoli-Moghaddam, R.: Addressing a non linear fixed-charge transportation using a spanning tree based genetic algorithm. Comput. Ind. Eng. **59**, 259–271 (2010)
18. Jo, J., Li, Y., Gen, M.: Nonlinear fixed charge transportation problem by spanning tree-based genetic algorithm. Comput. Ind. Eng. **53**, 290–298 (2007)
19. Kowalski, K., Lev, B.: On step fixed-charge transportation problem. OMEGA Int. J. Manag. Sci. **36**(5), 913–917 (2008)
20. Kowalski, K., Lev, B., Shen, W., Tu, Y.: A fast and simple branching algorithm for solving small scale fixed-charge transportation problem. Oper. Res. Perspect. **1**(1), 1–5 (2014)
21. Liu, S.-T.: The total cost bounds of transportation problem with varying demand and supply. OMEGA Int. J. Manag. Sci. **31**(4), 247–251 (2003)
22. Ma, H., Miao, Z., Lim, A., Rodrigues, B.: Cross docking distribution networks with setup cost and time window constraint. OMEGA Int. J. Manag. Sci. **39**(1), 64–72 (2011)
23. Murthy, M.S.: A bulk transportation problem. OPSEARCH **13**(3–4), 143–155 (1976)
24. Murty, K.G.: Solving the fixed charge problem by ranking the extreme points. Oper. Res. **16**(2), 268–279 (1968)
25. Ortega, F., Wolsey, L.: A branch and cut algorithm for the single-commodity, uncapacitated, fixed charge network flow problem. Networks **41**, 143–158 (2003)
26. Puri, M.C., Swarup, K.: A systematic extreme point enumeration procedure for fixed charge problem. Trab Estad y Investig Oper. **25**(1), 99–108 (1974)
27. Robers, P., Cooper, L.: A study of the fixed charge transportation problem. Comput. Math. Appl. **2**(2), 125–135 (1976)
28. Sadagopan, S., Ravindran, A.: A vertex ranking algorithm for the fixed-charge transportation problem. J. Optim. Theory Appl. **37**(2) (1982)
29. Srinivasan, V., Thompson, G.L.: An algorithm for assigning uses to sources in a special class of transportation problems. INFORMS **21**(1), 284–295 (1973)
30. Steinberg, D.I.: The fixed charge problem. Nav. Res. Logist. Q. **17**(2) (1970)
31. Thirwani, D.: A note on fixed charge Bi-Criterion transportation problem with enhanced flow. Indian J. Pure Appl. Math. **29**(5), 565–571 (1998)
32. Walker, W.: A heuristic adjacent extreme point algorithm for fixed charge problem. Manag. Sci. **22**, 587–596 (1976)
33. Warren, M.H, George, B.D.: The fixed charge problem. RAND Corp. **RM-1383**, 413–424 (1954)
34. Warren, M.H., Hoffman, A.J.: Extreme varieties, concave functions, and the fixed charge problem. Commun. Pure Appl. Math. **XIV**, 355–369(1961)

Reduction of Type-2 Lognormal Uncertain Variable and Its Application to a Two-Stage Solid Transportation Problem

Dipanjana Sengupta and Uttam Kumar Bera

Abstract The main focus of the paper is to develop a multi-objective solid transportation problem under uncertain environment, where transportation parameters are taken as type-2 lognormal uncertain variables. For reduction of the type-2 uncertain lognormal variables, expected value-based reduction method has been proposed. A two-stage solid transportation model has been also proposed here. Finally, an illustrative example with real-life data has been solved with the proposed expected value-based reduction method. A comparison has been shown between the result obtained using linear variable and lognormal variable. Lingo 13.0 optimization software has been used to find the optimal result.

Keywords Solid transportation problem · Type-2 lognormal uncertain variable
Expected value-based reduction method

1 Introduction

Logistic management manages the upward and downward movement of logistics from supply centres to demand centres to meet the customers' meet with less cost and more profitability. Initially, transportation problem was stated by Hitchcock [1]. Shell [2] stated the extension of transportation problem which is solid transportation problem (STP). The solution procedure of STP was described by [Haley] in the year 1962.

Uncertainty is the general part of transportation problem. Uncertainty arises due to different practical situation which arises in the everyday's life. It occurs in case of transportation of goods. When goods are transported from one place to another, this

D. Sengupta (✉) · U. K. Bera
Department of Mathematics, National Institute of Technology, Agartala, Barjala, Jirania, West Tripura 799046, India
e-mail: dipanjanasengupta09@gmail.com

U. K. Bera
e-mail: bera_uttam@yahoo.co.in

S. Kar et al. (eds.), *Operations Research and Optimization*, Springer Proceedings in Mathematics & Statistics 225, https://doi.org/10.1007/978-981-10-7814-9_23

transportation depends on road conditions, environmental condition, share market's condition, availability of fuels, availability of drivers, toll tax, insurance cost, etc. Due to uncertainty, different situations like fuzziness, randomness roughness may occur. In the literature, there are many researchers who have done many developments. For these the readers may refer [2–11]. Uncertainty theory which is the effective branch of mathematics was described by Liu [12–14].

To deal with those types of uncertainties in real-life situations, theories of probability, fuzzy set and rough set have been introduced. There are so many uncertain variables, namely linear, zigzag, normal, lognormal. All are practically a series of uncertain variables, which are represented by their corresponding series of uncertain distribution. In developing the different study of uncertainty theory, Liu et al. [15, 16] have a great contribution. So many researchers have been developed many models in this field. Cui and Sheng [5] presented the uncertain programming model for solid transportation problem. Sheng and Yao studied a transportation model with uncertain costs and demands in [17]. They also presented fixed charge transportation problem and its uncertain programming model in [18].

In this paper, our aim is to minimize transportation cost with the help of proposed expected value-based reduction method of type-2 lognormal uncertain variables by developing a two-stage solid transportation model. The major contribution of the paper is described below:

- Reduction method of type-2 lognormal uncertain variables with the use of expected value has been proposed.
- A multi-objective two-stage solid transportation problem has been proposed here.
- An illustrative example with real-life data is presented here.
- The model has been solved with the use of this expected value-based reduction method.
- A comparison has been shown between reduction methods of linear uncertain variable and lognormal uncertain variable.

With this introductory part, the rest of the paper is as follows, Sect. 2 describes some preliminaries, Sect. 3 represents the proposed expected value-based reduction method of type-2 lognormal uncertain variables. Section 4 presents the problem description and formulation. In Sect. 5, solution technique has been presented. An illustrative example using real-life data is presented in Sect. 6. Some managerial insights have been drawn in Sect. 7. In the last, conclusion and future scope have been given in Sect. 8.

2 Preliminaries

From mathematical point of view, uncertainty theory is essentially an alternative theory for measurement. Uncertainty theory should begin with a measurable space. Uncertainty theory depends on some certain sets like algebra, σ-algebra, measurable set, Borel-algebra, Borel set and measurable function. To learn uncertainty theory, we have to know about some definitions of uncertainty theory. In order to provide an axiom-based mathematical tool for describing and handling realistic inexactness, Liu introduced uncertainty theory and also successfully applied to handling and solving a variety of optimization problems. For completeness of the research, some basic concepts and definitions of uncertainty theory will be introduced in the following.

2.1 Definition

2.1.1 Uncertain Measure

Let Γ be a nonempty set. A collection $\mathcal{L}$ of subsets of Γ is called a σ-algebra if,

(a) $\Gamma \in \mathcal{L}$;
(b) if $\Lambda \in \mathcal{L}$, then $\Lambda^c \in \mathcal{L}$ and
(c) if $\Lambda_1, \Lambda_2, \ldots \in \mathcal{L}$, then $\Lambda_1 \cup \Lambda_2 \cup \cdots \in \mathcal{L}$.

Each element Λ in the σ-algebra $\mathcal{L}$ is called an event. Uncertain measure is a function from $\mathcal{L}$ to [0, 1]. In order to present an axiomatic definition of uncertain measure, it is necessary to assign each event Λ a number $\mathcal{M}\{\Lambda\}$ which indicates the belief degree that the event Λ will occur. In order to ensure that the number $\mathcal{M}\{\Lambda\}$ have certain mathematical properties, Liu [12] proposed the following three axioms:

Axiom 1. *(Normality Axiom)* $\mathcal{M}\{\Gamma\} = 1$ for the universal set Γ.

Axiom 2. *(Duality Axiom)* $\mathcal{M}\{\Lambda\} + \mathcal{M}\{\Lambda^c\} = 1$ for any event Λ.

Axiom 3. *(Subadditivity Axiom)* For every countable sequence of events Λ_1, Λ_2, …, we have

$$\mathcal{M}\left\{\bigcup_{i=1}^{\infty} \Lambda_i\right\} \le \sum_{i=1}^{\infty} \mathcal{M}\{\Lambda_i\}.$$

The set function $\mathcal{M}$ is called an uncertain measure if it satisfies the normality, duality and subadditivity axioms.

Example 2.1.1 Let $\Gamma = \{\gamma_1, \gamma_2, \gamma_3\}$. In this case, there are only 8 events. Define $\mathcal{M}\{\gamma_1\} = 0.6$, $\mathcal{M}\{\gamma_2\} = 0.3$, $\mathcal{M}\{\gamma_3\} = 0.2$, $\mathcal{M}\{\gamma_1, \gamma_2\} = 0.8$, $\mathcal{M}\{\gamma_1, \gamma_3\} = 0.7$, $\mathcal{M}\{\gamma_2, \gamma_3\} = 0.4$, $\mathcal{M}\{\emptyset\} = 0$, $\mathcal{M}\{\Gamma\} = 1$. Then $\mathcal{M}$ is an uncertain measure because it satisfies the three axioms.

2.1.2 Uncertain Variable (Liu [12])

An uncertain variable is a measurable function ξ from an uncertainty space $(\Gamma, \mathcal{L}, \mathcal{M})$ to the set of real numbers, i.e. for any Borel set B of real numbers, the set $\{\xi \in B\} = \{\gamma \in \Gamma | \xi(\gamma) \in B\}$ is an event.

Example 2.1.2 Take $(\Gamma, \mathcal{L}, \mathcal{M})$ to be $\{\gamma_1, \gamma_2\}$ with $\mathcal{M}\{\gamma_1\} = \mathcal{M}\{\gamma_2\} = 0.5$. Then the function

$$\xi(\gamma) = \begin{cases} 0, & \text{if } \gamma = \gamma_1 \\ 1, & \text{if } \gamma = \gamma_2 \end{cases}$$

is an uncertain variable.

2.1.3 Uncertainty Distribution (Liu [12])

The uncertainty distribution Φ of an uncertain variable ξ is defined by $\Phi(x) = \mathcal{M}\{\xi \le x\}$ for any real number x.

Example 2.1.3 The uncertain variable $\xi(\gamma) \equiv b$ on the uncertainty space $(\Gamma, \mathcal{L}, \mathcal{M})$ (i.e. a crisp number b) has an uncertainty distribution

$$\Phi(x) = \begin{cases} 0, & \text{if } x < b \\ 1, & \text{if } x \ge b \end{cases}.$$

2.1.4 Regular Uncertainty Distribution

An uncertainty distribution Φ is said to be regular if its inverse function $\Phi^{-1}(\alpha)$ exists and is unique for each $\alpha \in (0, 1)$.

For example, linear uncertainty distribution, zigzag uncertainty distribution, normal uncertainty distribution and lognormal uncertainty distribution are all regular.

2.1.5 Expected Value of Uncertain Variable (Liu [12])

Let ξ be an uncertain variable. Then the expected value of ξ is defined by $\mathrm{E}[\xi] = \int_0^{+\infty} \mathcal{M}\{\xi \ge r\}dr - \int_{-\infty}^0 \mathcal{M}\{\xi \le r\}dr$ Provided that at least one of the two integrals is finite. Let ξ be uncertain variable with uncertainty distribution Φ. If the expected value exists, then $\mathrm{E}[\xi] = \int_0^1 \Phi^{-1}(\alpha)d\alpha$.

2.2 Theorem

If $ln\xi = \mathrm{N}(e, \sigma; \theta_l, \theta_r)$ is a type-2 lognormal uncertain variable, then the distribution of the generated type-1 uncertainty distribution via expected value criterion is described below. For uncertainty distribution of optimistic and pessimistic value criterion, people can refer the paper of Yang [10].

$$\phi_{\xi}^{exp}(x) = \begin{cases} \left(1+\exp\left(\frac{\pi(e-lnx)}{\sqrt{3}\sigma}\right)\right)^{-1} - \frac{1}{2}(\theta_l-\theta_r)\left(1+\exp\left(\frac{\pi(e-lnx)}{\sqrt{3}\sigma}\right)\right)^{-1}, 0 \le x \le \exp(e)/3^{(\sqrt{3}\sigma)/\pi} \\ \left(1+\exp\left(\frac{\pi(e-lnx)}{\sqrt{3}\sigma}\right)\right)^{-1} - \frac{1}{2}(\theta_l-\theta_r)\left[\frac{1}{2}-\left(1+\exp\left(\frac{\pi(e-lnx)}{\sqrt{3}\sigma}\right)\right)^{-1}\right] \exp(e)/3^{\frac{\sqrt{3}\sigma}{\pi}} \le x \le \exp(e) \\ \left(1+\exp\left(\frac{\pi(e-lnx)}{\sqrt{3}\sigma}\right)\right)^{-1} - \frac{1}{2}(\theta_l-\theta_r)\left[\left(1+\exp\left(\frac{\pi(e-lnx)}{\sqrt{3}\sigma}\right)\right)^{-1} - \frac{1}{2}\right], \exp(e) \le x \le \exp(e).3^{(\sqrt{3}\sigma)/\pi} \\ \left(1+\exp\left(\frac{\pi(e-lnx)}{\sqrt{3}\sigma}\right)\right)^{-1} - \frac{1}{2}(\theta_l-\theta_r)\left[1-\left(1+\exp\left(\frac{\pi(e-lnx)}{\sqrt{3}\sigma}\right)\right)^{-1}\right], x \ge \exp(e).3^{(\sqrt{3}\sigma)/\pi} \end{cases}$$

For any $x \in (a, b) \cup (b, c)$.

3 Proposed Expected Value-Based Reduction Method of Type-2 Lognormal Uncertain Variables

Let ξ_i be the E reduction of the type-2 lognormal uncertain variable $ln\xi_i = N(e, \sigma; \theta_{l,i}, \theta_{r,i})$, $i = (1, 2, \ldots, n)$. Suppose $\xi_1, \xi_2, \xi_3, \ldots, \xi_n$ are mutually independent, and for $\theta_{r,1} - \theta_{l,1} \le \theta_{r,2} - \theta_{l,2} \le \ldots \theta_{r,n} - \theta_{l,n}$ for $i = 1, 2, \ldots n$.

(i) If $\alpha \in (0, 0.25]$, then $Cr\{\sum_{i=1}^{n} \xi_i k_i \le t\} \ge \alpha$ equivalent to

$$\sum_{i=1}^{n} exp\left[e - \frac{\sqrt{3}\sigma}{\pi}\{\ln(2(1-\alpha)+\theta_r-\theta_l) - ln2\alpha\}\right] k_i \le t$$

(ii) If $\alpha \in (0.25, 0.5]$, then $Cr\{\sum_{i=1}^{n} \xi_i k_i \le t\} \ge \alpha$ equivalent to

$$\sum_{i=1}^{n} \exp\left[e - \frac{\sqrt{3}\sigma}{\pi}\{\ln(4(1-\alpha)+\theta_l-\theta_r)\} - \ln(4\alpha+\theta_l-\theta_r)\right] k_i \le t$$

(iii) If $\alpha \in (0.5, 0.75]$, then $Cr\{\sum_{i=1}^{n} \xi_i k_i \leq t\} \geq \alpha$ equivalent to

$$\sum_{i=1}^{n} \exp\left[e - \frac{\sqrt{3}\sigma}{\pi}\{\ln(4(1-\alpha) - \theta_l + \theta_r)\} - \ln(4\alpha - \theta_l + \theta_r)\right] k_i \leq t$$

(iv) If $\alpha \in (0.75, 1]$, then $Cr\{\sum_{i=1}^{n} \xi_i k_i \leq t\} \geq \alpha$ equivalent to

$$\sum_{i=1}^{n} \exp\left[e - \frac{\sqrt{3}\sigma}{\pi}\{\ln[2(1-\alpha)] - \ln[2\alpha + \theta_l - \theta_r]\}\right] k_i \leq t$$

4 Problem Description and Formulation

4.1 Notation

The following are the lists of notation which have been used here for the formulation of the two-stage solid transportation problem:

- $\tilde{c}_{ijk}^{1}$: denotes the per unit type-2 lognormal uncertain transportation cost from ith supply centre to jth distribution centre using kth conveyance in stage-1.
- $\tilde{c}_{ijk}^{2}$: denotes the per unit type-2 lognormal uncertain transportation cost from jth distribution centre to rth retailer using kth conveyance in stage-2.
- f_j: denotes crisp per unit fixed cost including operating cost of jth distribution centre.
- M_k: denotes crisp per unit maintenance cost of kth conveyance.
- z_j: denotes the binary variable of fixed cost.

$$z_j = \begin{cases} 1, & \textit{if operating cost needed for } j\text{th } \textit{distribution centre.} \\ 0, & \textit{otherwise.} \end{cases}$$

- d_{jrk}: denotes distance between jth *distribution centre to* rth *retailer through* kth vehicle.
- a_k: denotes the binary variable of maintenance cost,

$$a_k = \begin{cases} 1, & \textit{if maintenance cost needed for } k\text{th } \textit{conveyance.} \\ 0, & \textit{otherwise.} \end{cases}$$

- $\tilde{a}_i^1, \tilde{b}_j^1$ *and* $\tilde{q}_k^1$: denotes source, demand and conveyance constraints, respectively, for stage-1.
- $\tilde{b}_r^2, \tilde{q}_k^2$: denotes demand and conveyance constraints, respectively, for stage-2.
- x_{ijk}^1: denotes quantity transported in the stage-1.
- x_{jrk}^2: denotes quantity transported in the stage-2.

Assumption:

- The model is an unbalanced problem.
- Here all vehicles are fully loaded. Partially loaded case is not considered here.
- The fixed cost and maintenance cost are taken only in stage-2.

The solid transportation problem focuses on decision making for transporting some products from *i*th source to *j*th destination using *k*th conveyance with minimum cost and maximum profit. Here the problem is a two-stage problem. In stage-1, supply centre is source point and distribution centre is demand point. In stage-2, distribution centre works as source point and retailer works as demand point. All the transportation is done with same type of vehicles.

$$\text{Min}\,Z = \sum_{i=1}^{m}\sum_{j=1}^{n}\sum_{k=1}^{l} \tilde{c}_{ijk}^1 x_{ijk}^1 + \sum_{j=1}^{n}\sum_{r=1}^{R}\sum_{k=1}^{l} \left(\tilde{c}_{jrk}^2 x_{jrk}^2 + f_j z_j + M_k d_{jrk} a_k\right) \tag{1}$$

$$\text{S.t} \quad \sum_{j=1}^{n}\sum_{k=1}^{l} x_{ijk}^1 \le \tilde{a}_i^1 \quad i = 1, 2, \ldots, m \tag{2}$$

$$\sum_{i=1}^{m}\sum_{k=1}^{l} x_{ijk}^1 \ge \tilde{b}_j^1 \quad j = 1, 2, \ldots, n \tag{3}$$

$$\sum_{i=1}^{m}\sum_{j=1}^{n} x_{ijk}^1 \le \tilde{q}_k^1 \quad k = 1, 2, \ldots, l \tag{4}$$

$$\sum_{r=1}^{R}\sum_{k=1}^{l} x_{jrk}^2 \le \sum_{i=1}^{m}\sum_{k=1}^{l} x_{ijk}^1 \quad j = 1, 2, \ldots, n \tag{5}$$

$$\sum_{j=1}^{n}\sum_{k=1}^{l} x_{ijk}^2 \ge \tilde{b}_r^2 \quad r = 1, 2, \ldots, R \tag{6}$$

$$\sum_{j=1}^{n}\sum_{r=1}^{R} x_{ijk}^2 \le \tilde{q}_k^2 \quad k = 1, 2, \ldots, l \tag{7}$$

where $x_{ijk}^1 \ge 0, x_{jrk}^2 \ge 0$, $\forall i, j$ and k and they denote the unknown quantity to be transport from supply centre to distribution centre and distribution centre to retailers, respectively.

Here, the objective function (1) denotes the minimization of total transportation cost of the proposed model. The transportation cost $\tilde{c}^1_{ijk}$ of stage-1 is multiplied with the transported amount x^1_{ijk}. The transportation cost c^2_{jrk} of stage-2 is multiplied with the transported amount x^2_{jrk}. Here in the model, fixed cost including operating cost and maintenance cost of kth conveyance is added with the transportation cost. Here, fixed cost and maintenance cost are included only in stage-2. Equations (2)–(4) denote the source, demand and conveyance constraints of stage. In stage-1, Eqs. (2) and (4) denote source and conveyance constraint, respectively, which should be less than or equal to capacity of source and demand. Equation (3) denotes demand constraint which should be greater than or equal to capacity of demand. Equations (5)–(7) denote the source, demand and conveyance constraints of second stage. In stage-2, Eq. (5) denotes that the transported amount of stage-2 is less than or equal to transported amount of stage-1. Equations (6) and (7) are the demand and conveyance constraint, respectively.

5 Solution Procedure

5.1 Crisp Conversion of Type-2 Lognormal Uncertain Variable Using Proposed Expected Value-Based Reduction Method

Here $\tilde{c}^1_{ijk}$, $\tilde{c}^2_{ijk}$ are type-2 lognormal uncertain variable. The deterministic form has shown below.

$$\text{Min}\,Z = \sum_{i=1}^{m}\sum_{j=1}^{n}\sum_{k=1}^{l} F_{c^1_{ijk}} x^1_{ijk} + \sum_{j=1}^{n}\sum_{r=1}^{R}\sum_{k=1}^{l}\left(F_{c^2_{jrk}} x^2_{jrk} + +f_j z_j + M_k d_{jrk} a_k\right) \tag{8}$$

$$\text{S.t}\quad \sum_{j=1}^{n}\sum_{k=1}^{l} x^1_{ijk} \le F_{a^1_i} \quad i = 1, 2, \ldots, m \tag{9}$$

$$\sum_{i=1}^{m}\sum_{k=1}^{l} x^1_{ijk} \ge F_{b^1_j} \quad j = 1, 2, \ldots, n \tag{10}$$

$$\sum_{i=1}^{m}\sum_{j=1}^{n} x^1_{ijk} \le F_{q^1_k} \quad k = 1, 2, \ldots, l \tag{11}$$

$$\sum_{r=1}^{R}\sum_{k=1}^{l} x^2_{jrk} \le \sum_{i=1}^{m}\sum_{k=1}^{l} x^1_{ijk} \quad j = 1, 2, \ldots, n \tag{12}$$

$$\sum_{j=1}^{n}\sum_{k=1}^{l} x_{ijk}^{2} \geq F_{b_r^2} \quad r = 1, 2, \ldots, R \tag{13}$$

$$\sum_{j=1}^{n}\sum_{r=1}^{R} x_{ijk}^{2} \leq F_{q_k^2} \quad k = 1, 2, \ldots, l \tag{14}$$

where $x_{ijk}^{1} \geq 0, x_{jrk}^{2} \geq 0, \forall i, j \text{ and } k$ and they denote the unknown quantity to be transport from supply centre to distribution centre and distribution centre to retailers, respectively.

Equations (9)–(11) denote the source, demand and conveyance constraints of first stage and Eqs. (12)–(14) denote the source, demand and conveyance constraints of second stage. Here, $F_{c_{ijk}^1}$, $F_{c_{jrk}^2}$, $F_{a_j^1}$, $F_{b_j^1}$, $F_{q_k^1}$, $F_{b_r^2}$ and $F_{q_k^2}$ are equivalent crisp form of type-2 uncertain lognormal variable.

There are four cases in deterministic form (0–25), (0.25–0.50), (0.50–0.75) and (0.75–1). Here we have taken only (0.25–0.50) case

(i) Case-1: when $\alpha \in (0.25–0.50)$, then the equivalent deterministic form of the model is as under

$$\begin{aligned}
&\text{Min}\,\bar{f}\\
&\text{S.t}\\
&\sum_{i=1}^{m}\sum_{j=1}^{n}\sum_{k=1}^{l} exp\left[e_{c_{ijk}^1} - \frac{\sqrt{3}\sigma_{c_{ijk}^1}}{\pi}\{ln4(1-\alpha)+\theta_l-\theta_r\} - \ln(4\alpha+\theta_l-\theta_r)\right] x_{ijk}^{1} + \\
&\sum_{j=1}^{n}\sum_{r=1}^{R}\sum_{k=1}^{l}\left(exp\left[e_{c_{ijk}^2} - \frac{\sqrt{3}\sigma_{c_{ijk}^2}}{\pi}\{ln4(1-\alpha)+\theta_l-\theta_r\} - \ln(4\alpha+\theta_l-\theta_r)\right] x_{jrk}^{2} + \right.\\
&\left. f_j z_j + M_k d_{jrk} a_k \right)
\end{aligned} \tag{15}$$

$$\text{S.t} \quad \sum_{j=1}^{n}\sum_{k=1}^{l} x_{ijk}^{1} \leq \tilde{a}_i^1 \quad i = 1, 2, \ldots, m \tag{16}$$

$$\sum_{i=1}^{m}\sum_{k=1}^{l} x_{ijk}^{1} \geq \tilde{b}_j^1 \quad j = 1, 2, \ldots, n \tag{17}$$

$$\sum_{i=1}^{m}\sum_{j=1}^{n} x_{ijk}^{1} \leq \tilde{q}_k^1 \quad k = 1, 2, \ldots, l \tag{18}$$

$$\sum_{r=1}^{R}\sum_{k=1}^{l} x_{jrk}^{2} \leq \sum_{i=1}^{m}\sum_{k=1}^{l} x_{ijk}^{1} \quad j = 1, 2, \ldots, n \tag{19}$$

$$\sum_{j=1}^{n}\sum_{k=1}^{l} x_{ijk}^2 \geq \tilde{b}_r^2 \quad r = 1, 2, \ldots, R \tag{20}$$

$$\sum_{j=1}^{n}\sum_{r=1}^{R} x_{ijk}^2 \leq \tilde{q}_k^2 \quad k = 1, 2, \ldots, l \tag{21}$$

where $x_{ijk}^1 \geq 0, x_{jrk}^2 \geq 0$, $\forall i, j \text{ and } k$ and they denote the unknown quantity to be transport from supply centre to distribution centre and distribution centre to retailers, respectively.

Equations (16)–(18) imply the source, demand and conveyance constraint of stage-1, respectively, and Eq. (19)–(21) denote source, demand and conveyance constraint of stage-2.

6 Numerical Example with Real-Life Data

For numerical experiment, we have considered a two-stage solid transportation problem model with two sources, two distribution centres, two retailers and two conveyances. Here the problem is a two-stage problem. In stage-1, supply centre is source point and distribution centre is demand point. In stage-2, distribution centre works as source point and retailer works as demand point. We have taken the real-life data from Tarasankar plastic industry and RFL plastic industry. Both are situated in the Bodhjungnagar, Special Economic Zone. They supply plastic products to the distribution centres located in different parts of Tripura and outside Tripura. From them, we have chosen two distribution centres located in Agartala and Udaipur. The distribution centres supply them to the retailer. From them, we have taken only two retailers. All the transportation is done with two same type of vehicles. Truck and Tripper are two vehicles which have been taken here. All the data are taken according to expert opinion. The respective inputs for objective function in type-2 lognormal uncertain variables are given in Table 1 (Table 2).

Here per unit crisp fixed cost $f_x = \text{Rs.}\,5$ and per unit crisp maintenance cost $M_k = \text{Rs.}\,4$ are same in all cases. The sources are $\tilde{a}_1^1 = (3, 0.9; 0.5, 0.6)$, $\tilde{a}_2^1 = (2.3, 1.2; 0.5, 0.6)$, demands are $\tilde{b}_1^1 = (2, 0.8; 0.5, 0.6)$, $\tilde{b}_2^1 = (2.4, 1.6; 0.5, 0.6)$, capacities of conveyances are $\tilde{q}_1^1 = (2.5, 0.8; 0.5, 0.6)$, $\tilde{q}_2^1 = (3.1, 0.7; 0.5, 0.6)$ in first stage. The demand and conveyance in case of second stage are $\tilde{b}_1^2 = (1.8, 1.3; 0.5, 0.6)$, $\tilde{b}_2^2 = (1.9, 1.4; 0.5, 0.6)$ and $\tilde{q}_1^2 = (2.1, 0.6; 0.5, 0.6)$, $\tilde{q}_2^2 = (1.8, 0.9; 0.5, 0.6)$. Here credibility level is taken as 0.35. When we solve this by LINGO 13.0 software the result is as below:

Table 1 Inputs for stage-1 cost parameters $\ln \xi_i = N(e, \sigma; \theta_{l,i}, \theta_{r,i})$

$\tilde{c}^1_{111} = (3, 1.0; 0.8, 1.0)$	$\tilde{c}^1_{112} = (1.2, 1.5; 0.5, 1.5)$	$\tilde{c}^1_{121} = (1, 0.5; 0.6, 1.3)$	$\tilde{c}^1_{122} = (1.2, 2.0; 0.9, 1.9)$
$\tilde{c}^1_{211} = (1.5, 1.5; 0.5, 1.0)$	$\tilde{c}^1_{212} = (1.5, 1.0; 1.4, 2.0)$	$\tilde{c}^1_{221} = (1.7, 1.0; 0.5, 1.5)$	$\tilde{c}^1_{222} = (2.7, 1.0; 1.2, 1.8)$

Table 2 Inputs for stage-2 cost parameters $\ln \xi_i = N(e, \sigma; \theta_{l,i}, \theta_{r,i})$

$\tilde{c}^2_{111} = (2.3, 1.0; 0.8, 1.0)$	$\tilde{c}^2_{112} = (1.1, 1.5; 0.5, 1.5)$	$\tilde{c}^2_{121} = (0.8, 0.5; 0.6, 1.3)$	$\tilde{c}^2_{122} = (1.1, 2.0; 0.9, 1.9)$
$\tilde{c}^2_{211} = (1.2, 1.5; 0.5, 1.0)$	$\tilde{c}^2_{212} = (1.3, 1.0; 1.4, 2.0)$	$\tilde{c}^2_{221} = (1.5, 1.0; 0.5, 1.5)$	$\tilde{c}^2_{222} = (2.4, 1.0; 1.2, 1.8)$

7 Discussion and Managerial Insights

From Table 3, we can find that the minimum transportation cost is Rs. 1676.42 and Rs. 1662.05, respectively, which have been found by two different reduction methods using linear type-2 and lognormal type-2 uncertain variables, respectively. The results are somehow almost same. So both the reduction methods are helpful. But lognormal uncertain variables are used when the complexity is much higher. The table shows that the amount transported in stage-1 is more than amount transported in stage-2. It occurs because in stage-2 distribution centre may not transfer the whole amount to the retailer. Some amount may store in the warehouse. This type of problem occurs due to uncertain conditions. So type-2 uncertain lognormal variables are used with the help of expert opinion. The above model is helpful for a manager as he can decide how to transport goods in two-stage transportation problem in minimum cost. The type-2 uncertain lognormal and linear variables are used according to expert opinion.

Table 3 Result of the STP model obtained by comparison of proposed solution methods using type-2 lognormal and linear uncertain variables

Solution using expected value-based reduction method using type-2 lognormal	Solution using expected value-based reduction method using type-2 linear
Total cost = 1676.42	Total cost = 1662.045
Solution of first stage	Solution of first stage
$x^1_{111} = 0$, $x^1_{112} = 9$, $x^1_{121} = 0$, $x^1_{122} = 23$, $x^1_{211} = 32$, $x^1_{212} = 0$, $x^1_{221} = 0$, $x^1_{222} = 0$	$x^1_{111} = 8.5$, $x^1_{112} = 0.5$, $x^1_{121} = 23$, $x^1_{122} = 0$, $x^1_{211} = 0$, $x^1_{212} = 32$, $x^1_{221} = 0$, $x^1_{222} = 0$
Solution of second stage	Solution of second stage
$x^2_{111} = 0$, $x^2_{112} = 3$, $x^2_{121} = 0$, $x^2_{122} = 38$, $x^2_{211} = 2.71$, $x^2_{212} = 0$, $x^2_{221} = 1.35$, $x^2_{222} = 0$	$x^2_{111} = 0$, $x^2_{112} = 3$, $x^2_{121} = 0$, $x^2_{122} = 38$, $x^2_{211} = 2.71$, $x^2_{212} = 9$, $x^2_{221} = 1.35$, $x^2_{222} = 0$

8 Conclusion and Future Scope

In this paper, a cost minimization problem with the help of two-stage solid transportation has been proposed. An expected value-based reduction method has been proposed to solve the proposed model. Linear and lognormal uncertain type-2 variables have been used to solve the proposed model. Lingo 13.0 optimization software has been used here to solve the model. One comparison has been made between reduction methods of two types of type-2 uncertain variables. In future, also we can solve this type of problem with the help of different uncertain type-2 variables, viz. zigzag, normal uncertain variables.

References

1. Hitchcock, F.L.: The distribution of a product from several sources to numerous localities. J. Math. Phy. **20**, 224–230 (1941)
2. Shell, E.: Distribution of a product by several properties, Directorate of Management Analysis. In: Proceedings of 2nd Symposium on Linear programming, DCs/ Comptroller, vol. 2, pp. 615–642. H.Q. U.S.A.F., Washington, DC (1955)
3. Baidya, A., Bera, U.K., Maiti, M.: Solution of multi-item interval valued solid transportation problem with safety measure using different methods. Opsearch (Springer) **51**(1), 1–22 (2014)
4. Das, A., Bera, U.K., Maiti, M.: A profit maximizing solid transportation model under rough interval approach. IEEE Trans. Fuzzy Syst. **25**(3), 485–498
5. Cui, Q., Sheng, Y.: Uncertain programming model for solid transportation problem. Information **15**(12), 342–348 (2012)
6. Jimenez, F., Verdegay, J.L.: Uncertain solid transportation problems. Fuzzy Sets Syst. **100**, 45–57 (1998)
7. Li, Y., Ida, K., Gen, M., Kobuchi, R.: Neural network approach for multicriteria solid transportation problem. Comput. Ind. Eng. **33**, 465–468 (1997)
8. Liu, B.: Why is there a need for uncertainty theory? J. Uncertain Syst. **6**(1), 3–10 (2012)
9. Liu, B.: Uncertainty Theory: A Branch of Mathematics for Modeling Human Uncertainty. Springer, Heidelberg, Germany (2010)
10. Yang, L., Liu, P., Li, S., Gao, Y., Ralescu, Dan A.: Reduction method of type-2 uncertain variables and their application to solid transportation problem. Inf. Sci. **291**, 204–237 (2015)
11. Yang, L., Feng, Y.: A bi-criteria solid transportation problem with fixed charge under stochastic environment. Appl. Math. Model. **31**, 2668–2683 (2007)
12. Liu, B.: Uncertainty Theory, 2nd edn. Springer, Berlin (2007)
13. Liu, Y., Ha, M.: Expected value of function of uncertain variables. J. Uncertain Syst. **4**, 181–186 (2010)
14. Liu, B.: Theory and Practice of Uncertain Programming, 2nd edn. Springer, Berlin (2009)
15. Liu, B.: Some Research Problems in Uncertainty Theory. J. Uncertain. Syst. **3**(1), 3–10 (2009)
16. Liu, B., Chen, X.W.: Uncertain multi-objective programming and uncertain goal programming. Technical Report (2013)
17. Sheng, Y., Yao, K.: A transportation model with uncertain costs and demands. Information **15**, 3179–3186 (2012)
18. Sheng, Y., Yao, K.: Fixed charge transportation problem in uncertain environment. Ind. Eng. Manag. Syst. **11**, 183–187 (2012)

19. Das, A., Bera, U.K., Das, B.: A solid transportation problem with mixed constraint in different environment. J. Appl. Anal. Comput. SCI **6**(1), 179–195 (2016)
20. Gen, M., Ida, K., Li, Y., Kubota, E.: Solving bi-criteria solid transportation problem with fuzzy numbers by a genetic algorithm. Comput. Ind. Eng. **29**, 537–541 (1995)
21. Haley, K.B.: The solid transportation problem. Oper. Res. Int. J. **11**, 446–448 (1962)

Performance Evaluation of Green Supply Chain Management Using the Grey DEMATEL–ARAS Model

Kajal Chatterjee, Edmundas Kazimieras Zavadskas, Jagannath Roy and Samarjit Kar

Abstract Under stakeholder pressure and more strict regulations, firms need to enhance green supply chain management (GSCM) practice using multidimensional approaches. In view of these facts, a multi-criteria decision-making (MCDM) technique can be implemented while evaluating GSCM performance of alternative suppliers based on a set of criteria to deal with vagueness of human perceptions. The grey set theory is used to interpret the linguistic preference in accordance with the subjective evaluation. The cause–effect relationships among GSCM criteria, as well as their weights are considered using the grey DEMATEL approach. The grey ARAS method is also applied, using the weights obtained, for evaluating and ranking the GSCM performance of alternative suppliers. A sensitivity analysis conducted to ensure the reliability of solutions is described and the comparison of the applied technique with other MCDM methods such as grey TOPSIS and grey COPRAS is provided.

Keywords Green supply chain management (GSCM) · DEMATEL · Grey set ARAS · Multi-criteria decision-making (MCDM)

K. Chatterjee (✉) · J. Roy · S. Kar
Department of Mathematics, National Institute of Technology, Durgapur 713209, India
e-mail: chatterjeekajal7@gmail.com

J. Roy
e-mail: jaksformath@gmail.com

S. Kar
e-mail: kar_s_k@yahoo.com

E. K. Zavadskas
Department of Construction Technology and Management,
Vilnius Gediminas Technical University, Sauletekio al. 11, 10223 Vilnius, Lithuania
e-mail: edmundas.zavadskas@vgtu.lt

S. Kar et al. (eds.), *Operations Research and Optimization*, Springer Proceedings in Mathematics & Statistics 225, https://doi.org/10.1007/978-981-10-7814-9_24

1 Introduction

Under the conditions of stricter governmental regulations and rising public awareness of the environmental protection problems, many firms are now undertaking major initiatives to make their supply chains greener by enhancing technological innovation and improving green activities [23]. As environmental awareness increases, firms purchase products from suppliers that provide a high quality, low-cost product, while displaying high environmental responsibility [8]. Since a green partner is expected to achieve green product design and life cycle analysis, companies need rigorous partner selection and performance evaluation techniques. Thus, firms should systematically embrace an evaluation model of supplier selection in determining potential and appropriate partners to maintain a competitive advantage in the globalization trend [5]. In view of the above considerations, the selection of proper green suppliers, which is a critical issue for organizations, can be modelled as a multi-criteria decision-making (MCDM) problem that can handle various and conflicting criteria for making a selection among predetermined alternatives. With structural relationships among the criteria constructed for supplier selection, firms obtain a clear understanding of the cause–effect relationships for facilitating suppliers reflecting more realistic results among decision attributes and alternatives [3]. The DEMATEL method is considered due to its ability to confirm interdependence of the considered factors and can help identify key criteria to improve performance and provide decision-making information [11]. However, it is unable to deal with uncertain situations, the lack of information and conflict resolution among experts and cannot express ambiguous values around a given discrete value [2]. Although some fuzzy methods offset the shortcomings of the crisp methods, they suffer from limitations of mapping the memberships functions. Grey theory, being superior in the mathematical analysis of systems with uncertain information, can deal flexibly with the fuzziness situation [9]. The grey set theory studies the relationships among various attributes (criteria) in an MCDM problem, proving to be an effective approach to theoretical analysis of systems with imprecise information and incomplete samples [22]. Grey theory can be successfully amalgamated with any of the decision-making process, so as to improve the exactitude of judgments, and grey numbers are easily convertible into crisp numbers using various methods [22].

Thus, in this paper, the grey DEMATEL method is used to identify key influencing criteria in forecasting and selecting suppliers that help companies solve decision-making problems in an imprecise environment. Particularly, this method can also be successfully used to divide a set of complex factors (criteria) into cause and effect groups through a causal diagram. Thus, the complexity of a problem is easier to be captured and profound decisions can be made. Furthermore, the additive ratio assessment method (ARAS) expressed in intervals in the grey theory [18] is employed to aggregate the performance values for selecting the best supplier for the GSCM based on the weights of criteria and using the grey DEMATEL method. The method is particularly helpful when decision-makers may hold diverse opinions and preferences due to incomplete information and knowledge or some inherent conflict between various departments.

The rest of the paper is structured as follows. Section 2 discusses the scenario and proposes the criteria for GSCM. Section 3 deals with the proposed methodology for ranking the alternatives based on the critical risk criteria. The discussion of results and the sensitivity analysis is presented in Sect. 4. The conclusion part is given in Sect. 5.

2 Problem Discussion

2.1 Background Scenario

As companies in developing countries are lagging behind in adopting the GSCM concept, the understanding of the drivers and risk criteria for adopting GSCM practices in their business strategies, is crucial [13]. The study is based on the analysis of a electronics manufacturing company in Taiwan, adopted from Liou et al. [10], for which five green supplier alternatives should be selected based on the proposed green crucial criteria, determined from the extensive literature sources (discussed in Sect. 2.2). The hierarchal structure of the GSCM initiatives is as follows. The top level of the hierarchy represents the final goal of the problem solution, while the second level consists of ten main risk criteria (Table 1) influencing the GSCM practices. Finally, the bottom level of the hierarchy represents the supplier companies as the alternatives selected for evaluation. As the company chooses the green supplier partners, the last level should represent the alternative suppliers that must be prioritized according to their greenness.

2.2 The Criteria Proposed for GSCM

GSCM has been defined as the incorporation of the ecological component into the supply chain management [17]. Specifically, risks to the green supply chain are unforeseen events that might affect the green or environmentally friendly material movement, and even disturb the proposed flow of green materials and products from their point of origin to the point of consumption in business (Yang and Li 2010). Mangla et al. [12] identified six categories of risks and thirteen specific risks, associated with the GSCM on the basis of the literature and inputs from industrial experts. Mathiyazhagan et al. [14] identified twenty-six barriers based on consultation with industrial experts and academicians and applied the ISM-based approach to barrier analysis in implementing GSCM. Wu and Chang [19] used four dimensions and twenty criteria to identify critical dimensions and factors to construct a digraph, showing causal relationships among them in GSCM. Govindan and Chaudhuri [7]

Table 1 GSC specific risk-based drivers/criteria

No.	Risk drivers/ criteria	References
1	Process design and planning risks (C_1)	Chan and Wang [4]
2	Lack of technical expertise (C_2)	Deleris and Erhun [6]
3	Non-availability of fund to encourage green products (C_3)	Mathiyazhagan et al. [14]
4	Lack of new technologies, materials and processes (C_4)	Perron [15]
5	Inflation and currency exchange rate (C_5)	Yang and Li [20]
6	Market dynamics and bullwhip effect risk (C_6)	Mangla et al. [12]
7	Procurement costs risks (C_7)	Yang and Li [20]
8	High cost of hazardous waste disposal (C_8)	Mathiyazhagan et al. [14]
9	IT and information sharing risks (C_9)	Selviaridis et al. [16]
10	Supplier quality issues (C_{10})	Mangla et al. [12]

analysed the interrelationships between risks faced by the third-party logistics service providers (3PLs) with respect to one of its customers, using DEMATEL. Based on the above papers, we identified ten (mainly risk driven) criteria for prioritizing the green supplier alternatives (see Table 1).

3 The Methodology Proposed for Ranking the Alternatives Based on Criteria Weight

After setting the decision goal, construct a committee of experts with K members for determining the evaluation criteria and the alternatives. The evaluation criteria are already discussed in Table 1. Based on the interviews with academicians and industry experts, the computational steps of the above hybrid MCDM framework are implemented.

3.1 The Grey DEMATEL Method for Criteria Weight

The aim is to investigate the major relationships among the criteria (Table 1) that have been identified in the literature. The analysis is performed based on the method

Table 2 The grey linguistic scale of respondent's assessments

Linguistic terms	Grey numbers	Normal values
No influence (N)	[0, 0]	0
Very low influence (VL)	[0, 0.25]	1
Low influence (L)	[0.25, 0.5]	2
High influence (H)	[0.5, 0.75]	3
Very high influence (VH)	[0.75, 1]	4

guidelines and steps presented in Zhang et al. [22], Dalalah et al. [5], Bai and Sarkis [2]. Based on the above papers, grey DEMATEL algorithmic methodology is applied here. The program code for computation is written along in MATLAB, following the steps given below:

Step 1. Develop a complete grey direct-relation matrix

Step 1.1: Defining the grey influence comparison linguistic scale. Firstly, for the purpose of measuring the relationships, it is required to define a comparison scale. The different degree of influence among criteria are expressed with 5-level grey linguistic scale: 0 = no influence, 1 = very low influence, 2 = low influence, 3 = high influence, 4 = very high influence. The grey linguistic scale for respondents evaluations is defined in Table 2.

Step 1.2: Acquire the initial grey direct-relation matrix.
To show the relationships among the set of the criteria ($C_i | i = 1, 2, \ldots, n$), a group of K experts developed $n \times n$ pair-wise comparison grey decision matrices $Z^1, Z^2, \ldots,$ Z^K (using Eq. 1), with the principal diagonal elements initially set to the grey value of zero.

$$Z^k = [\otimes z_{ij}^k]_{n \times n}, k = 1, 2, \ldots, K \tag{1}$$

$\otimes z_{ij}^k = [z_{ij\alpha}^k, z_{ij\alpha}^k]$ is a grey number for the influence of criterion on criterion for expert. The matrices constructed for four experts ($K = 4$) and ten criteria used in the case study are given in Table 3.

Step 1.3: Combine all grey direct-relation matrices grey into an aggregate matrix as:

$$Z = \frac{\sum_{i=1}^{K} Z^k}{K} \tag{2}$$

Step 2: Calculate the normalized grey direct-relation matrix (N)
The normalized grey direct-relation matrix $N = [\otimes n_{ij}]_{n \times n}$ can be obtained using Eqs. (3)–(5). ($i, j = 1, 2, \ldots, n$)

$$N = \otimes s \cdot Z, \quad where \tag{3}$$

$$\otimes n_{ij} = [s_\alpha \cdot z_{ij\alpha}, s_\beta \cdot z_{ij\beta}] \tag{4}$$

Table 3 The pair-wise direct-relation matrix

Expert 1	C1	C2	C3	C4	C5	C6	C7	C8	C9	C10	Expert 2	C1	C2	C3	C4	C5	C6	C7	C8	C9	C10
C1	0	2	0	1	2	1	1	4	1	1	C1	0	2	1	0	1	1	0	2	1	1
C2	2	0	1	1	2	1	2	0	2	2	C2	3	0	1	0	2	2	1	1	0	1
C3	1	2	0	2	1	1	3	4	3	3	C3	3	2	0	1	3	3	2	3	0	0
C4	3	2	0	0	4	1	4	0	3	1	C4	2	2	1	0	1	1	1	1	0	1
C5	4	1	2	1	0	2	2	0	0	4	C5	3	2	2	3	0	2	3	2	3	2
C6	2	1	3	2	1	0	1	3	1	2	C6	2	1	2	2	4	0	3	3	1	3
C7	0	3	4	3	4	3	0	4	0	1	C7	1	3	1	1	3	1	0	3	3	0
C8	1	4	1	3	0	1	3	0	4	3	C8	1	3	1	1	2	2	3	0	4	0
C9	3	2	3	3	1	4	1	3	0	1	C9	2	1	2	0	1	2	4	4	0	1
C10	2	3	1	1	0	1	2	3	1	0	C10	4	1	3	1	0	1	1	3	2	0
Expert 3	C1	C2	C3	C4	C5	C6	C7	C8	C9	C10	Expert 4	C1	C2	C3	C4	C5	C6	C7	C8	C9	C10
C1	0	1	2	1	1	4	0	2	1	2	C1	0	2	2	3	3	3	4	1	2	4
C2	3	0	3	2	0	1	2	1	2	1	C2	4	0	3	3	3	3	3	2	3	3
C3	1	4	0	2	1	4	2	2	3	2	C3	3	3	0	1	1	1	1	3	2	1
C4	4	1	1	0	2	1	2	3	4	3	C4	1	2	3	0	3	2	3	1	3	2
C5	3	3	2	0	0	2	3	4	1	1	C5	3	4	2	4	0	2	2	1	4	3
C6	2	1	2	1	3	0	1	1	2	2	C6	3	3	2	3	4	0	4	1	2	3
C7	1	1	1	1	2	4	0	2	1	3	C7	4	2	2	3	1	3	0	3	2	3
C8	2	2	3	1	1	3	2	0	3	4	C8	2	1	1	1	2	2	3	0	1	1
C9	0	0	2	3	0	1	1	3	0	1	C9	1	1	2	2	1	1	2	4	0	2
C10	1	0	2	0	1	1	0	0	1	0	C10	0	2	1	2	3	1	3	2	2	0

$$\otimes s = [s_\alpha, s_\beta] = \frac{1}{\sum_{j=1}^{n} \otimes z_{ij}} \tag{5}$$

Step 3: Calculate the total relationship matrix (T)
The total relationship matrix $T = (t_{ij})_{n\times n}$ is found by expression (6), where I represents the $n \times n$ identity matrix.

$$T = N + N^2 + N^3 + \cdots = \sum_{i=1}^{\infty} N_i = N(I - N)^{-1} \tag{6}$$

Step 4: Develop the causal influence
Sub-step 4.1: Calculate the sum of row ($\otimes R_i$) and column ($\otimes D_j$) for each row i and j from the total relation matrix T as follows:

$$\otimes R_i = \sum_{j=1}^{n} \otimes t_{ij}, \forall i = 1, 2, \ldots, n \tag{7}$$

$$\otimes D_j = \sum_{i=1}^{n} \otimes t_{ij}, \forall j = 1, 2, \ldots, n \tag{8}$$

The row values $\otimes R_i$ present the sum of direct and indirect influence of the criterion i on the other criteria. Similarly, the column values $\otimes D_j$ present the sum of direct and indirect influence that factor j is receiving from other. The grey numbers $\otimes R_i = [R_{i\alpha}, R_{i\beta}]$ and $\otimes D_j = [D_{j\alpha}, D_{j\beta}]$ are transformed into white numbers using Eqs. (9) and (10), suggested by Zhang et al. [22], as shown in Table 4.

$$R_i = \rho_i \times R_{i\alpha} + (1 - \rho_i) \times R_{i\beta} \tag{9}$$

$$D_j = \rho_j \times D_{j\alpha} + (1 - \rho_j) \times D_{j\beta} \tag{10}$$

where ρ_i is the orientation coefficient of the grey numbers R_i and D_i, $i = 1, 2, \ldots, n$.

Sub-step 4.2: Determine the prominence P_i and net effect E_i of the criterion i using the expressions:

$$P_i = \{R_i + D_j | i = j\} \tag{11}$$

$$E_i = \{R_i - D_j | i = j\} \tag{12}$$

The values P_i show the index representing the total cause and effect. In this case, larger the value of P_i, the greater the overall prominence (*visibility/importance/influence*) of the criterion i with respect to others criteria. The values E_i shows the net effect or cause of the criterion i. If $E_i > 0$, the criterion i represents the net cause, and if $E_i < 0$, then, the criterion i represents the net effect. The result shown in Table 4.

Table 4 The degree of prominence and net cause/effect of criteria

Criteria	$\otimes D_i$		$\otimes R_i$		D_i	R_i	Prominence	Net effect
	$D_{i\alpha}$	$D_{i\beta}$	$R_{i\alpha}$	$R_{i\beta}$			$R_i + D_i$	$R_i - D_i$
C1	0.711	3.119	0.472	2.542	1.915	1.507	3.422	−0.408
C2	0.607	2.923	0.584	2.793	1.765	1.689	3.454	−0.077
C3	0.537	2.771	0.688	3.140	1.654	1.914	3.568	0.260
C4	0.470	2.475	0.624	2.912	1.473	1.768	3.240	0.295
C5	0.567	2.701	0.810	3.365	1.634	2.088	3.721	0.454
C6	0.581	2.926	0.695	3.779	1.754	2.237	3.991	0.483
C7	0.722	3.143	0.764	3.294	1.933	2.029	3.962	0.096
C8	0.811	3.334	0.653	3.042	2.072	1.847	3.920	−0.225
C9	0.638	2.878	0.578	2.800	1.758	1.689	3.447	−0.069
C10	0.608	3.589	0.385	2.193	2.099	1.289	3.387	−0.810

Table 5 Non-normalized and normalized weight of green supplier criteria

Criteria	Weights	Normalized weight	Ranking
C_1	5.288	0.817	8
C_2	5.437	0.84	6
C_3	5.726	0.885	5
C_4	5.22	0.807	9
C_5	6.034	0.933	4
C_6	6.470	1	1
C_7	6.295	0.973	2
C_8	6.128	0.947	3
C_9	5.428	0.839	7
C_{10}	5.126	0.792	10

Step 5: Calculate the weights of criteria

Inspired by Dalalah et al. [5], we proposed a formula to determine the weight W_i and normalized weight w_i of criteria given in Eqs. (13) and (14) respectively. The calculation is presented in Table 5.

$$W_i = \left\{(P_i)^2 + (E_i)^2\right\}^{\frac{1}{2}} \tag{13}$$

$$w_i = \frac{W_i}{\max_{1 \le j \le n}(W_i)}, \forall i = 1, 2, \ldots, n \tag{14}$$

3.2 Grey Value-Based ARAS for Alternative Selection

Grey ARAS technique adapted from Turskis and Zavadskas [18] uses a utility function value to determine complex relative efficiency of a reasonable alternative that is directly proportional to the relative weights of the main criteria.

The steps fused in the methodology are given for ten risk-based criteria $C_j(j = 1, 2, \ldots, 10)$ and five alternatives suppliers $A_i(i = 1, 2, \ldots, 5)$. Taking the weights derived from grey DEMATEL of criteria $C_j(j = 1, 2, \ldots, 10)$, we can prioritize the alternative $A_i(i = 1, 2, \ldots, 5)$. Applying the data from Table 6, the decision-makers express to determine the initial direct relationship matrix.

Step 1: Form a grey decision-making matrix
Construct the decision matrix X^k of the kth decision maker and the average decision matrix X according to Eqs. (15)–(16). In multi-criteria decision-making problem related to discrete optimization, any problem to be solved is represented by the following decision-making method (DMM) of preferences for m reasonable alternatives (rows) rated on n criteria (columns):

$$X^k = [\otimes x_{ij}^k] \tag{15}$$

where $\otimes x_{ij}^k = [x_{ij\alpha}^k, x_{ij\beta}^k], \forall i = 0, 1, \ldots, m; j = 1, 2, \ldots, n,$
and

$$X = [\otimes x_{ij}] \tag{16}$$

where $\otimes x_{ij} = [x_{ij\alpha}, x_{ij\beta}], \forall i = 0, 1, \ldots, m; j = 1, 2, \ldots, n$
and $x_{ij} = \left(\frac{x_{ij}^1 \oplus x_{ij}^2 \oplus \cdots \oplus x_{ij}^K}{K}\right)$.
Here, m is the alternatives, n is the number of criteria describing each alternative, $\otimes x_{ij}$ represents the performance value of the ith alternative in terms of jth criterion, $\otimes x_{0j}$ is an optimal value of jth criterion.
If the optimal value of jth criterion is unknown, then

Table 6 Linguistic variables and grey numbers for evaluating the alternatives

Scale	Grey numbers
Very Poor (VP)	[0, 1]
Poor (P)	[1, 3]
Medium Poor (MP)	[3, 4]
Fair (F)	[4, 5]
Medium Good (MG)	[5, 6]
Good (G)	[6, 9]
Very Good (VG)	[9, 10]

$$\begin{cases} \otimes x_{0i} = max_i \otimes x_{ij}, \text{ if } max_i \otimes x_{ij} \text{ is preferable} \\ \otimes x_{0j} = min_i \otimes x^*_{ij}, \text{ if } min_i \otimes x^*_{ij} \text{ is preferable} \end{cases} \tag{17}$$

Usually, the performance values $\otimes x_{ij}$ and the criteria weights w_j are viewed as the entries of a DMM. When the dimensionless value of the criteria are known, all the criteria, originally having different dimensions, can be compared. Based on the data given in Table 7 with ($i = 0, 1, \ldots, 5; j = 1, 2, \ldots, 10$), the initial grey decision-making matrix is formed.

Step 2: Calculate the normalized decision-making matrix
In the second step, the initial values of all criteria are normalized by defining the values $\otimes \bar{x}_{ij}$ of normalized decision-making matrix $\otimes \bar{X}$.

$$\otimes \bar{X} = [\otimes \bar{x}_{ij}]_{m \times n} = [\bar{x}_{ij\alpha}, \bar{x}_{ij\alpha}]_{m \times n}; \forall i = 0, 1, \ldots, m; j = 1, 2, \ldots, n \tag{18}$$

The criteria, whose preferable values are maxima, are normalized as follows:

$$\otimes \bar{x}_{ij} = \frac{\otimes x_{ij}}{\sum_{i=0}^{m} \otimes x_{ij}} \tag{19}$$

The criteria, whose preferable values are minima, are normalized by applying two-stage procedure as follows:

$$\otimes x_{ij} = \frac{1}{\otimes x^*_{ij}}; \otimes \bar{x}_{ij} = \frac{\otimes x_{ij}}{\sum_{i=0}^{m} \otimes x_{ij}} \tag{20}$$

Step 3: Define the weighted normalized decision-making matrix $\otimes \hat{X}$
The criteria can be evaluated using weights $0 < w_j \leq 1$. Only well-founded weights should be used because weights are always subjective and influence the solution. The weighted normalized decision matrix $\otimes \hat{X}$ is calculated as follows:

$$\otimes \hat{X} = [\otimes \hat{x}_{ij}]_{m \times n}; \forall i = 0, 1, \ldots, m; j = 1, 2, \ldots, n \tag{21}$$

$$\otimes \hat{x}_{ij} = w_{ij} \times \bar{x}_{ij} \tag{22}$$

where w_{ij} is the weight (*importance*) of the *j*th criterion and $\bar{x}_{ij}$ is the normalized rating of the *j*th criterion.

Step 4: Determine the values of optimality function $\otimes S_i$

$$\otimes S_i = \sum_{j=1}^{n} \otimes \hat{x}_{ij}; \forall i = 0, 1, \ldots, m \tag{23}$$

Step 5: Calculate the utility degree
The result of grey decision-making for each alternative is optimal function $\otimes S_i$ expressed in grey number. The grey number $\otimes S_i = [S_{i\alpha}, S_{i\beta}]$ is transformed into the

Table 7 The rating scale of suppliers alternative with respect to the criteria used

Expert 1	C1	C2	C3	C4	C5	C6	C7	C8	C9	C10	Expert 3	C1	C2	C3	C4	C5	C6	C7	C8	C9	C10
A1	G	MG	G	G	G	MG	VP	VP	VP	VP	A1	VP	MG	VP	G	VP	G	G	G	G	G
A2	MG	G	F	MG	F	G	G	G	G	G	A2	MG	G	G	G	G	G	F	F	G	F
A3	G	MG	F	F	MP	F	MP	P	G	MP	A3	VP	F	VP	F	F	F	VP	VP	MP	VG
A4	P	F	VG	MG	G	MG	G	F	MG	VG	A4	VG	VG	G	F	VG	VP	F	F	F	VP
A5	MG	MG	F	G	MG	MG	F	G	G	P	A5	G	F	F	F	G	F	VG	VP	F	P
Expert 2	C1	C2	C3	C4	C5	C6	C7	C8	C9	C10	Expert 4	C1	C2	C3	C4	C5	C6	C7	C8	C9	C10
A1	F	G	G	G	VG	MG	F	G	P	MP	A1	MP	MP	P	VG	G	G	VP	P	F	VP
A2	MG	G	G	G	G	F	P	MP	MP	MP	A2	MG	G	G	F	F	F	G	VG	MG	F
A3	G	VG	VG	VG	VG	G	G	G	G	G	A3	VG	G	F	G	P	PP	F	VG	VP	G
A4	MP	MP	MP	G	VG	G	VG	MP	P	MP	A4	MG	MP	VG	G	MP	MG	VG	VG	MG	MG
A5	VG	VG	VG	MG	MG	MG	MG	MG	G	P	A5	VG	G	G	VG	MG	G	G	MG	G	F

Table 8 The values of optimality function and ranking of the alternatives

Alternatives A_i	Optimality function $\otimes S_i$		Crisp value S_i	Utility degree R_i	Ranking
	$S_{i\alpha}$	$S_{i\beta}$			
A_0	1.9293	2.4311	2.1802	1	*
A_1	0.9391	1.6721	1.3056	0.5988	5
A_2	1.124	1.8668	1.4954	0.6859	2
A_3	1.0436	1.8507	1.4471	0.6638	3
A_4	1.0126	1.6609	1.3367	0.6131	4
A_5	1.1453	1.8745	1.5099	0.6925	1

white number S_i using Eq. (24) proposed by Zhang et al. [22].

$$S_i = \rho_i \times S_{i\alpha} + (1 - \rho_i) \times S_{i\beta} \tag{24}$$

The degree of the alternative utility is determined by a comparison of the variant analysed with the ideally best one S_0. Equation (25) used for calculating the utility degree S_0 for alternative A_i is given below:

$$K_i = \frac{S_i}{S_0}; \forall i = 0, 1, \ldots, m \tag{25}$$

S_i and S_0 are optimality criterion values, ordered in the increasing order. The details are given in Table 8. The complex relative efficiency of the reasonable alternative can be determined according to the utility function values.

4 Result Discussion

4.1 The Position of the Considered Criteria and Alternatives

The results of MCDM analysis (Table 8) suggest that supplier A_5, with the weight 0.6925, should be considered the best in terms of green criteria followed by A_2 (0.6859) and A_3 (0.6638). The result also show that four dominant criteria based on normalized weights (Table 5) are as follows: Market dynamics and bullwhip effect risk (C_6) with weight (1.000) make the first priority. Procurement costs risks (C_7) with the weight (0.973) make the second priority. High cost of hazardous wastes disposal (C_8), with the weight of (0.947), makes the third priority. Inflation and currency exchange rate (C_5), with the weight of (0.933), make the fourth priority. Supplier quality issues (C_{10}) with the weight of (0.792) are considered to be of the lowest importance.

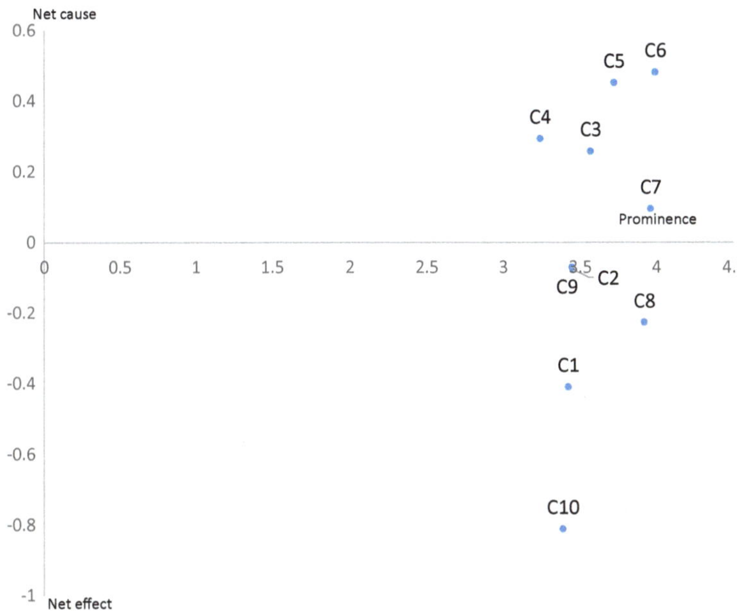

Fig. 1 DEMATEL-yielded prominence causal graphs

4.2 A Description of the Causal Diagram

Using Table 5, we can identify the most important (prominent) criteria (factors) and most important relationships among the green supply chain key criteria. The four most prominent criteria are market dynamics and bullwhip effect risk (C_6), procurement costs risks (C_7), high cost of disposing hazardous wastes disposal (C_8), and inflation and currency exchange rate (C_5) (see Fig. 1).

Prominence includes the integration of the criteria from both cause (influencing) and effect (resulting) perspectives. This analysis provides a temporal perspective, showing us what factors should be initiated in place and which ones should be added in the near future. Now, each of these considered relationships will be evaluated. The most important influencing criteria, producing a strong effect on GSCM, are given the highest score: $(R - D)$. These criteria not only play a significant role in the evaluation, but also affect other critical criteria. In Fig. 1, five key net cause criteria, with net effect scores over 0, can be identified. These are lack of new technology, materials and processes (C_4), Inflation and currency exchange rate (C_5), market dynamics and bullwhip effect risk (C_6). The remaining similar net cause criteria are non-availability of fund to encourage green products (C_3) and procurement costs risks (C_7) but they are not as prevalent in terms of their relationships with others. The key net effect green criteria denote the most strongly influencing resulting criteria that are the last ones to be addressed. From Fig. 1, we can identify five key net effect

Table 9 Sensitivity analysis of ranking based on the variation of risks criteria weight

Scenarios	Priority based criteria			Green supplier alternatives				
				A_1	A_2	A_3	A_4	A_5
1	C_6	C_7	C_8	5	2	3	4	1
2	C_3	C_8	C_7	4	3	2	5	1
3	C_1	C_9	C_{10}	4	1	3	5	2
4	C_5	C_3	C_{10}	4	3	2	5	1
5	C_{10}	C_1	C_7	4	1	3	2	5
6	C_7	C_3	C_8	3	4	1	5	2
7	C_3	C_8	C_5	3	5	2	4	1

criteria as follows: IT and information sharing risks (C_9), Lack of technical expertise (C_2), High cost of disposing hazardous wastes disposal (C_8), Process design and planning risks (C_1) and Supplier quality issues (C_{10}), with net effect scores below 0.

4.3 Sensitivity Analysis

A sensitivity analysis of the alternatives ranking results based on various weights of criteria was performed and the obtained data are presented in Table 9. Six scenarios were considered to prioritize the criteria. The change in ranking indicates the robustness of the method. The criteria C_3 and C_8 have a strong effect on the ranking of the alternatives.

4.4 The Comparison of the Results with Those Yielded by Other MCDM Methods

When analysing multiple criteria evaluation of alternatives, one can observe that relative significances of alternatives and change in the priority order of alternatives, established by applying several MCDM methods. The aim of presented case study is to determine priorities as well as produce some recommendations concerning rational redevelopment of buildings. A comparative analysis was performed using five alternatives as shown in Table 10. For the weights based on the grey DEMATEL, the results show that the ranking data for the alternatives A_1, A_3, and A_4 are the same for grey ARAS and grey COPRAS [21], while there is change of position for A_2 and A_5 can be observed for grey TOPSIS [24] and grey ARAS, and only for A_1 they have similar ranking order.

Table 10 Comparison of ranking data based on using different MCDM methods

Green supplier alternatives	Grey ARAS [18]	Grey TOPSIS [24]	Grey COPRAS [21]
	As per utility degree K_i	As per closeness coefficient CC_i	As per utility degree N_i
A_1	5	5	5
A_2	2	1	1
A_3	3	4	3
A_4	4	3	4
A_5	1	2	2

4.5 Reliability Test of Alternative Ranking Using Spearman's Rank Correlation

In Sect. 4.4, multiple criteria analysis of different methods are performed (Table 10) and results are compared. Spearman's correlation coefficient analysis is performed among the ranking results of alternatives, for every possible pair of MCDM methods in uncertain domain, to show the reliability of ranking [1]. The current coefficient best fits the aim of the presented research as it compares ranks of alternative decisions obtained in a process of multiple criteria analysis when applying different MCDM methods. First of all, Spearman's rank correlation coefficients are calculated for three pairs of ranking results, viz. (ARAS and COPRAS), (ARAS and TOPSIS), and (COPRAS and TOPSIS), respectively in grey number domain. Priorities of alternatives computed by grey-based ARAS provide significant relations with grey-based COPRAS ranking results and grey-based TOPSIS. The value of Spearman's rank correlation coefficient is high enough for ARAS and COPRAS ranking results (0.9), for ARAS and TOPSIS ranking (0.8), and COPRAS and TOPSIS (0.9), respectively, showing the reliability of the ranking results.

5 Conclusions

Green supply chains (GSCs) are becoming ever more complex and exposed to risks due to the increased globalization and vertical integrations. The present research has attempted to identify the key criteria of green supply chain risk mitigation. The plotted cause–effect relationships can help managers to identify primary causal criteria for addressing the vulnerability issues of supply chain. Managers can also plan the future direction of the implementation of strategies by determining how the particular criteria influence the other criteria, adding more criteria at the cost of complexity. The proposed method also reveals the strengths and weaknesses of the alternative companies from the viewpoint of their greenness. Furthermore, the suggested approach can be used to benchmark, improve, and develop better products and green processes. The results can guide managers in their choice of the best partner among the candidates for future cooperation. The limitation of this method is that it depends on experience and quality of statements from experts review. Investigating other hybrid techniques of DEMATEL method with other MCDM methods would be interesting for future study.

References

1. Antucheviciene, J., Zakarevicius, A., Zavadskas, E.: Measuring congruence of ranking results applying particular MCDM methods. INFORMATICA. **22**(3), 319–338 (2011)
2. Bai, C., Sarkis, J.: A grey-based DEMATEL model for evaluating business process management critical success factors. Int. J. Prod. Econ. **146**, 281–292 (2013)
3. Bykzkan, G., Cifci, G.: A novel hybrid MCDM approach based on fuzzy DEMATEL, fuzzy ANP and fuzzy TOPSIS to evaluate green suppliers. Expert Syst. Appl. **39**(3), 3000–3011 (2012)
4. Chan, H., Wang, X.: An integrated fuzzy approach for aggregative supplier risk assessment. In: Fuzzy Hierarchical Model for Risk Assessment, pp. 45–69. Springer, London (2013)
5. Dalalah, D., Hayajneh, M., Batieha, F.: A fuzzy multi-criteria decision making model for supplier selection. Expert Syst. Appl. **38**, 8384–8391 (2011)
6. Deleris, L., Erhun, F.: Quantitative risk assessment in supply chains: a case study based on engineering risk analysis concepts. In: Planning Production and Inventories in the Extended Enterprise, pp. 105–131. Springer, New York (2011)
7. Govindan, K., Chaudhuri, A.: Interrelationships of risks faced by third party logistics service providers: a DEMATEL based approach. Trans. Res. Part E **90**, 177–195 (2016)
8. Lee, A., Kang, H., Hsu, C., Hung, H.: A green supplier selection model for high-tech industry. Expert Syst. Appl. **36**, 791727 (2009)
9. Li, G., Yamaguchi, D., Nagai, M.: A grey-based decision-making approach to the supplier selection problem. Math. Comput. Model. **46**, 573–581 (2007)
10. Liou, J., Tamosaitiene, J., Zavadskas, E., Tzeng, G.: New hybrid COPRAS-G MADM model for improving and selecting suppliers in green supply chain management. Int. J. Prod. Res. **54**(1), 114–134 (2016)
11. Lin, C., Chen, S., Tzeng, G.: Constructing a cognition map of alternative fuel vehicles using the DEMATEL method. J. Multi-Crit. Decis. Anal. **16**, 5–19 (2009)
12. Mangla, S., Kumar, P., Barua, M.: An integrated methodology of FTA and fuzzy AHP for risk assessment in green supply chain. Int. J. Oper. Res. (2015) (in press)
13. Mathiyazhagan, K., Ali, D., Abbas, A., Xu, L.: Application of analytical hierarchy process to evaluate pressures to implement green supply chain management. J. Clean. Prod. **107**, 229–236 (2015)
14. Mathiyazhagan, K., Govindan, K., NoorulHaq, A., Geng, Y.: An ISM approach for barrier analysis in implementing green supply chain management. J. Clean. Prod. **47**, 283–297 (2013)
15. Perron, G.M.: Barriers to Environmental Performance Improvements in Canadian SMEs. Dalhousie University, Canada (2005)
16. Selviaridis, K., Spring, M., Profillidis, V., Botzoris, G.: Benefits, risks, selection criteria and success factors for third-party logistics services. Marit. Econ. Logist. **10**(4), 380–392 (2008)
17. Srivastava, S.: Green supply-chain management: a state-of-the-art literature review. Int. J. Manag. Rev. **9**(1), 53–80 (2007)
18. Turskis, Z., Zavadskas, E.: A novel method for multiple criteria analysis: grey additive ratio as sessment (ARAS-G) method. Informatica **21**, 597–610 (2010)

19. Wu, H., Chang, S.: A case study of using DEMATEL method to identify critical factors in green supply chain management. Appl. Math. Comput. **256**, 394–403 (2015)
20. Yang, Z., Li, J.: Assessment of green supply chain risk based on circular economy. In: IEEE 17th International conference on industrial engineering and engineering management (IE&EM), pp. 1276–1280 (2010). https://doi.org/10.1109/ICIEEM.2010.5645996
21. Zavadskas, E., Kaklauskas, A., Turskis, Z., Tamoaitien, J.: Selection of the effective dwelling house walls by applying attributes values determined at intervals. J. Civ. Eng. Manag. **14**(2), 85–93 (2008)
22. Zhang, W., Zhang, X., Fu, X., Liu, Y.: A grey analytic network process (ANP) model to identify storm tide risk. GSIS 582–587 (2009)
23. Zhu, Q., Sarkis, J., Lai, K.: Institutional-based antecedents and performance outcomes of internal and external green supply chain management practices. J. Purch. Supply Manag. **19**, 106–117 (2013)
24. Zolfani, S., Antucheviciene, J.: Team member selecting based on AHP and TOPSIS grey. Inz. Ekon.—Eng. Econ. **23**(4), 425–434 (2012)

Performance Evaluation of Management Faculty Using Hybrid Model of Logic—AHP

Anupama Chanda, R. N. Mukherjee and Bijan Sarkar

Abstract The main objective of this paper is to show how the two approaches of Boolean logic and analytical hierarchy process (AHP) can be utilized to solve management faculty selection problem. The problem is solved in two phases. In phase I, we use logic to find the minimal criteria required for management faculty selection. In phase II, two different methods have been used separately: logical analysis and AHP for final selection from the shortlisted candidates to arrive at a decision.

Keywords LAD · Propositional logic · Computational complexity AHP

1 Introduction

Logical analysis of data (LAD) is a data analysis technique which integrates principles of combinatorics, optimization, and Boolean functions; the idea was first described by Hammer [1]. It is applicable for analysis of binary data sets whose attributes take only two values (0–1). It is very much applicable to real-life problem including real values; hence, 'binarization' method was developed [2]. Binarization consists of introduction of several binary attributes associated with each of numerical attributes; each binary attribute is supposed to take value 1 or 0 if the corresponding numerical attribute associated takes value which is above or below a certain threshold value. LAD has been applied to various disciplines, e.g., medicine, economics, and business (see [3, 4] and [5]).

A. Chanda (✉)
Burdwan University, Bardhaman, India
e-mail: anupama.chanda@gmail.com

R. N. Mukherjee · B. Sarkar
Jadavpur University, Kolkata, India
e-mail: bsarkar@production.jdvu.ac.in

S. Kar et al. (eds.), *Operations Research and Optimization*, Springer Proceedings in Mathematics & Statistics 225, https://doi.org/10.1007/978-981-10-7814-9_25

LAD is used in finding minimal size (or economical) logical formula which is highly desirable in many problems from the view of cost or utility. The computational problem in logic arises as the size of the data increases which leads to exponential complexity; i.e., the computation time increases exponentially with the size of the database. LAD has been applied to various medical problems particularly in disease diagnosis [6]. Crama et al. [7] introduced a classification rule which was later developed into a rich classification framework named as logical analysis of data or LAD [2, 4, 8]. Here, they discussed a rule-based classifier called 'justifiability' in order to classify new observations. Important areas are electrical circuit design, credit scoring, disease diagnosis, pattern recognition, etc.

When there are multiple criteria important to a decision-maker, it may pose difficulty in taking decision. A mathematical technique such as analytical hierarchy process (AHP) developed by Saaty [9] is a powerful technique which is used to choose between various alternatives. The two approaches, viz. Boolean logic and AHP are used in management faculty selection problem.

In Sect. 2, logic is used to determine the selection criteria for management faculty. Here, computational logic is used to determine a minimal size (or economical) logical formula [10] that can be used to predict a result from a given input data. This minimal logical representation which consists of minimal number of literals is based on the method of Quine's procedure [11, 12]. It consists of repeatedly applying resolution and absorption, with appending the resolvent clauses and deleting the absorbed clauses, until no further simplification is possible. The method of resolution [13, 14] is normally applied to a logical statement in conjunctive normal form (CNF). This ends with phase I.

In Sect. 3, the final selection of the candidate based on the short list obtained in phase I is determined by (i) data analysis technique using logic and (ii) mathematical technique particularly AHP [15].

2 Minimal Criteria for Faculty Selection Process

The process of selection of minimal (optimal) criteria is deduced through a process [6] developed by H.P. William.

Example 1 Obtain a economical (or minimum size) logical formulae which can be used to find optimal selection criteria for faculty selection process:-

Here, five criteria are considered: age, qualification, experience, publication, and faculty development program (FDP). There are four experts for evaluation.

The data are converted into a finite set of True (T) or False (F) observations. 'T' and 'F' denote respectively that a particular criteria is either important or unimportant for an expert while '_' indicates irrelevance of the selection critera.

To get the logical formula for job selection criteria, let us first use the notations:

X_i: Satisfies the faculty selection criteria
$\bar{X}_i$: Does not satisfy the faculty selection criteria

Table 1 is expressed in disjunctive form as follows:

$$(X_1 \cdot X_2 \cdot X_5) \vee (X_1 \cdot \bar{X}_2 \cdot X_4 \cdot \bar{X}_5) \vee (X_1 \cdot \bar{X}_2 \cdot X_3 \cdot X_5) \vee (X_1 \cdot X_3 \cdot \bar{X}_5) \quad (1)$$

Equation (1) indicates the necessary selection criteria for management faculty. Here, the logical connectives '.', '$\vee$', and '_' are 'and', 'or', and 'not', respectively. Here, X_1, X_2 are atomic propositions which can be either true 'T' or false 'F'. X_1 or its negation $\bar{X}_1$ is called literal. When these literals are connected by '.', for example $X_1 \cdot X_3 \cdot \bar{X}_5$, they form a clause. These clauses are connected by '$\vee$' disjunctions to give a disjunctive form (DF).

It is important to obtain a logically equivalent form to (1) which contains a minimum number of literals. It is highly desirable to obtain a minimum (number of criteria) expression from the point of view of time or cost. It is achieved by applying the Quine's procedure [11, 12] which consists of applying consensus and subsumption alternatively to obtain the complete disjunction of prime implicants. For example, $X_1 \cdot \bar{X}_2 \cdot X_3 \cdot X_5$ subsumes $X_1 \cdot X_3 \cdot X_5$; hence, the subsuming clause can be dropped. Similarly, $X_1 \cdot X_3 \cdot X_5$ is consensus of $X_1 \cdot \bar{X}_2 \cdot X_3 \cdot X_5$ and $X_1 \cdot X_2 \cdot X_5$ and the resultant clause can be appended in the problem.

After successively applying these procedures, we get

$$(X_1 \cdot X_2 \cdot X_5) \vee (X_1 \cdot \bar{X}_2 \cdot X_4 \cdot \bar{X}_5) \vee (X_1 \cdot X_3) \quad (2)$$

This is the logically equivalent form to (1) with least number of clauses. Using truth table, it can be shown that none of the prime implicants are redundant.

Table 2 is represented in DNF as follows:

$$(X_1 \cdot X_2 \cdot \bar{X}_4) \vee (X_1 \cdot X_4) \vee (X_1 \cdot X_2 \cdot X_4 \cdot \bar{X}_5) \vee (X_3 \cdot \bar{X}_5) \vee (X_1 \cdot X_4 \cdot X_5) \quad (3)$$

Table 1 Criteria for faculty selection process

Selection criteria	S_1	S_2	S_3	S_4	S_5
Expert 1	T	T	–	–	T
Expert 2	T	F	–	T	F
Expert 3	T	F	T	–	T
Expert 4	T	–	T	–	F

Table 2 Non-selection criteria for management faculty

Selection criteria	S_1	S_2	S_3	S_4	S_5
Expert 1	T	T	–	F	–
Expert 2	T	–	–	T	–
Expert 3	T	T	–	T	F
Expert 4	–	–	T	–	F
Expert 5	T	–	–	T	T

Table 3 New set of criteria for faculty selection process

Selection criteria	S_1	S_2	S_3	S_4	S_5
Expert 1	T	T	–	–	T
Expert 2	T	F	–	T	F
Expert 3	T	–	T	–	–

Now this statement would be false if job applicant is selected for the interview. Here, we use De Morgan's theorem to negate the statement resulting in CNF form of the same statement.

$$(\bar{X}_1 \vee \bar{X}_2 \vee X_4) \cdot (\bar{X}_1 \vee \bar{X}_4) \cdot (\bar{X}_1 \vee \bar{X}_2 \vee \bar{X}_4 \vee X_5) \cdot (\bar{X}_3 \vee X_5) \cdot (\bar{X}_1 \vee \bar{X}_4 \vee \bar{X}_5) \quad (4)$$

This statement is true if the job applicant is selected for the job interview. This statement must be consistent with (1).

After simplification, the minimal logical expression in DNF is given by

$$(X_1 \cdot X_2) \vee (X_1 \cdot X_4) \vee (X_3 \cdot \bar{X}_5) \vee (X_2 \cdot X_3) \quad (5)$$

The conjunction of (2) with (5) must be consistent; i.e., it must give satisfiable set of values.

Equation (2) can be used as new (smaller) set of selection criteria for job interview. This is given in Table 3.

Table 3 gives smaller set of selection criteria for management faculty.

3 Final Selection Procedure

Once we have determined the minimum criteria for faculty selection, we will now use two different approaches namely data analysis using logic and mathematical technique of analytical hierarchy process (AHP) to arrive at final decision. Figure 1 gives a comprehensive framework of faculty selection problem using a hybrid model of logic and AHP.

3.1 Logical Analysis of Data

We take a small example which is similar to a real-life situation where a simple scoring method [3] is used to select a suitable candidate.

Let us say there are three candidates who have been shortlisted based on the minimum criteria set. Now based on the three experts' opinion, one candidate will be selected. In Table 4, there are three candidates A, B, and C with the available information.

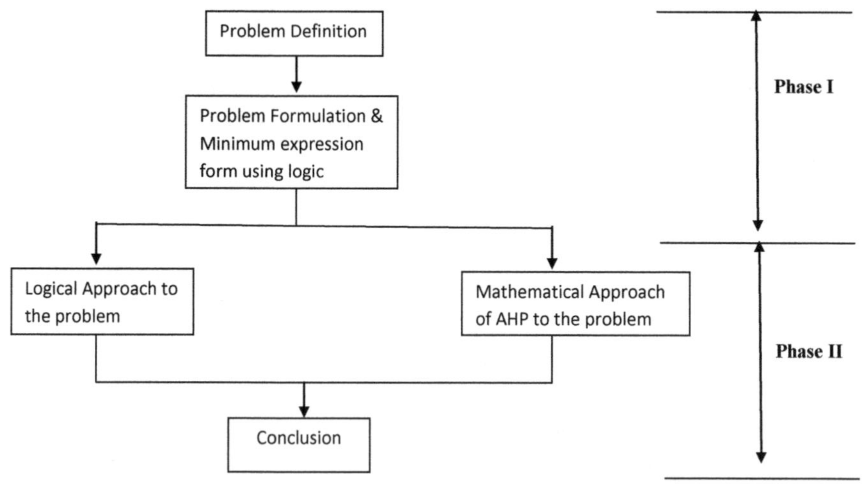

Fig. 1 A comprehensive framework for analysis of human resource problem

Table 4 Candidates' job profile selection criteria

Selection criteria					
Candidate	S_1	S_2	S_3	S_4	S_5
A	1	0	1	1	1
B	1	1	0	0	1
C	1	0	0	1	0

Now suppose there are three experts who develop their own selection rule. Let expert P say that four criteria out of five are important for selection process. Hence, expert P considers a candidate suitable for job if the vector $X = (X_1, X_2, X_3, X_4, X_5)$ is

$$X_1 + X_2 + X_3 + X_4 + X_5 \geq 4 \tag{6}$$

According to this selection rule, candidate A is selected and B and C are rejected. The expert P selection rule can be modeled as disjunction of five basic rules namely:

$$\begin{aligned} P(x) = &(X_1 \cdot X_2 \cdot X_3 \cdot X_4) \vee (X_1 \cdot X_2 \cdot X_4 \cdot X_5) \vee (X_1 \cdot X_3 \cdot X_4 \cdot X_5) \\ &\vee (X_1 \cdot X_2 \cdot X_3 \cdot X_5) \vee (X_2 \cdot X_3 \cdot X_4 \cdot X_5) \end{aligned} \tag{7}$$

Expert Q has a different opinion: He considers criteria X_3 as thrice more important than other ones and regards X_2 as irrelevant. Hence, a candidate is selected if

$$X_1 + 3X_3 + X_4 + X_5 \geq 4 \tag{8}$$

Here, candidate A is selected and B and C are rejected according to expert Q. A disjunction of three rules can be used to describe the selection criteria given by expert Q.

$$Q(x) = (X_1 \cdot X_3 \cdot X_4) \vee (X_1 \cdot X_3 \cdot X_5) \vee (X_3 \cdot X_4 \cdot X_5) \tag{9}$$

Expert R believes criteria X_3 as twice more important than other ones and X_2 and X_5 as irrelevant.

$$X_1 + 2X_3 + X_4 \geq 3 \tag{10}$$

Similarly, expert R rule can be described by disjunction of two basic rules. His rule is consistent with the other two experts' view and given by:

$$R(x) = (X_1 \cdot X_3) \vee (X_3 \cdot X_4) \tag{11}$$

It can be seen easily that all the three experts have given a perfect selection rule; i.e., in this small data set, all the three candidates are correctly judged by these rules. Thus, the classification rules given by the three experts are well grounded in the given data sets.

However, the selection rule adopted by the three experts is not identical in the sense that it may give contradictory results in some possible cases. To explain this point, let us say a candidate D has the following characteristics: X_D = 10110.

Expert P will reject the candidate, thus leading to taking a candidate who is not suitable for the job, and it may lead to inefficiency and later realizing that it was a wrong decision. However, according to experts Q and R, the candidate D is selected. Later, the management may wish to know the reason why expert P rejected candidate D. Expert P found the absence of first rule, i.e., simultaneous occurrence of criteria 1, 2, 3, and 4.

Another situation may arise shown in candidate E with characteristics: X_E = 11011. According to expert P, candidate E is selected for the job. Later, he may

Table 5 Classification results given by the three classifiers: experts P, Q, and R

Selection criteria						Classification by		
Candidate	S_1	S_2	S_3	S_4	S_5	Expert P	Expert Q	Expert R
A	1	0	1	1	1	1	1	1
B	1	1	0	0	1	0	0	0
C	1	0	0	1	0	0	0	0
D	1	0	1	1	0	0	1	1
E	1	1	0	1	1	1	0	0

find that the candidate is not doing his work efficiently and is not suitable for the job. Experts Q and R reject the candidate. Then, the question comes on what basis expert P justifies his decision? The only reason why he selected candidate E is found in his second rule, i.e., the co-occurrence of 1, 2, 4, and 5. However, there is no supporting evidence in the initial data to justify this rule, since this combination of characteristics was never observed in the data set (Table 5).

3.2 Analytical Hierarchy Process

Let us solve the same problem using a mathematical approach of analytical hierarchy process (AHP) [9]. AHP is a decision-making framework used for multi-criteria decision analysis. The basic principle underlying AHP involves the pair-wise comparison of various alternatives of which the best is chosen. It helps in capturing both subjective and objective aspects of a decision. In this section, the AHP technique will be discussed to show how it helps in management faculty selection problem.

The pair-wise comparison matrix (PWCM) A for 'm' criteria is given below:

$$A = \begin{bmatrix} a_{11} & a_{12} & \cdots & a_{1m} \\ a_{21} & a_{22} & \cdots & a_{2m} \\ \vdots & & & \\ a_{m1} & a_{m2} & \cdots & a_{mm} \end{bmatrix}$$

where aij indicates how much more important the ith selection criteria is than the jth one for a particular candidate for finding the best candidate.

For all i and j, it is necessary that $a_{ii} = 1$ $a_{ij} = 1/a_{ji}$. The pair-wise comparison matrix is constructed based on the nine-point scale. The possible assignment value of a_{ij} with the corresponding interpretation is shown in Table 6.

Table 6 Assessment of a_{ij}

Value of a_{ij}	Interpretation
1	Criteria i and j are of equal importance
3	Criteria i is weakly more important than criteria j
5	Criteria i is strongly more important than criteria j
7	Criteria i is very strongly more important than criteria j
9	Criteria i is absolutely more important than criteria j
2, 4, 6, 8	Intermediate values

Here, the objective is to select a candidate based on five criteria: age, qualification, experience, publication, and FDP and there are three candidates: A, B, and C (Table 7, 8, 9).

The composite weight of the three candidates is given in Table 10.

Since the composite weight of candidate A is higher than others, hence the best alternative is to select candidate A.

Table 7 Pair-wise comparison matrix of factors (selection criteria) for expert P

A=	X_1	X_2	X_3	X_4	X_5	GM	PV
X_1	1	1/3	1/5	1/3	1/2	0.407	0.068
X_2	3	1	3/2	2	3	1.933	0.324
X_3	5	2/3	1	2	4	1.928	0.323
X_4	3	1/2	1/2	1	2	1.084	0.182
X_5	2	1/3	1/4	1/2	1	0.608	0.102
						$\sum$ GM=5.96	$\sum$ PV=1.000

GM Geometric Mean *PWCM* Pair-wise Comparison Matrix

Consistency Ratio CR is found to be less than 0.1.

Table 8 Priority vector of factors (selection criteria) for expert P, expert Q, and expert R

	X_1	X_2	X_3	X_4	X_5	
Expert P	0.068	0.324	0.323	0.18	0.102	
Expert Q	0.088	0.218	0.372	0.197	0.126	
Expert R	0.118	0.299	0.307	0.184	0.092	
	GM_1	GM_2	GM_3	GM_4	GM_5	Grand Total
	(0.089)	(0.276)	(0.333)	(0.187)	(0.106)	(0.991)

$X_1 = \frac{GM_1}{\text{Grand Total}} = 0.089$ $X_2 = 0.279$ $X_3 = 0.336$ $X_4 = 0.189$ $X_5 = 0.107$

Table 9 PWCM of alternatives (candidates) w.r.t. each factor (criteria) for three experts

	A				B				C			
	Exp. P	Exp. Q	Exp. R	PV	Exp. P	Exp. Q	Exp. R	PV	Exp. P	Exp. Q	Exp. R	PV
X_1	0.424	0.353	0.39	0.212	0.344	0.294	0.368	0.225	0.232	0.353	0.238	0.189
X_2	0.471	0.464	0.406	0.244	0.239	0.251	0.25	0.166	0.29	0.285	0.344	0.176
X_3	0.417	0.417	0.454	0.235	0.378	0.378	0.33	0.243	0.205	0.205	0.216	0.12
X_4	0.353	0.351	0.393	0.199	0.218	0.227	0.205	0.145	0.429	0.422	0.402	0.241
X_5	0.222	0.197	0.181	0.109	0.327	0.319	0.34	0.221	0.451	0.484	0.48	0.272

Table 10 Individual composite score

Candidate	Composite weight
A	0.216
B	0.199
C	0.182

4 Conclusion

In this paper, we have considered to solve a faculty selection problem through application of two methods. One method is qualitative, i.e., the Boolean logic, and the other is quantitative, i.e., the analytical hierarchy process (AHP). It is seen that the results obtained by the two approaches are identical. As the candidate 'A' is selected by the two different set of experts through two different approaches and only one post is to be filled up, we conclude that candidate 'A' is therefore selected. The said procedure in selection of candidate has thus helped the management in taking a correct decision.

It may also be noted that the above combined method can be made applicable for evaluation and problem solving in different areas of management by suitably altering and modifying for its application.

References

1. Hammer, P.L.: The logic of cause-effect relationship. In: Lecture at International Conference on Multi Attribute Decision Making via Operations Research based expert systems, Passau, Germany (1986)
2. Boros, E., Hammer, P.L., Ibarki, T., Kogan, A.: Logical analysis of numerical data. Math. Program. **79**, 163–190 (2000)
3. Boros, E., Crama, Y., Hammer, P.L., Ibaraki, T., Kogan, A., Makino, K.: Logical analysis of data: classification with justification. DIMACS Technical report 2009-02, February (2009)
4. Boros, E., Hammer, P.L., Ibarki, T., Kogan, A., Mayoraz, E., Muchnik, I.: An implementation of logical analysis of data. IEEE Trans. Knowl. Data Eng. **12**(2), 292–306 (2000)
5. Hammer, A., Hammer, P.L., Muchnik, I.: Logical analysis of Chineese Labor productivity patterns. Ann. Oper. Res. **87**, 165–176 (1999)
6. Williams, H.P.: Logic and Integer Programming. International Series in Operations Research and Management Science (2009)
7. Crama, Y., Hammer, P.L., Ibaraki, T.: Cause-effect relationships and partially defined boolean functions. Ann. Oper. Res. **16**, 299–326 (1988)
8. Anthony, M., Hammer, P.L.: A boolean measure of similarity. Discrete Appl. Math **154**, 2242–2246 (2006)
9. Saaty, T.L.: How to make a decision: the analytic hierarchy process. Eur. J. Oper. Res. **48**, 9–26 (1990)
10. Chanda, A., Sarkar, B., Mukherjee, R.N.: Logical deduction for an integer programming problem and integer programming (IP) formulation for a logical problem. Int. J. Oper. Res. Nepal **4**, 33–40 (2015)
11. Quine, W.V.: The problem of simplifying truth functions. Am. Math. Monthly **59**, 521–531 (1952)

12. Quine, W.V.: A way to simplify truth functions. Am. Math. Mon **62**, 627–631 (1955)
13. Robinson, J.A.: A machine oriented logic based on the resolution principles. J. ACM **12**, 23–41 (1965)
14. Robinson, J.A.: A generalized resolution principle. Mach. Intell. **3**, 77–93 (1968)
15. Taha, H.A.: Operations Research: An Introduction. Pearson Education (2007)

A Multi-item Inventory Model with Fuzzy Rough Coefficients via Fuzzy Rough Expectation

Totan Garai, Dipankar Chakraborty and Tapan Kumar Roy

Abstract In this paper, we concentrated on developing a multi-item inventory model under fuzzy rough environment. Here, demand and holding cost rates are assumed as the functions of stock level. Fuzzy rough expectation method is used to transform the present fuzzy rough inventory model into its equivalent crisp model. A numerical example is provided to illustrate the proposed model. To show the validity of the proposed models, few sensitivity analyses are also presented under the major parameter, and the results are illustrated numerically and graphically.

Keywords Multi-item inventory · Trapezoidal fuzzy rough variable · Fuzzy rough expectation

1 Introduction

Uncertainty is common to all real-life problems, for example fuzziness, roughness and randomness. Since Zadeh [1] introduced the fuzzy set in 1965, fuzzy set theory has been well developed and employed to an extensive variety of real problems [2, 3]. Fuzziness and roughness play a significant role in among types of uncertain problems. Fuzzy rough variable, which is a combination of rough variable and fuzzy variable, can characterize both roughness and fuzziness in real problems. The concept of fuzzy rough sets was first introduced by Dubois and Prade [4] and then investigated by Morsi and Yakout [5], who defined the upper and lower approximations of the fuzzy set with respect to a fuzzy min–similarity relationship. Additionally, some researchers [5, 6] generalized above definitions of the fuzzy rough set

T. Garai (✉) · T. K. Roy
Department of Mathematics, Indian Institute of Engineering Science and Technology, Shibpur, Howrah 711103, West Bengal, India
e-mail: garaitotan@gmail.com

D. Chakraborty
Department of Mathematics, Heritage Institute of Technology, Anandapur, Kolkata 700107, West Bengal, India
e-mail: dipankar1920@gmail.com

S. Kar et al. (eds.), *Operations Research and Optimization*, Springer Proceedings in Mathematics & Statistics 225, https://doi.org/10.1007/978-981-10-7814-9_26

to a more general case. Liu [7] presented the some definitions and discussed some valuable properties of the fuzzy rough variable. At present, using these approaches, some researches [8–12] modelled different practical problems where both fuzziness and roughness exist simultaneously.

In numerous cases, it is established that the parameters of some inventory problems are considered fuzzy and rough uncertainties. For example, production cost, set-up cost, holding cost, repairing cost relies on various factors such as inflation, labour travail wages, wear and tear cost, bank interest which are uncertain in fuzzy rough sense. To be more specific, set-up cost depends on the total quantity to be produced in a scheduling period, and the inventory holding cost for an item is supposed to be dependent on the amount of storage. Moreover, because in the inventory, the total quantity to be produced in a scheduling period and the amount storage may be uncertain, and range on an interval, uncertainties may in fuzzy environment. In these circumstances, fuzzy rough can be used for the formulation of inventory problems. In the literature, very few researchers [9, 13–15] developed and solved inventory or production-inventory problems with the fuzzy rough environment.

For the inventory problem, the classical inventory decision-making models have deliberated a single item. However, single item inventories rarely occur, whereas multi-item inventories are common in real-life circumstances. Many researchers (cf. Balkhi and Foul [16], Hartly [17], Lee and Yao [18], Taleizadeh et al. [19]) investigated the multi-item inventory models under resource constraints. Dutta et al. [20] studied an inventory model for single-period products with reordering opportunities under fuzzy demand. An EOQ model with deteriorating items under inflation when supplier credits linked to order quantity was proposed by Chang [21]. Shi et al. [22] considered a probability maximization model based on rough approximation and its application to the inventory problem. Recently, Wang [23], Taleizadeh et al. [24], Kazemi et al. [25], Bazan et al. [26] and Das et al. [27] have investigated the inventory models with the imprecise fuzzy environment. However, no attempt has been made that includes all replenishment cost, set-up cost, purchasing cost and shortage cost, which are fuzzy rough variables, are considered within the multi-item inventory model with demand is a power function and holding cost is a nonlinear function of the stock level. Therefore, we have developed and solved a multi-item inventory model with fuzzy rough coefficients via expected value approaches.

The rest of the paper is organized as follows: In Sect. 2, we present some basic knowledge of fuzzy rough theory and optimization theory. Section 3 provides the notations and assumptions which are used throughout the paper. In Sect. 4, a multi-item inventory model has been developed in the fuzzy rough environment and discusses its solution method. Numerical example to illustrate the model is provided in Sect. 5. In Sect. 6, the result of the change of different parameters is discussed graphically. Finally, the conclusion and scope of the future work plan have been made in Sect. 7.

2 Preliminaries and Deductions

Definition 1 Let Λ be a non-empty [9] set, $\mathscr{A}$ a σ algebra of subsets of Λ, and Δ an element in $\mathscr{A}$ and π a trust measure, Then, $(\Lambda, \Delta, \mathscr{A}, \pi)$ is called a rough space.

Definition 2 Let $(\Lambda, \Delta, \mathscr{A}, \pi)$ be rough space. A rough variable ζ is a measurable function [9] from the rough space $(\Lambda, \Delta, \mathscr{A}, \pi)$ to the set of real numbers $\mathbb{R}$. That is, for every Borel set $\mathbb{B}$ of $\mathbb{R}$, we have

$$\{\eta \in \Lambda : \zeta(\eta) \in \mathbb{B}\} \in \mathscr{A}$$

The upper $(\overline{\zeta})$ and lower $(\underline{\zeta})$ approximations of the rough variable ζ are defined as follows:

$$\overline{\zeta} = \{\zeta(\eta) : \eta \in \Lambda\} \qquad \underline{\zeta} = \{\zeta(\eta) : \eta \in \Delta\}$$

Definition 3 Let $(\Lambda, \Delta, \mathscr{A}, \pi)$ be a rough space. The trust measure [9] of the event A is defined by

$$Tr\{A\} = \frac{1}{2}(\underline{Tr}\{A\} + \overline{Tr}\{A\})$$

where the upper trust measure $\overline{Tr}\{A\} = \dfrac{\pi\{A\}}{\pi\{\Lambda\}}$ and lower trust measure $\underline{Tr}\{A\} = \dfrac{\pi\{A \cap \Delta\}}{\pi\{\Delta\}}$. When the enough information about the measure π is not given, for this case, the measure π may be treated as the Lebesgue measure.

Example 1 Let $\zeta = ([a_1, a_2]\,[b_1, b_2])$ be a rough variable with $b_1 \le a_1 \le a_2 \le b_2$ representing the identity function $\zeta(\eta) = \eta$ from the rough space $(\Lambda, \Delta, \mathscr{A}, \pi)$ to the set of real numbers $\mathbb{R}$, where $\Lambda = \{\eta : b_1 \le \eta \le b_2\}$, $\Delta = \{\eta : a_1 \le \eta \le a_2\}$, $\mathscr{A}$ is the σ-algebra on Λ, and π is the Lebesgue measure.

According to the Definitions 2 and 3, we can obtain the trust measure of the event $\{\zeta \ge x\}$(cf. Fig. 1) and and $\{\zeta \le x\}$(cf. Fig. 1) as follows:

$$Tr(\zeta \ge x) = \begin{cases} 0 & \text{if } b_2 \le x \\ \dfrac{b_2 - x}{2(b_2 - b_1)} & \text{if } a_2 \le x \le b_2 \\ \dfrac{1}{2}\left(\dfrac{b_2 - x}{b_2 - b_1} + \dfrac{a_2 - x}{a_2 - a_1}\right) & \text{if } a_1 \le x \le a_2 \\ \dfrac{1}{2}\left(\dfrac{b_2 - x}{b_2 - b_1} + 1\right) & \text{if } b_1 \le x \le a_1 \\ 1 & \text{if } x \le b_1 \end{cases} \qquad (1)$$

and

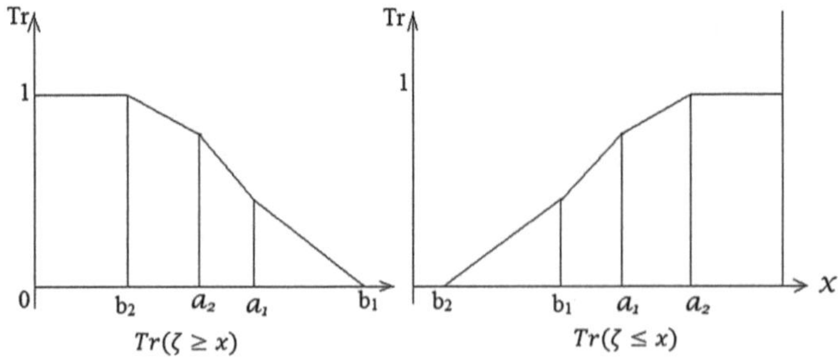

Fig. 1 Function curve of rough variable ζ

$$Tr(\zeta \le x) = \begin{cases} 0 & \text{if } r \le b_1 \\ \dfrac{x - b_1}{2(b_2 - b_1)} & \text{if } b_1 \le x \le a_1 \\ \dfrac{1}{2}\left(\dfrac{x - b_1}{b_2 - b_1} + \dfrac{x - a_1}{a_2 - a_1}\right) & \text{if } a_1 \le x \le a_2 \\ \dfrac{1}{2}\left(\dfrac{x - b_1}{b_2 - b_1} + 1\right) & \text{if } a_2 \le x \le b_2 \\ 1 & \text{if } b_2 \le x \end{cases} \tag{2}$$

Example 2 Let's consider the trapezoidal fuzzy variable $\zeta = (r_1, r_2, r_3, r_4)$ with the following membership function

$$\mu_\zeta(x) = \begin{cases} \dfrac{x - r_1}{r_2 - r_1} & \text{if } r_1 \le x < r_2 \\ 1 & \text{if } r_2 \le x \le r_3 \\ \dfrac{r_4 - x}{r_4 - r_3} & \text{if } r_3 < x \le r_4 \\ 0 & \text{otherwise} \end{cases} \tag{3}$$

where $r_i \vdash ([a_1, a_2]\,[b_1, b_2])$ for $i = 1, 2, 3, 4$. Then, ζ is called trapezoidal fuzzy rough variable.

Example 3 Let $\zeta = (\rho - 1, \rho - 2, \rho + 2.5, \rho + 3)$ with $\rho = ([3, 5]\,[2, 7])$, where the quadruple (r_1, r_2, r_3, r_4) with $r_1 \le r_2 \le r_3 \le r_4$ denotes a trapezoidal fuzzy variable and ρ is a rough variable, then ζ is a fuzzy rough variable.

Definition 4 Let $\widehat{\zeta}$ be a rough variable. The expected [8] value of the rough variable $\widehat{\zeta}$ is denoted by $E_r[\widehat{\zeta}]$ and defined by

$$E_r[\widehat{\zeta}] = \int_0^\infty Tr\{\widehat{\zeta} \ge x\}dx - \int_{-\infty}^0 Tr\{\widehat{\zeta} \le x\}dx \tag{4}$$

provided that at least one of the two integrals is finite.

Example 4 Let $\widehat{\zeta} = ([a_1, a_2]\,[b_1, b_2])$ be a rough variable. We then have

$$E_r[\widehat{\zeta}] = \frac{1}{4}(a_1 + a_2 + b_1 + b_2)$$

Particularly, when $[a_1, a_2] = [b_1, b_2]$, the rough variable $\widehat{\zeta}$ degenerates to an interval number $[a_1, a_2]$. Then we have

$$E_r[\widehat{\zeta}] = \frac{1}{2}(a_1 + a_2)$$

Theorem 1 *Let $\widehat{\zeta}_1$ and $\widehat{\zeta}_2$ be two rough variables [28] with finite expected values. Then, for any two real numbers a and b, we have*

$$E_r[a\widehat{\zeta}_1 + b\widehat{\zeta}_2] = aE_r[\widehat{\zeta}_1] + bE_r[\widehat{\zeta}_2]$$

Theorem 2 *Let $\widehat{\tilde{\zeta}}$ be a trapezoidal fuzzy [28] rough variable $\widehat{\tilde{\zeta}} = (\widehat{r}_1, \widehat{r}_2, \widehat{r}_3, \widehat{r}_4)$, where $\widehat{r}_1, \widehat{r}_2, \widehat{r}_3, \widehat{r}_4$ are rough variables defined on $(\Lambda, \Delta, \mathscr{A}, \pi)$, and*

$$\begin{aligned}
\widehat{r}_1 &= ([m_2, m_3]\,[m_1, m_4]), 0 < m_1 \le m_2 < m_3 \le m_4\\
\widehat{r}_2 &= ([n_2, n_3]\,[n_1, n_4]), 0 < n_1 \le n_2 < n_3 \le n_4\\
\widehat{r}_3 &= ([p_2, p_3]\,[p_1, p_4]), 0 < p_1 \le p_2 < p_3 \le p_4\\
\widehat{r}_4 &= ([q_2, q_3]\,[q_1, q_4]), 0 < q_1 \le q_2 < q_3 \le q_4
\end{aligned}$$

Then, the expected value of $\widehat{\tilde{\zeta}}$ is

$$E_r[\widehat{\tilde{\zeta}}] = \frac{1}{16}\sum_{i=1}^{4}(m_i + n_i + p_i + q_i)$$

2.1 Single-Objective Fuzzy Rough Expected Value Model

Let us consider the following single-objective model with trapezoidal fuzzy rough coefficients:

$$\begin{aligned}
&\text{Max} \sum_{j=1}^{n} \widehat{\tilde{c}}_{ij} x_j, \; i = 1, 2, \ldots, m\\
&\text{subject to} \sum_{j=1}^{n} \widehat{\tilde{a}}_{jr} x_j \le \widehat{\tilde{b}}_r; \quad r = 1, 2, \ldots, p \qquad (5)\\
&x_j \ge 0;
\end{aligned}$$

where $\widehat{\tilde{c}}, \widehat{\tilde{a}}, \widehat{\tilde{b}}$ are the trapezoidal fuzzy rough variables.

In order to solve the uncertain model with fuzzy rough coefficients, we must transform into a deterministic model. The technique of computing the expected value is a proficient method and is easily perceived. Consequently, the above problem is equivalent to

$$\begin{aligned} &\text{Max } E_r\left[\sum_{j=1}^{n} \widehat{\tilde{c}}_{ij} x_j,\ i = 1, 2, \ldots, m\right] \\ &\text{subject to } E_r\left[\sum_{j=1}^{n} \widehat{\tilde{a}}_{jr} x_j\right] \le E_r\left[\widehat{\tilde{b}}_r\right];\quad r = 1, 2, \ldots, p \\ &x_j \ge 0; \end{aligned} \tag{6}$$

where E_r denote the fuzzy rough expected value operator.

Theorem 3 *If the trapezoidal fuzzy rough [28] variables* $\widehat{\tilde{c}}_{ij}$ *are defined as* $\widehat{\tilde{c}}_{ij}(\eta) = (\widehat{c}_{ij1}, \widehat{c}_{ij2}, \widehat{c}_{ij3}, \widehat{c}_{ij4})$, *with* $\widehat{\tilde{c}}_{ijt} \vdash ([c_{ijt1}, c_{ijt2}]\,[c_{ijt3}, c_{ijt4}])$, *for* $i = 1, 2, \ldots, m$, $j = 1, 2, \ldots, n$, $t = 1, 2, 3, 4$, $x = (x_1, x_2, \ldots, x_m)$, $0 \le c_{ijt3} \le c_{ijt1} \le c_{ijt2} \le c_{ijt4}$. *Then,* $E_r[\widehat{\tilde{c}}_1^T x]$, $E_r[\widehat{\tilde{c}}_2^T x], \ldots, E_r[\widehat{\tilde{c}}_m^T x]$ *is equivalent to*

$$\frac{1}{16}\sum_{j=1}^{n}\sum_{t=1}^{4}\sum_{k=1}^{4} c_{1jtk} x_j,\ \frac{1}{16}\sum_{j=1}^{n}\sum_{t=1}^{4}\sum_{k=1}^{4} c_{2jtk} x_j, \ldots, \frac{1}{16}\sum_{j=1}^{n}\sum_{t=1}^{4}\sum_{k=1}^{4} c_{mjtk} x_j$$

Theorem 4 *If the trapezoidal fuzzy rough [28] variables* $\widehat{\tilde{a}}_{rj}$, $\widehat{\tilde{b}}_r$ *are defined as follows:*

$\widehat{\tilde{a}}_{rj}(\eta) = (\widehat{a}_{rj1}, \widehat{a}_{rj2}, \widehat{a}_{rj3}, \widehat{a}_{rj4})$, *with* $\widehat{a}_{rjt} \vdash ([a_{rjt1}, a_{rjt2}]\,[a_{rjt3}, a_{rjt4}])$,
$\widehat{\tilde{b}}_r(\eta) = (\widehat{b}_{r1}, \widehat{b}_{r2}, \widehat{b}_{r3}, \widehat{b}_{r4})$, *with* $\widehat{b}_{rt} \vdash ([b_{rt1}, b_{rt2}]\,[b_{rt3}, b_{rt4}])$,

for $j = 1, 2, \ldots, n$, $r = 1, 2, \ldots p$, $t = 1, 2, 3, 4$, $0 \le a_{rjt3} \le a_{rjt1} \le a_{rjt2} \le a_{rjt4}, 0 \le b_{rt3} \le b_{rt1} \le b_{rt2} \le b_{rt4}$. *Then* $E_r[\widehat{\tilde{a}}_{rj}^T x] \le E_r[\widehat{\tilde{b}}_{rj}]$, $r = 1, 2, \ldots p$ *is equivalent to*

$$\frac{1}{16}\sum_{j=1}^{n}\sum_{t=1}^{4}\sum_{k=1}^{4} a_{rjtk} x_j \le \sum_{t=1}^{4}\sum_{k=1}^{4} b_{rtk},\quad r = 1, 2, \ldots, p.$$

3 Notation and Assumptions

To develop the mathematical model of inventory replenishment intention, the notation affected in this paper is as below:

3.1 Notation

(i) Q_i = the ordering quantity per cycle for ith item
(ii) A_i = the replenishment cost per order of ith item
(iii) c_i = purchasing cost of each product of the ith item
(iv) c_{1i} = shortage cost per unit time for ith item
(v) c_{2i} =set-up cost for ith item
(vi) c_{3i} = the cost of lost sales per unit of ith item
(vii) S_i = shortage level for the ith item
(viii) $D_i(t)$ = demand rate of ith item, which is a function of inventory level at time t
(ix) t_{1i} = the time at which the inventory level reach zero for ith item (a decision variable)
(x) t_{2i} = the length of period during which are allowed for ith item
(xi) T_i = the length of the inventory cycle; hence, $T_i = t_{1i} + t_{2i}$ (a decision variable)
(xii) $H_i[q_i(t)]$ = holding cost for the ith item, which is function of inventory level at time t
(xiv) h_i = scaling constant for holding cost
(xv) w_i = storage space per unit quantity for the ith item
(xvi) B = budget available for replenishment
(xvii) F = available storage space in the inventory system

In addition, the following assumptions are instated.

3.2 Assumption

(i) The replenishment rate is infinite and the lead time zero.
(ii) The time horizon of the inventory system is infinite.
(iii) Shortage is allowed; during the stock-out period, a fraction $\frac{1}{1+\delta_i x}$ of the demand will be back order, and the remaining fraction $(1 - \frac{1}{1+\delta_i x})$ will be lost, where x is the waiting time up to the next replenishment and δ_i is a positive constant.
(iv) The demand rate function $D_i(t)$ is deterministic and a power function of instantaneous stock level $q_i(t)$ at time t; that is:

$$D_i(t) = D_i[q_i(t)] = \begin{cases} \lambda_i[q_i(t)]^\beta, & \text{if } 0 \le t \le t_{1i},\ q_i(t) \ge 0; \\ \lambda_i, & \text{if } t_{1i} < t \le T_i,\ q_i(t) > 0; \end{cases}$$

where $\lambda_i > 0$ and $0 < \beta < 1$.

(v) The holding cost is nonlinear function of the stock level $q_i(t)$ at time t and is given as $H_i(t) = H_i[q_i(t)] = h_i[q_i(t)]^\gamma$, where $h_i > 0$ and $0 < \gamma < 1$.

4 Model Formulation

Using the above assumption, the inventory level follows the pattern depicted in Fig. 1. The depletion of the inventory happens due to the effect of demand in $[0, t_{1i}]$ and the demand backlogged in $[t_{1i}, T_i]$. Now, in the interval $[0, T_i]$, the inventory level gradually decreases to meet the demands. By this process, the inventory level reaches zero at $t = t_{1i}$, and shortages are allowed to occur in $[t_{1i}, T_i]$. Hence, the variety of inventory level $q_i(t)$ with respect to time t can be described by the following differential equations:

$$\frac{dq_{1i}(t)}{dt} = -\lambda_i[q_i(t)]^\beta, \quad 0 \le t \le t_{1i} \tag{7}$$

with conditions $q_{1i}(0) = R_i(= Q_i - S_i)$ and $q_{1i}(t_{1i}) = 0$.

The solution of Eq. (1) is (Fig. 2).

$$q_{1i}(t) = \left[R_i^{1-\beta} - (1-\beta)\lambda_i t\right]^{\dfrac{1}{1-\beta}} \tag{8}$$

By the interval $[t_{1i}, T_i]$, the inventory level only build on demand, and some demand is lost while a fraction $\dfrac{1}{1+\delta_i(T_i - t)}$ of the demand is backlogged, where $t \in [t_{1i}, T_i]$. The inventory level is controlled by the following differential equation:

$$\frac{dq_{2i}}{dt} = -\frac{\lambda_i}{1+\delta_i(T_i - t)}, \quad t_{1i} \le t \le T_i \tag{9}$$

with conditions $q_{2i}(T_i) = -S_i$ and $q_{2i}(t_{1i}) = 0$.

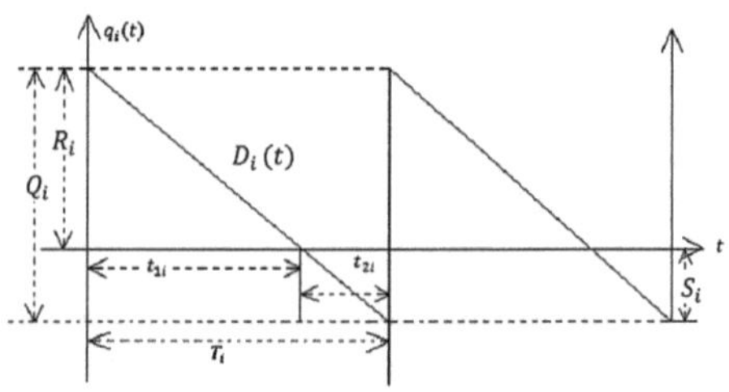

Fig. 2 Inventory model for the ith item

The solution of Eq. (3) is

$$q_{2i}(t) = -\frac{\lambda_i}{\delta_i}\left\{\ln\left[1+\delta_i(T_i - t_{1i})\right] - \ln\left[1+\delta_i(T_i - t)\right]\right\} \tag{10}$$

Considering the continuity of $q_{1i}(t)$ and $q_{2i}(t)$ at point $t = t_{1i}$, i.e., $q_{1i}(t_{1i}) = q_{2i}(t_{2i}) = 0$, we have:

$$R_i = \left[(1-\beta)\lambda_i t_{1i}\right]^{\frac{1}{1-\beta}} \quad S_i = \frac{\lambda_i}{\delta_i}\ln\left\{1+\delta_i(T_i - t_{1i})\right\} \tag{11}$$

Therefore, the ordering quantity over the replenishment cycle for the ith item can be determined as

$$Q_i = q_{1i}(0) - q_{2i}(T_i) = \left[(1-\beta)\lambda_i t_{1i}\right]^{\frac{1}{1-\beta}} + \frac{\lambda_i}{\delta_i}\ln\left\{1+\delta_i(T_i - t_{1i})\right\} \tag{12}$$

Based on Eqs. (8), (10) and (12), the total inventory cost per cycle consists of the following elements.

The ordering cost per cycle for ith item is A_i.

The inventory holding cost per cycle for the ith item is given by

$$\begin{aligned} & h_i \int_0^{t_{1i}} [q_{1i}(t)]^{\gamma} dt \\ &= h_i \int_0^{t_{1i}} \left[R_i^{1-\beta} - (1-\beta)\lambda_i t\right]^{\frac{\gamma}{1-\beta}} dt \\ &= \frac{h_i}{(\gamma+1-\beta)\lambda_i} R_i^{\gamma+1-\beta} \\ &= \frac{h_i}{(\gamma+1-\beta)\lambda_i}\left[(1-\beta)\lambda_i t_{1i}\right]^{\frac{\gamma+1-\beta}{1-\beta}} \end{aligned} \tag{13}$$

Purchase cost of ith item per cycle is

$$\begin{aligned} & c_i Q_i \\ &= c_i\left[(1-\beta)\lambda_i t_{1i}\right]^{\frac{1}{1-\beta}} + \frac{c_i\lambda_i}{\delta_i}\ln\left\{1+\delta_i(T_i - t_{1i})\right\} \end{aligned} \tag{14}$$

The opportunity cost due to lost sales for ith item is

$$\begin{aligned} & c_{3i}\lambda_i \int_{t_{1i}}^{T_i} \left(1 - \frac{1}{1+\delta_i(T_i - t)}\right) dt \\ &= \frac{c_{3i}\lambda_i}{\delta_i}\left\{(T_i - t_{1i}) - \ln\left[1+\delta_i(T_i - t_{1i})\right]\right\} \end{aligned} \tag{15}$$

Shortage cost for the ith item is given by

$$c_{1i}\int_{t_{1i}}^{T_i} -q_{2i}(t)dt$$

$$= \frac{c_{1i}\lambda_i}{\delta_i}\int_{t_{1i}}^{T_i}\left\{\ln\left[1+\delta_i(T_i-t_{1i})\right]-\ln\left[1+\delta_i(T_i-t)\right]\right\}dt$$

$$= \frac{c_{1i}\lambda_i}{\delta_i^2}\left\{(T_i-t_{1i})-\ln\left[1+\delta_i(T_i-t_{1i})\right]\right\} \quad (16)$$

In conjunction with the pertinent costs mentioned above, we can then simplify the total average cost per inventory cycle as follows:

$$TAC(t_{1i},T_i) = \frac{1}{T_i}[holding\ cost + ordering\ cost + purchasing\ cost + set\ up\ cost + shortage\ cost + opportunity\ cost]$$

$$TAC(t_{1i},T_i) = \sum_{i=1}^{n}\frac{1}{T_i}\left[\frac{h_i}{(\gamma+1-\beta)\lambda_i}\left[(1-\beta)\lambda_i t_{1i}\right]^{\frac{\gamma+1-\beta}{1-\beta}} + A_i + c_{2i} + c_i\left[(1-\beta)\lambda_i t_{1i}\right]^{\frac{1}{1-\beta}} + \frac{c_i\lambda_i}{\delta_i}\ln\left\{1+\delta_i(T_i-t_{1i})\right\} + \left(\frac{c_{3i}\lambda_i}{\delta_i}+\frac{c_{1i}\lambda_i}{\delta_i^2}\right)\left\{(T_i-t_{1i})-\ln\left[1+\delta_i(T_i-t_{1i})\right]\right\}\right] \quad (17)$$

Our problem is to minimize the total average cost under two subjects to constraints, such as one budget constraint and another space constraint. Hence, the multi-item crisp inventory problem is given by

$$\text{Min } TAC(t_{1i},T_i) = \sum_{i=1}^{n}\frac{1}{T_i}\left[\frac{h_i}{(\gamma+1-\beta)\lambda_i}\left[(1-\beta)\lambda_i t_{1i}\right]^{\frac{\gamma+1-\beta}{1-\beta}} + A_i + c_{2i} + c_i\left[(1-\beta)\lambda_i t_{1i}\right]^{\frac{1}{1-\beta}} + \frac{c_i\lambda_i}{\delta_i}\ln\left\{1+\delta_i(T_i-t_{1i})\right\} + \left(\frac{c_{3i}\lambda_i}{\delta_i}+\frac{c_{1i}\lambda_i}{\delta_i^2}\right)\left\{(T_i-t_{1i})-\ln\left[1+\delta_i(T_i-t_{1i})\right]\right\}\right]$$

$$\text{subject to } \sum_{i=1}^{n} c_i \left[\left[(1-\beta)\lambda_i t_{1i}\right]^{\frac{1}{1-\beta}} + \frac{\lambda_i}{\delta_i} \ln\left\{1+\delta_i(T_i - t_{1i})\right\}\right] \le B; \tag{18}$$

$$\sum_{i=1}^{n} w_i \left[\left[(1-\beta)\lambda_i t_{1i}\right]^{\frac{1}{1-\beta}} + \frac{\lambda_i}{\delta_i} \ln\left\{1+\delta_i(T_i - t_{1i})\right\}\right] \le F;$$

$$t_1 \ge 0,\ T \ge 0$$

where $t_1 = (t_{11}, t_{12}, \ldots, t_{1n})^{\mathrm{T}}$ and $T = (T_1, T_2, \ldots, T_n)^{\mathrm{T}}$ are decision variables.

When $c_i, c_{1i}, c_{2i}, c_{3i}, h_i, w_i, A_i, B$ and F become fuzzy rough variables, the above problem (18) can be formulated by the following model:

$$\text{Min } \widehat{T\tilde{A}C}(t_{1i}, T_i) = \sum_{i=1}^{n} \frac{1}{T_i} \left[\frac{\widehat{\tilde{h}}_i}{(\gamma+1-\beta)\lambda_i} \left[(1-\beta)\lambda_i t_{1i}\right]^{\frac{\gamma+1-\beta}{1-\beta}} \right.$$

$$+ \widehat{\tilde{A}}_i + \widehat{\tilde{c}}_{2i} + \widehat{\tilde{c}}_i \left[(1-\beta)\lambda_i t_{1i}\right]^{\frac{1}{1-\beta}} + \frac{\widehat{\tilde{c}}_i \lambda_i}{\delta_i}$$

$$\ln\left\{1+\delta_i(T_i - t_{1i})\right\} + \left(\frac{\widehat{\tilde{c}}_{3i}\lambda_i}{\delta_i} + \frac{\widehat{\tilde{c}}_{1i}\lambda_i}{\delta_i^2} \right)$$

$$\left. \left\{ (T_i - t_{1i}) - \ln\left[1+\delta_i(T_i - t_{1i})\right] \right\} \right]$$

$$\text{subject to } \sum_{i=1}^{n} \widehat{\tilde{c}}_i \left[\left[(1-\beta)\lambda_i t_{1i}\right]^{\frac{1}{1-\beta}} + \frac{\lambda_i}{\delta_i} \ln\left\{1+\delta_i(T_i - t_{1i})\right\}\right] \le \widehat{\tilde{B}}; \tag{19}$$

$$\sum_{i=1}^{n} \widehat{\tilde{w}}_i \left[\left[(1-\beta)\lambda_i t_{1i}\right]^{\frac{1}{1-\beta}} + \frac{\lambda_i}{\delta_i} \ln\left\{1+\delta_i(T_i - t_{1i})\right\}\right] \le \widehat{\tilde{F}};$$

$$t_1 \ge 0,\ T \ge 0$$

where $t_1 = (t_{11}, t_{12}, \ldots, t_{1n})^{\mathrm{T}}$ and $T = (T_1, T_2, \ldots, T_n)^{\mathrm{T}}$ are decision variables.

4.1 *Solution Procedure of the Proposed Inventory Model*

To solve the above multi-items fuzzy rough inventory model, we transform the multi-items fuzzy rough inventory model (19) into the deterministic problem considering expected value operator and solved it by using soft computing technique generalized reduced gradient (GRG) method (Lingo-14.0). The deterministic model of the multi-item fuzzy rough inventory problem (19) is given by

$$
\begin{aligned}
&\text{Min}\, E_r\left[\widehat{\widetilde{TAC}}(t_{1i}, T_i)\right] \\
&\text{subject to}\, E_r\left[\sum_{i=1}^{n}\left(\widehat{\widetilde{c}}_i\left[(1-\beta)\lambda_i t_{1i}\right]^{\frac{1}{1-\beta}} + \frac{\widehat{\widetilde{c}}_i\lambda_i}{\delta_i}\ln\left\{1+\delta_i(T_i - t_{1i})\right\}\right)\right] \le E_r\left[\widehat{\widetilde{B}}\right]; \qquad (20)\\
&E_r\left[\sum_{i=1}^{n}\left(\widehat{\widetilde{w}}_i\left[(1-\beta)\lambda_i t_{1i}\right]^{\frac{1}{1-\beta}} + \frac{\widehat{\widetilde{w}}_i\lambda_i}{\delta_i}\ln\left\{1+\delta_i(T_i - t_{1i})\right\}\right)\right] \le E_r\left[\widehat{\widetilde{F}}\right];\\
&t_1 \ge 0,\ T \ge 0
\end{aligned}
$$

where $t_1 = (t_{11}, t_{12}, \ldots, t_{1n})^{\mathrm{T}}$ and $T = (T_1, T_2, \ldots, T_n)^{\mathrm{T}}$ are decision variables.

5 Numerical Example

We have considered a numerical example to illustrate the expected value approach for solving the multi-item fuzzy rough inventory model. A multinational manufacturing company produces soft drinks. It is given that the purchasing cost(c_i), set-up cost(c_{2i}), the cost of lost sales(c_{3i}), shortage cost(c_{1i}), ordering cost(A_i) and storage area per unit item(w_i). The total available storage area and available budget cost are $\widehat{\widetilde{F}} = (\widehat{F} - 600, \widehat{F} - 700, \widehat{F} + 900, \widehat{F} + 1000)$ m^2 and $\widehat{\widetilde{B}} = \$(\widehat{B} - 1100, \widehat{B} - 1200, \widehat{B} + 1800, \widehat{B} + 2200)$, where $\widehat{F} \in ([1600, 1800]\,[1000, 2360])$, $\widehat{B} \in ([2400, 3200]\,[2200, 3620])$. The company wants to optimize the total average cost under the limitations of budget cost and storage area.

So, the problem is to minimize total average inventory costs with the limitation of space capacity, total budget cost. The costs and parameters of the inventory problem are given in Tables 1 and 2. Table 3 and Table 4 represent the optimum results of the above problem for different values of γ and β, respectively.

Table 1 Input crisp parameters

Item	I	II	III
λ_i	11.5	12.5	9.40
δ_i	0.80	0.75	0.70

Table 2 Input fuzzy rough cost parameters

Item	I	II	III
$\hat{\hat{h}}_i$	$(\hat{h}_1 - 0.5, \hat{h}_1 - 1.5, \hat{h}_1 + 2.5, \hat{h}_1 + 3.5)$ $\hat{h}_1 \in ([13.0, 14.5]\,[12.0, 17.3])$	$(\hat{h}_2 - 1.0, \hat{h}_2 - 1.5, \hat{h}_2 + 2.5, \hat{h}_2 + 4.0)$ $\hat{h}_2 \in ([10.5, 12.0]\,[08.5, 16.2])$	$(\hat{h}_3 - 1.0, \hat{h}_3 - 2.0, \hat{h}_3 + 3.0, \hat{h}_3 + 4.0)$ $\hat{h}_3 \in ([17.0, 19.0]\,[15.5, 22.9])$
$\hat{\hat{c}}_i$	$(\hat{c}_1 - 0.3, \hat{c}_1 - 1.7, \hat{c}_1 + 2.0, \hat{c}_1 + 3.0)$ $\hat{c}_1 \in ([16.0, 17.5]\,[14.0, 21.5])$	$(\hat{c}_2 - 0.4, \hat{c}_2 - 1.5, \hat{c}_2 + 2.5, \hat{c}_2 + 2.8)$ $\hat{c}_2 \in ([18.5, 20.0]\,[15.0, 23.1])$	$(\hat{c}_3 - 0.8, \hat{c}_3 - 1.2, \hat{c}_3 + 3.0, \hat{c}_3 + 4.5)$ $\hat{c}_3 \in ([15.5, 16.0]\,[14.5, 17.3])$
$\hat{\hat{c}}_{1i}$	$(\hat{c}_{11} - 0.5, \hat{c}_{11} - 1.8, \hat{c}_{11} + 2.8, \hat{c}_{11} + 3.5)$ $\hat{c}_{11} \in ([13.5, 16.0]\,[12.0, 18.5])$	$(\hat{c}_{12} - 0.8, \hat{c}_{12} - 1.6, \hat{c}_{12} + 2.6, \hat{c}_{12} + 3.8)$ $\hat{c}_{12} \in ([15.5, 17.4]\,[14.2, 20.9])$	$(\hat{c}_{13} - 1.2, \hat{c}_{13} - 2.2, \hat{c}_{13} + 3.0, \hat{c}_{13} + 4.0)$ $\hat{c}_{13} \in ([16.2, 18.6]\,[15.0, 23.4])$
$\hat{\hat{c}}_{2i}$	$(\hat{c}_{21} - 14, \hat{c}_{21} - 16, \hat{c}_{21} + 38, \hat{c}_{21} + 42)$ $\hat{c}_{21} \in ([98.5, 115.6]\,[90.0, 125.9])$	$(\hat{c}_{22} - 15, \hat{c}_{22} - 18, \hat{c}_{22} + 40, \hat{c}_{22} + 46)$ $\hat{c}_{22} \in ([125.0, 135.0]\,[105.0, 142.0])$	$(\hat{c}_{23} - 18, \hat{c}_{23} - 20, \hat{c}_{23} + 36, \hat{c}_{23} + 48)$ $\hat{c}_{23} \in ([128.0, 145.0]\,[110.0, 147.0])$
$\hat{\hat{c}}_{3i}$	$(\hat{c}_{31} - 7.0, \hat{c}_{31} - 8.5, \hat{c}_{31} + 12.0, \hat{c}_{31} + 16.0)$ $\hat{c}_{31} \in ([34.0, 40.0]\,[30.0, 43.5])$	$(\hat{c}_{32} - 7.5, \hat{c}_{32} - 9.0, \hat{c}_{32} + 13, \hat{c}_{32} + 17)$ $\hat{c}_{32} \in ([38.0, 44.0]\,[34.0, 46.0])$	$(\hat{c}_{33} - 8.0, \hat{c}_{33} - 8.5, \hat{c}_{33} + 14, \hat{c}_{33} + 18)$ $\hat{c}_{33} \in ([38.0, 48.0]\,[34.8, 55.2])$
$\hat{\hat{A}}_i$	$(\hat{A}_1 - 20, \hat{A}_1 - 25, \hat{A}_1 + 45, \hat{A}_1 + 65)$ $\hat{A}_1 \in ([155.0, 280.0]\,[120.0, 260.0])$	$(\hat{A}_2 - 18, \hat{A}_2 - 26, \hat{A}_2 + 42, \hat{A}_2 + 60)$ $\hat{A}_2 \in ([152.0, 258.0]\,[118.0, 278.0])$	$(\hat{A}_3 - 22, \hat{A}_2 - 26, \hat{A}_3 + 46, \hat{A}_3 + 68)$ $\hat{A}_3 \in ([156.0, 170.0]\,[123.0, 281.0])$
$\hat{\hat{w}}_i$	$(\hat{w}_1 - 3.0, \hat{w}_1 - 4.0, \hat{w}_1 + 5.0, \hat{w}_1 + 6.0)$ $\hat{w}_1 \in ([6.8, 8.0]\,[4.0, 10.0])$	$(\hat{w}_2 - 2.5, \hat{w}_2 - 3.5, \hat{w}_2 + 4.0, \hat{w}_2 + 5.0)$ $\hat{w}_2 \in ([7.5, 9.5]\,[6.0, 14.0])$	$(\hat{w}_3 - 3.5, \hat{w}_3 - 4.0, \hat{w}_3 + 6.0, \hat{w}_3 + 8.5)$ $\hat{w}_3 \in ([10.5, 11.0]\,[9.0, 13.7])$

6 Sensitivity Analysis

Here, sensitivity analysis is performed for minimization of the total average cost (TAC) and order quantity(Q_i) with respect to the parameter γ. From Table 3 and 4, following decisions can be constructed. These are also depicted in Figs. 3 and 6.

(i) For fixed value of β, the total average cost(TAC) increases with the increase in the value of parameter γ (cf. Fig. 4).
(ii) For fixed value of β, the order quantity(Q_i) decreases with increase in the value of parameter γ (cf. Fig. 3).
(iii) For fixed value of γ, the total average cost(TAC) increases with increase in the value of parameter γ (cf. Fig. 5).
(iv) For fixed value of γ, the order quantity(Q_i) decreases with increase in the value of parameter γ (cf. Fig. 6).

Table 3 Optimal result for different values of γ and $\beta = 0.12$

Item	γ	t_{1i}	t_{2i}	T_i	Q_i	Min TAC
I	0.35	3.2491	0.8946	4.1438	60.703	314.441
II		2.8977	1.0243	3.9220	56.985	
III		3.4303	0.7654	4.1957	62.720	
I	0.40	3.1958	0.9599	4.1557	60.148	318.757
II		2.8959	1.0853	3.9812	56.677	
III		3.4122	0.8279	4.2401	62.112	
I	0.45	3.1079	1.0374	4.1454	59.023	323.715
II		2.8831	1.1600	4.0432	56.652	
III		3.4141	0.9055	4.3196	62.060	
I	0.50	3.0015	1.1295	4.1310	57.636	329.372
II		2.8674	1.2526	4.1201	56.574	
III		3.4177	1.0027	4.4204	61.216	
I	0.55	2.6855	1.2090	3.8946	52.724	335.474
II		2.6066	1.3263	3.9329	52.724	
III		3.0403	1.0860	4.1263	56.687	
I	0.60	2.3739	1.2881	3.6621	47.243	341.456
II		2.3295	1.3981	3.7277	48.228	
III		2.6605	1.1668	3.8274	50.204	
I	0.65	2.1016	1.3689	3.4706	42.905	347.196
II		2.0805	1.4721	3.5527	44.299	
III		2.3382	1.2484	3.5867	44.863	

Table 4 Optimal result for different values of β and $\gamma = 0.50$

Item	β	t_{1i}	t_{2i}	T_i	Q_i	Min *TAC*
I	0.05	3.9334	0.6515	4.5580	58.409	290.234
II		4.0312	0.6545	4.6857	60.406	
III		4.0596	0.6727	4.7324	59.331	
I	0.07	3.6462	0.7704	4.4167	58.281	301.601
II		3.6803	0.8029	4.4832	59.751	
III		3.8530	0.7567	4.6089	60.227	
I	0.10	3.2702	0.9768	4.2471	58.358	318.455
II		3.1971	1.0613	4.2585	58.575	
III		3.5578	0.8965	4.4544	61.513	
I	0.12	3.0015	1.1295	4.1310	57.636	329.372
II		2.8674	1.2526	4.1201	56.574	
III		3.4177	1.0027	4.4204	61.216	
I	0.15	2.3675	1.3395	3.7070	50.759	344.475
II		2.1592	1.5083	3.6675	48.764	
III		2.8911	1.1488	4.0400	58.890	
I	0.18	1.8378	1.5414	3.3795	43.965	357.138
II		1.6069	1.7397	3.3466	41.436	
III		2.3807	1.2935	3.6743	53.104	

Fig. 3 γ versus Min Q_i

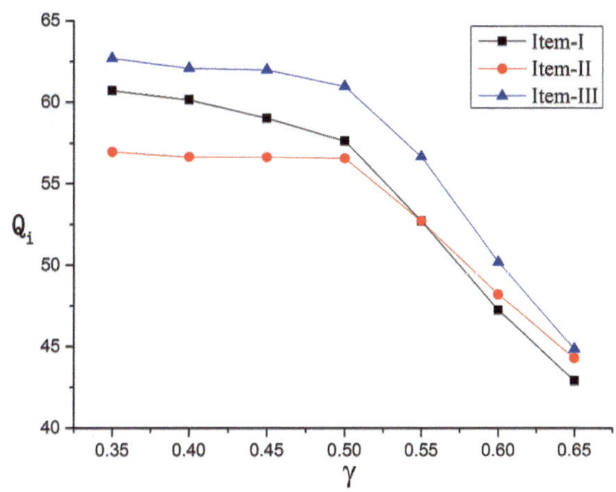

Fig. 4 γ versus Min TAC

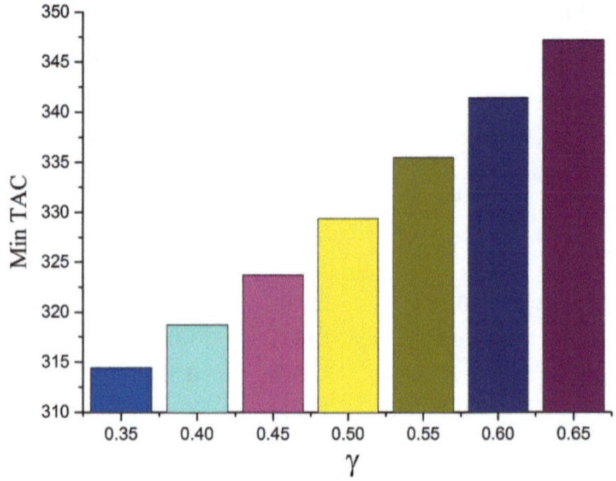

Fig. 5 β versus Min TAC

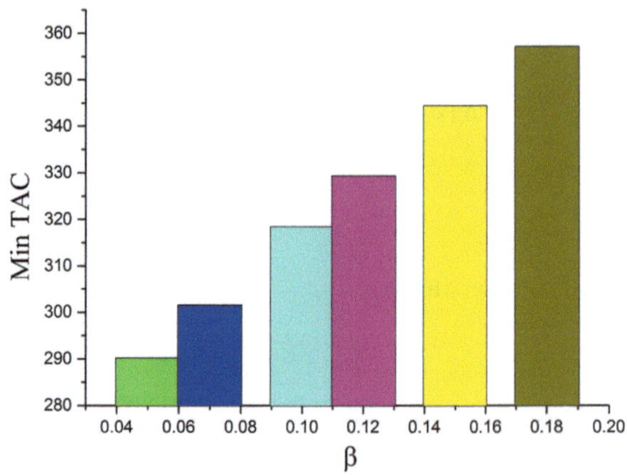

Fig. 6 β versus Q_i

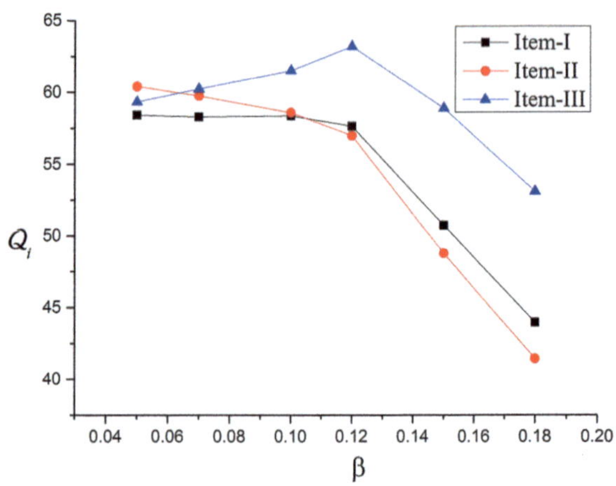

7 Conclusion

This paper develops a multi-item inventory model by considering demand as a power function of instantaneous stock level and holding cost as a nonlinear function of stock level. The time horizon and replenishment rate of the inventory system are considered as infinite. In this model, shortages are allowed and partially backlogged. To capture the real-life business situations, the cost and other parameters are considered under fuzzy rough environment. In Table 3 and Table 4, we have shown the optimal solutions of the problem (18) for different value of γ and β, respectively, and the solution's sensitivity has been analysed. The proposed model can be further extended in several ways like assuming fuzzy demand, quantity discount, variable rate of reworking and others.

References

1. Zadeh, L.A.: Fuzzy sets. Inf. Control **8**, 338–353 (1965)
2. Zadeh, L.A.: Fuzzy sets as a basis for a theory of possibility. Fuzzy Sets Syst. **1**, 3–28 (1978)
3. Ishii, H., Konno, T.: A stochastic inventory problem with fuzzy shortage cost. Eur. J. Oper. Res. **106**, 90–94 (1998)
4. Dubois, D., Prade, H.: Rough fuzzy sets and fuzzy rough sets. Int. J. Gen. Syst. **17**, 191–208 (1990)
5. Morsi, N.N., Yakout, M.M.: Axiomatics for fuzzy rough sets. Fuzzy Sets Syst. **100**, 327–342 (1998)
6. Radzikowska, M.A., Kerre, E.E.: A comparative study of rough sets. Fuzzy Sets Syst. **126**, 137–155 (2002)
7. Liu, B.: Theory and Practice of Uncertain Programming. Physica-Verlag, Heidelberg (2002)
8. Mondal, M., Maity, K.A., Maiti, K.M., Maiti, M.: A production-repairing inventory model with fuzzy rough coefficients under inflation and time value of many. Appl. Math. Model. **37**, 3200–3215 (2013)
9. Xu, J., Zhao, L.: A multi-objective decision-making model with fuzzy rough coefficients and its application to the inventory problem. Inf. Sci. **180**, 679–696 (2010)
10. Maiti, M.K., Maiti, M.: Production policy for damageable items with variable cost function in an imperfect production process via genetic algorithm. Math. Comput. Model. **42**, 977–990 (2005)
11. Khouja, M.: The economic production lot size model under volume flexibility. Comput. Oper. Res. **22**, 515–525 (1995)
12. Maity, K.A.: One machine multiple-product problem with production-inventory system under fuzzy inequality constraint. Appl. Soft Comput. **11**, 1549–1555 (2011)
13. Xu, J., Zaho, L.: A class of fuzzy rough expected value multi-objective decision making model and its application to inventory problems. Comput. Math. Appl. **56**, 2107–2119 (2008)
14. Lushu, S., Nair, K.P.K.: Fuzzy models for single-period inventory model. Fuzzy Sets Syst. **132**, 273–289 (2002)
15. Li, F.D.: An approach to fuzzy multi-attribute decision-making under uncertainty. Inf. Sci. **169**, 97–112 (2005)
16. Balkhi, Z.T., Foul, A.: A multi-item production lot size inventory model with cycle dependent parameters. Int. J. Math. Model. Methods Appl. Sci. **3**, 94–104 (2009)
17. Hartley, R.: An existence and uniqueness theorem for an optimal inventory problem with forecasting. J. Math. Anal. Appl. **66**, 346–353 (1978)

18. Lee, H., Yao, J.S.: Economic production quantity for fuzzy demand quantity and fuzzy production quantity. Eur. J. Oper. Res. **109**, 203–211 (1998)
19. Taleizadeh, A.A., Sadjadi, S.J., Niaki, S.T.A.: Multi-product EPQ model with single machine, back-ordering and immediate rework process. Eur. J. Ind. Eng. **5**, 388–411 (2011)
20. Dutta, P., Chakraborty, D., Roy, R.A.: An inventory model for single-period products with reordering opportunities under fuzzy demand. **53**, 1502–1517 (2007)
21. Chang, T.C.: An EOQ model with deteriorating items under inflation when supplier credits linked to order quantity. Int. J. Prod. Econ. **88**, 6159–6167 (2004)
22. Shi, Y., Yao, L., Xu, J.: A probability maximization model based on rough approximation and its application to the inventory problem. Int. J. Approx. Reason. **52**, 261–280 (2011)
23. Wang, Y.: Mining stock price using fuzzy rough set system. Expert Syst. Appl. **24**, 13–23 (2003)
24. Taleizadeh, A.A., Wee, M.H., Jolai, F.: Revisiting a fuzzy rough economic order quantity model for deteriorating items considering quantity discount and prepayment. Math. Comput. Model. **57**, 1466–1479 (2013)
25. Kazemi, N., Olugu, U.E., Rashid, H.S., Ghazilla, R.A.R.: A fuzzy EOQ model with back orders and forgetting effect on fuzzy parameters: an empirical study. Comput. Ind. Eng. **96**, 140–148 (2016)
26. Bazan, E., Jaber, Y.M., Zanoni, S.: A review of mathematical inventory models for reverse logistics and the future of its modelling: an environmental perspective. Appl. Math. Model. **40**, 4151–4178 (2016)
27. Das, C.B., Das, B., Mondal, K.S.: An integrated production inventory model under interactive fuzzy credit period for deteriorating item with several markets. Appl. Soft Comput. **28**, 453–465 (2015)
28. Xu, J., Zaho, L.: Fuzzy Link Multiple-Object Decision Making. Springer, Berlin (2009)

MIX
Papier aus verantwortungsvollen Quellen
Paper from responsible sources
FSC® C105338

If you have any concerns about our products,
you can contact us on
ProductSafety@springernature.com

In case Publisher is established outside the EU,
the EU authorized representative is:
Springer Nature Customer Service Center GmbH
Europaplatz 3, 69115 Heidelberg, Germany

Printed by Libri Plureos GmbH
in Hamburg, Germany